Lecture Notes in Computer Science   3469

Commenced Publication in 1973
Founding and Former Series Editors:
Gerhard Goos, Juris Hartmanis, and Jan van Leeuwen

Alden H. Wright   Michael D. Vose
Kenneth A. De Jong   Lothar M. Schmitt (Eds.)

Foundations
of Genetic
Algorithms

8th International Workshop, FOGA 2005
Aizu-Wakamatsu City, Japan, January 5-9, 2005
Revised Selected Papers

 Springer

Volume Editors

Alden H. Wright
University of Montana
Computer Science Department, Missoula, MT 59812, USA
E-mail: alden.wright@umontana.edu

Michael D. Vose
University of Tennessee
Computer Science Department, 203 Claxton Complex, 1122 Volunteer Blvd.
Knoxville, TN 37996-3450, USA
E-mail: vose@cs.utk.edu

Kenneth A. De Jong
George Mason University
Computer Science Department, 4400 University Drive, Fairfax, VA 22030, USA
E-mail: kdejong@cs.gmu.edu

Lothar M. Schmitt
University of Aizu
Aizu-Wakamatsu, Fukushima 965-8580, Japan
E-mail: LMSchmitt@yahoo.com

Library of Congress Control Number: 2005928446

CR Subject Classification (1998): F.1-2, I.2, I.2.6, I.2.8, D.2.2

ISSN   0302-9743
ISBN-10   3-540-27237-2 Springer Berlin Heidelberg New York
ISBN-13   978-3-540-27237-3 Springer Berlin Heidelberg New York

Springer is a part of Springer Science+Business Media

springeronline.com

© Springer-Verlag Berlin Heidelberg 2005
Printed in Germany

Typesetting: Camera-ready by author, data conversion by Olgun Computergrafik
Printed on acid-free paper   SPIN: 11513575   06/3142   5 4 3 2 1 0

Preface

The eighth workshop on the foundations of genetic algorithms, FOGA-8, was held at the University of Aizu in Aizu-Wakamatsu City, Japan, January 5-9, 2005. This series of workshops was initiated in 1990 to encourage further research in the theoretical aspects of genetic algorithms, and have been held biennually ever since. The papers presented at these workshops are revised, edited and published as a volume during the year following each workshop. This series of (now eight) volumes provides an outstanding source of reference for the theoretical work in this field.

At the same time this series of volumes provides a clear picture of how the theoretical research has grown and matured along with the field to encompass many evolutionary computation paradigms including evolution strategies (ES), evolutionary programming (EP), and genetic programming (GP), as well as the continuing growth in interactions with other fields such as mathematics, physics, and biology.

A tradition of these workshops is organize them in such a way as to encourage lots of interaction and discussion by restricting the number of papers presented and the number of attendees, and by holding the workshop in a relaxed and informal setting. This year's workshop was no exception. Thirty two researchers met for three days to present and discuss sixteen papers. The local organizer was Lothar Schmitt who, together with help and support from his University, provided the workshop facilities.

After the workshop was over, the authors were given the opportunity to revise their papers based on the feedback they received from the other participants. It is these revised papers that are included in this volume and follow the order in which they were presented at the workshop. In addition to these sixteen papers, there were two invited talks: an opening presentation by Alden Wright and a closing presentation by Kenneth De Jong. These slides-only presentations are not included in this volume, but can be obtained from the authors upon request. A brief summary of these presentations is provided here.

Alden Wright opened the workshop with a presentation titled "Can Evolutionary Computation Theory have Significance Outside of EC?" and subtitled "Can EC theory help us to understand evolution?". The field of artificial life has been successful in reaching wide audience with claims that artificial life experiments can give insight into natural evolution. Wright asked if EC theory can do the same? He proposed that EC theory might be relevant to some challenges in evolutionary research[1]. These included:

- Analysis of the evolution of rates of mutation and recombination. Do "optimal" rates evolve?
- Analysis of the evolution of the information content of genomes.
- Analysis of genic selection and of conflict within genomes (e.g., segregation distortion, evolution of gene expression, etc.).
- How does evolution maintain the great complexity of organisms while aso allowing "rapid" evolution in some areas?

---

[1] Some of these challenges came from the website:
http://evonet.sdsc.edu/evoscisociety/chall_and_oppors_in_e_res.htm

- How is it possible that phenotypic variations do not destroy brittle interactions between the subsystems of an organism while still allowing for the variablity that allows for evolutionary innovations?

Wright suggested that investigation of the genotype-phenotype map might give insight into the last two challenges.

Kenneth De Jong closed the workshop with a presentation titled "Unifying EC Theory". In this presentation he argued that developing a more unified framework for EC theory was important for further progress in the field. This was based on the observation that historically the field has evolved around a number of EA demes (GAs, ES, etc.), resulting in deme-specific terminology and theory. We now have deme-independent EC toolkits that provide creative mixing and matching of EA components, but we have no theory to guide EA design at this level.

De Jong outlined a strategy for developing such a theory. He suggested that we need to clearly distinguish between EAs as dynamical systems and EAs as problem solvers. Adopting a dynamical systems view allows us to answer questions about trajectories, fixed points, etc., and makes contact with a large body of existing theoretical work in evolutionary biology, evolutionary game theory, and dynamical systems theory. Adopting a problem solving view allows us to answer questions about the effectiveness of EAs for optimization, search, machine learning, etc., and makes contact with a large body of existing theoretical work from computer science, operations research, and artificial intelligence.

De Jong argued that in both of these areas it is important to find a middle ground between theories that are too abstract to be helpful and too specific to be be applicable to new situations. He gave several examples of how that might be done using a top-down strategy. He concluded by noting that several of the papers presented were nice examples of this middle theoretical ground, and express the hope that he would see more of them at the next FOGA.

In between those two presentations sixteen papers were presented on a wide range of theoretical evolutionary computation topics. We hope that you find them as interesting and provocative as we have. We fully expect that these papers will serve as a catalyst for further progress to be reported at the next FOGA workshop in 2007.

March 2005

Alden Wright
Michael Vose
Kenneth De Jong
Lothar Schmitt

Organization

Program Co-Chairs

Alden Wright (The University of Montana Missoula, USA)
Michael Vose (University of Tennessee, USA)
Kenneth De Jong (George Mason University, USA)
Lothar Schmitt (The University of Aizu, Japan)

Reviewers

Lee Altenberg
Anne Auger
Thomas Baeck
Wolfgang Banzhaf
Hans-Georg Beyer
Jürgen Branke
Anthony Bucci
Forbes Burkowski
Kenneth De Jong
Rolf Drechsler
Stefan Droste
Anton Eremeev
Larry Eshelman
James Foster
Marcus Gallagher
Oliver Giel
Jens Gottlieb
Walter Gutjahr
Jonathan Hallam
Nikolaus Hansen
William Hart
Jun He
Robert Heckendorn

Jeff Horn
Christian Igel
Yaochu Jin
Jens Jägersküpper
Thomas Jansen
Tim Kovacs
William Langdon
Anthony Liekens
Sean Luke
Evelyne Lutton
Keith Matthias
Nicholas McPhee
Peter Merz
Martin Middendorf
Heinz Mühlenbein
Bart Naudts
Silja Meyer-Nieberg
Una-May O'Reilly
Riccardo Poli
Adam Prugel-Bennett
Colin Reeves
Franz Rothlauf
Jonathan Rowe

Günter Rudolph
Dave Schaffer
Lothar Schmitt
Marc Schoenauer
Hans-Paul Schwefel
Jonathan Shapiro
Jim Smith
William Spears
Peter Stadler
Chris Stephens
Matthew Streeter
Dirk Thierens
Marc Toussaint
Ingo Wegener
Karsten Weicker
Nicole Weicker
Darrell Whitley
Paul Wiegand
Carsten Witt
Alden Wright
Annie Wu
Xin Yao

Sponsoring Institutions

International Society for Genetic and Evolutionary Computation (ISGEC)
University of Aizu, Japan

University of Aizu Institutional Support

President T. Ikegami, Vice-President S. Tsunoyama
Professor R.H. Fujii, Professor L. Pichl
K. Ishikawa, Y. Nagashima, M. Nanaumi
K. Doi, S. Ito, M. Oouchi, K. Takeyasu

FOGA 2005 Poster

FOGA 2005 Participants

Table of Contents

Genetic Algorithms for the Variable Ordering Problem of Binary Decision Diagrams

Wolfgang Lenders* and Christel Baier

Universität Bonn, Institut für Informatik I, Römerstrasse 164, 53117 Bonn, Germany
wolfgang.lenders@web.de, baier@cs.uni-bonn.de

Abstract. Ordered binary decision diagrams (BDDs) yield a data structure for switching functions that has been proven to be very useful in many areas of computer science. The major problem with BDD-based calculations is the variable ordering problem which addresses the question of finding an ordering of the input variables which minimizes the size of the BDD-representation. In this paper, we discuss the use of genetic algorithms to improve the variable ordering of a given BDD. First, we explain the main features of an implementation and report on experimental studies. In this context, we present a new crossover technique that turned out to be very useful in combination with sifting as hybridization technique. Second, we provide a definition of a distance graph which can serve as formal framework for efficient schemes for the fitness evaluation.

1 Introduction

Ordered binary decision diagrams (BDDs for short) are data structures to represent switching functions that rely on a compactification of binary decision trees. More general, using appropriate binary encodings, BDDs can serve to represent discrete functions with a finite domain. They were first introduced by Lee [28] and Akers [1]. In the meantime, various variants of BDDs have been suggested in the literature and applied successfully in many areas of computer science. Most popular are Bryant's *(reduced) ordered binary decisions diagrams* [8] that require a fixed variable ordering on any path. They have been proven to be very useful for the verification of reactive systems, often called *symbolic model checking* [10, 32]. Other application areas of BDDs include VLSI design, graph algorithms, complexity theory, matrix-operations, data bases, artificial intelligence, and many more. See e.g. the text books [18, 25, 33, 36, 48].

The crucial point with ordered BDD-based computations is the *variable ordering problem*. For a wide range of switching functions, there are polynomial-sized BDDs for "good" variable orderings, while the BDDs under "bad" variable orderings have exponential size. Unfortunately, the problem of finding an optimal variable ordering is NP-complete [6, 45]. However, there are many reordering algorithms that improve the ordering of a given BDD. Most popular are Rudell's sifting algorithm [41] and the window permutation algorithm [21]. A first attempt to use genetic algorithms for the variable ordering problem for BDDs was presented by Drechsler, Becker and Göckel [15] where the main genetic operations are partially-mapped crossover and mutation.

---

* The paper is based on material of the diploma thesis by the first author Wolfgang Lenders which he submitted in August 2004 at the Department of Computer Science, Universität Bonn.

A.H. Wright et al. (Eds.): FOGA 2005, LNCS 3469, pp. 1–20, 2005.

A related approach using simulated annealing was suggested by Bollig, Löbbing and Wegener [5]. In experimental studies it turned out that these methods yield better results (smaller BDDs) than other dynamic reordering techniques, but they are comparably slow, see e.g. [42]. To speed up the computations, several approaches have been suggested, including advanced tricks for the parameter setting and treating sifting as a genetic operation that replaces crossover techniques [16, 46], evolutionary algorithms with learning heuristics [17], the use of computed tables and approximate fitness values [24] or parallel genetic algorithms [12].

The goal of our paper is orthogonal to the above mentioned strategies by presenting alternative techniques to improve the efficiency and quality of genetic reordering algorithms for BDDs, while still retaining the concept of crossover (in contrast to the approaches of [16, 46]). We concentrate here on the purely genetic part of such reordering algorithms. However, the techniques suggested here can easily be combined with other (non-genetic) methods to increase the efficiency, e.g. by using "ordinary" sifting as in [16, 46].

Unlike [16, 46] which uses inversion as the only genetic recombination technique, we discuss several crossover techniques and present a new one, called *alternating crossover* which attempts to maximize the benefits of hybridization, i.e., the combination of a deterministic search algorithm with a genetic algorithm. The idea in the context of BDD minimization relies in generating an interleaving of the parent's variable orderings (alternating crossover) and moving the variables with the sifting-technique to the next local optimum after (the hybridization step). Our experimental results show that alternating crossover outperforms other recombination techniques such as order, partially matched or cycle crossover and inversion, by means of the BDD-sizes, while no significant differences in the runtime could be observed.

The second contribution is a formal framework to speed up the calculation of the fitness values for the newly generated individuals. In fact, for the variable ordering problem, calculating the BDD-size under a given variable ordering is a time-consuming step. It is typically realized by a sequence of local (level-wise) reorganizations of the BDD, the so called *swap*-operator (see e.g. [48]). Even when the final BDD is smaller than the original one, an exponential blow-up for the intermediate BDDs is possible. Thus, strategies that support the fitness calculation of the new population are highly desirable. We introduce a formal notion of a *distance graph*, a weighted graph where the nodes are orderings and the edges are labeled with the minimal number of swaps necessary to transform one ordering into another one. Using (variants of) heuristic algorithms for the traveling salesperson problem a "short" tour in the distance graph through the newly generated orderings, for which the fitness values (BDD-sizes in our case) are required, yields an appropriate scheme for the fitness evaluation. The distance graph can also serve as formal framework for other techniques that support the fitness calculation as suggested in [24]. Moreover, the fitness computation via our visiting strategy can easily be modified to weaken the drawback of crossover operations that might lead to unfeasible BDD-sizes, e.g., if they generate individuals that are far from both parents and combine the bad attributes of the parents.

Throughout the paper, we concentrate on the use of our algorithm for the minimization of ordinary BDDs, but our methods are also applicable to other types of decision diagrams, such as zero suppressed BDDs [36] algebraic decision diagrams, [2, 11] and their normalized version [39], and other DD-variants.

Organization of the paper. The basic concepts of binary decision diagrams and notations used in this paper are summarized in Section 2. Section 3 explains the main concepts of our genetic algorithm and its implementation we used for the experimental studies. Section 4 is concerned with alternating crossover. Our graph-based technique to reduce the runtime for the fitness calculation are described in Section 5. In Section 6, we report on experimental results. Section 7 concludes the paper.

2 Binary Decision Diagrams

In the remainder of this paper, we fix a finite set $\mathcal{Z} = \{z_1, \ldots, z_n\}$ of boolean variables and often refer to the variables by their indices (i.e., we identify index i with variable z_i). An evaluation for $\mathcal{Z}$ denotes a function that assigns a boolean value (0 or 1) to any variable $z_i \in \mathcal{Z}$. By a *switching function* over $\mathcal{Z}$, we mean a function f which maps any evaluation for $\mathcal{Z}$ to 0 or 1. If $z \in \mathcal{Z}$ then $f|_{z=0}$ and $f|_{z=1}$ denote the *cofactors* of f which arise by fixing the assignment $z \mapsto 0$ and $z \mapsto 1$ respectively. For instance, if $f = z_1 \wedge (z_2 \vee z_3)$ then $f|_{z_1=0} = 0$ and $f|_{z_1=1} = z_2 \vee z_3$.

The fact that there is no data structure for switching functions that is efficient for all switching functions becomes clear from the observation that the number of switching functions over $\mathcal{Z}$ grows double exponentially in the size of $\mathcal{Z}$. An *explicit* representation of switching functions using truth tables seems coherent, but a truth table for a switching function with n variables consists of 2^n lines and consequently its space complexity grows exponentially in the number of variables. *Implicit* descriptions, like propositional logic formulas and binary decision diagrams can be much more efficient.

Binary decision diagrams are a graph based representation of switching functions which rely on the decomposition of switching functions in their cofactors according to the *Shannon expansion* $f = (\neg z \wedge f|_{z=0}) \vee (z \wedge f|_{z=1})$. Formally, a BDD is an acyclic rooted directed graph where every inner node v is labeled with a variable and has two children, called the 0-successor and 1-successor. The terminal nodes are labeled with one of the truth values 0 or 1. In ordered BDDs (OBDD) [8], there is a variable ordering $\pi = (z_{i_1}, \ldots, z_{i_n})$ which is preserved on any path from the root to a terminal node. That is, if v is an inner node labeled with variable z_{i_ℓ} and w a child of v which is non-terminal and labeled with variable z_{i_r} then z_{i_ℓ} appears in π before z_{i_r}, i.e., $i_\ell < i_r$. In the sequel, we shall use the notation π-OBDD to denote an OBDD relying on the ordering π and we refer to any inner node labeled with variable z as a z-node.

The switching function represented by a terminal node agrees with the corresponding constant 0 or 1. The switching function of a z-node v with 0-successor w_0 and 1-successor w_1 is $f_v = (\neg z \wedge f_{w_0}) \vee (z \wedge f_{w_1})$. The switching function $f_{\mathcal{B}}$ represented by an OBDD $\mathcal{B}$ agrees with the switching function for its root node. Thus, given an evaluation for $\mathcal{Z}$, the truth value under $f_{\mathcal{B}}$ is obtained by traversing $\mathcal{B}$ starting in its root and branching in any inner node according to the given evaluation. Figure 1 depicts two π-OBDDs with the variable ordering $\pi = (z_1, z_2, z_3)$ for the function $f = (z_1 \wedge \neg z_2 \wedge z_3) \vee (\neg z_1 \wedge z_3 \wedge z_2)$. In the OBDD on the left, both z_3-nodes represent

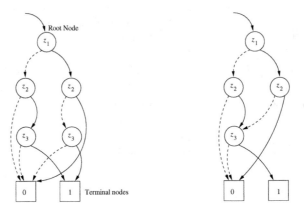

Fig. 1. OBDD and ROBDD

the same cofactor, namely $f|_{z_1=0,z_2=1} = f|_{z_1=1,z_2=0} = z_3$. Thus, a further reduction of the shown OBDD is possible by identifying the two z_3-nodes which yields the *reduced OBDD* (ROBDD) shown on the right. Intuitively, A OBDD is called reduced if it does not contain any redundancies. Formally, an ROBDD $\mathcal{B}$ denotes an OBDD such that $f_v \neq f_w$ for all nodes v, w in $\mathcal{B}$ with $v \neq w$. Given an π-OBDD, an equivalent π-ROBDD is obtained by identifying terminal nodes with the same value, identifying z-nodes with the same successors and eliminating all inner nodes where the 0- and 1-successor agree.

π-ROBDDs yield a *universal* representation for switching functions. (This follows from the fact that the above reduction procedure applied to the decision tree for a switching function with ordering π yields an π-ROBDD.) Moreover, the representation by π-ROBDDs is *canonical* up to isomorphism because the node-set of a π-ROBDD stands in one-to-one correspondence to the set of cofactors $f|_{z_{i_1}=b_1,\ldots,z_{i_k}=b_k}$ that can be obtained from f by assigning values to the "first" variables of π [1]. (Here, the range for k is $0, 1, \ldots, n$, and $b_1, \ldots, b_k \in \{0, 1\}$.)

ROBDDs yield a minimized OBDD-representation for a given switching function, provided the variable ordering is viewed to be fixed. However, by varying the ordering π the size of the BDD can be influenced. Figure 2 illustrates this observation by displaying two ROBDDs for the same switching function $f = (x_1 \wedge x_2) \vee (y_1 \wedge y_2) \vee (z_1 \wedge z_2)$ using different variable orderings. In the worst case, a ROBDD can have exponential size according to the number of variables n. There are functions, e.g. the middle bit of multiplication, whose ROBDD representation has exponential size for every variable ordering. Other functions, e.g. the most significant bit of addition, can vary between linear and exponential size depending on the chosen variable ordering while the number of any ROBDD for symmetric functions (e.g. n-ary disjunction or the parity function) is at most quadratic. See [9] and e.g. the text books [33, 48] for a detailed discussion of the complexity of ROBDDs.

Shared BDDs. Most BDD-packages follow the approach of [35] who suggested the simultaneous representation of several switching functions in one reduced graph (called

---

[1] Some of these cofactors might agree in which case they are represented by the same node

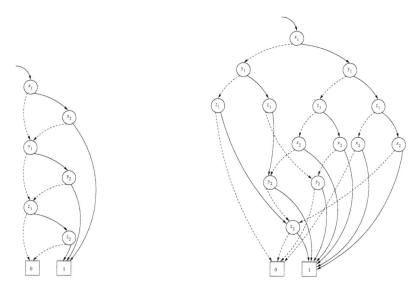

Fig. 2. Two BDDs for the same switching function using different variable orderings

shared or *multi-rooted BDD*) where the ROBDDs of the represented functions are realized as subgraphs and share the nodes for common cofactors. With several additional
implementation tricks (appropriate hash tables, the ITE-operator to treat all boolean
connectives, negated edges, etc.) the manipulation of switching functions and other
BDD-based calculations can be realized efficiently, such as checking equivalence of
switching functions in constant time or performing boolean combinations in time polynomial in the sizes of the ROBDDs for the arguments.

Throughout the paper the term BDD will refer to a shared BDD with negative edges.
(This also applies for the number of BDD-nodes in the experimental results.)

The variable ordering problem. For the wide range of functions where the BDD-sizes
range from polynomial to exponential, the variable ordering has an immense importance
for BDD-applications, not only for reasons of memory requirement but also for the runtime of BDD manipulation operations. Beside some heuristics that compute a variable
ordering from a given circuit description there is a wide range of *dynamic reordering*
algorithms that attempt to improve the given variable ordering. The problem of finding
an optimal variable ordering for a given BDD is known to be NP-complete [6, 45]. The
best known algorithm that determines an optimal variable ordering requires exponential
time [20]. However, there are several Greedy-heuristics that might return a suboptimal
ordering. All these reordering algorithms are based on sequentially exchanging pairs
of neighboring variables. This basic *swap* operation induces only local changes to the
involved variables and can be carried out in constant time for each node that has to be
handled. Thus, the running time of the operation $swap(z, z')$ on the BDD $\mathcal{B}$ with ordering π, where z and z' are adjacent in π, is linear in the number of z-nodes and the
number of their incoming edges in $\mathcal{B}$. Using appropriate sequences of swap operations,
any variable ordering can be transformed into another one.

One of the most commonly used deterministic heuristics for BDD minimization is Rudell's *sifting* algorithm [41] . The basic idea of sifting is to move each variable successively through the whole variable ordering and eventually leave it at the position that yields the best BDD size. This procedure can be repeated as long as the variable ordering changes (*iterated sifting*). Several additional heuristics can be used to improve the efficiency of the sifting algorithm. Most popular is the use of a *maximum growth factor* c which stops the movement of a variable in one direction if the BDD-size becomes c-times larger than the original one. In our genetic algorithm, we shall use (non-iterated) sifting as hybridization technique with small maximum growth factors c. With such a choice for c, the sifting procedure is quite fast and searches the local optimum for any variable in its neighborhood. In fact, we made good experience with a *local search* that we obtain by choosing max growth factor $c = 1$.

Genetic algorithms for the variable ordering problem rely on a representation of the variable orderings in permutation form. The main genetic operations used in the algorithm proposed in [15] are (i) *partially matched crossover* (PMX) [22] which selects a matching section between two cutpoints and uses exchange operations to make one parent's matching section assimilate the other's, (ii) *mutation* which exchanges the positions of two variables, and (iii) *inversion* [26] which selects at random two cutpoints and reverses the ordering in the enclosed segment. To improve the efficiency, [16, 46] suggest to skip crossover techniques and use sifting as a "normal" operation instead[2], while [12] deals with a parallel genetic algorithms with PMX and mutation as main operations. Other additional techniques to achieve a speed-up are proposed in e.g. [24].

Our approach where sifting serves as hybridization technique should be contrasted to the approach of [16, 46] where sifting serves as a "normal" operation which is chosen with a probability of 50% and executed with the maxgrowth factor $c = 2$. In our setting, we deal with a minimized version of sifting that only serves for a local search in the surrounding of an offspring generated by a crossover operation. In fact, by choosing the maxgrowth factor $c = 1$ we only look for the nearest local optimum of any variable which makes the sifting-phase much faster than with higher maxgrowth factors (such as $c = 2$).

3 A Genetic Algorithm for the Variable Ordering Problem

In this section, we summarize the main features of our implementation of a genetic algorithm for the BDD minimization problem. We realized the standard schema for evolutionary algorithms with hybridization, sketched in Figure 3, using several genetic operations. We adapted several techniques for evolutionary algorithms suggested somewhere else in the literature and developed a new crossover technique (see Section 4) as well as a graph-algorithmic approach for the design of an efficient schema for the fitness computation (see Section 5).

---

[2] More precisely, the main "proper" genetic operation in [16, 46] is inversion, but they skip the crossover techniques, and use mutation only if the offspring is equal to the parent element. In [16] some additional problem-specific recombination and mutation operators have been used for incompletely specified boolean functions. As we shrink our attention to completely specified function these techniques are not applicable in our setting

Genetic Algorithm with Hybridization
Input: Population p as a collection of individuals   **Output:** Individual i with "good" fitness

initialize(p)
 evaluateFitness(p)
 $i = $ fittestElementOf(p)
 REPEAT
   selectParents(p)
   recombination(p)   (* crossover and inversion *)
   mutation(p)
   evaluateFitness(p)   (* see section 5 *)
   hybridization()   (* sifting with maxgrowth $c = 1$ *)
   $i = $ fittestElementOf(p)
 UNTIL(i was not improved)

Fig. 3. A hybrid genetic algorithm

The population size is parametric in our implementation. Even for large circuits, we made good experience with small population sizes, such as 8 individuals per population (see Section 6). The initial population is chosen at random. Techniques that derive a promising ordering from the topology of a circuit description (e.g. the fanin heuristic [30] or weight heuristic [35]) could be used in addition. Also an improvement of the initial population with deterministic reordering algorithms (such as sifting or window permutation) could be integrated, as e.g. in [15].

Recombination. Beside the partially matched crossover (PMX) [22], which is also used in [15] and [12], we consider three other crossover techniques. *Order crossover* [13] chooses $n/2$ pairwise different positions and copies the genes at the selected positions to the offspring, and finally, fills up the gaps using the missing genes in the order they are found in the second parent. In general, the offspring under order crossover assimilates the first parent more than the second. Another version of order crossover incorporates cutpoints instead of randomly selected positions. Every element between the two cutpoints is copied from the first parent, the elements outside the cutpoints are filled up with the missing elements, preserving the second parents' order. This variant has the benefit of being less disruptive. *Cycle crossover* [38] attempts to retain the original position of genes in their parents. This is achieved by continuous copying of genes from one parent until the end of a cycle is reached, then switching and continuing from the other parent. In rare occasions the offspring can be equal to one of its parents. This case has to be combined with forced mutation to achieve a modification in the next generation. In addition, we implemented *alternating crossover*, that will be explained in Section 4, and the *inversion* operator [26], which reverses the fragment of a given variable ordering between randomly chosen cutpoints, as an asexual recombination technique.

Mutation. Mutation of a permutation means the exchange of the positions of two variables by appropriate swap-operations. The approach we have chosen in our implementation first takes a general decision whether a given offspring is to be mutated or not.

If so, a level of mutation is chosen and expressed as a number of variable exchanges to be executed. The positions of the variables to be exchanged are picked randomly, also multiple selection of the same variable is possible. This approach is efficient in implementation and execution, and it resembles the original mutation scheme. A forced mutation in case a crossover does not generate (enough) differences between offspring and parents is available. For measuring "differences", a *distance* is defined in Section 5.

Fitness scaling. Choosing the BDD size as a natural measure for the fitness of a variable ordering seems straightforward. Nevertheless the fitness values will be "negated", conducted by setting $fitness(\pi) = max\_bdd\_size\_found - bdd\_size(\pi)$, for implementation reasons, which also retains the comfort of speaking of a *higher* fitness as a better one, whereas a higher BDD size would imply a worse variable ordering. In Section 5, we will explain our new scheme to minimize the number of swaps necessary for fitness calculation by a distance minimizing strategy.

To handle the problem of premature convergence[3] or the problem of fitness values that are too close to each other (which can happen in "late" populations, also in the non-premature case, in particular for small population sizes), we adapt the approach of Goldberg [23] and use a *linear scaling mechanism*. That is, we replace the original fitness function f by the scaling function $f' = af + b$ by first fixing f' (avg) to f (avg), which ensures that each not less than average individual obtains a scaled value ≥ 1 and is therefore guaranteed a mating opportunity in a subsequent remainder selection scheme. Toward the end of a GA's run, the population has largely converged. In this environment, the maximum fitness is generally close to the average fitness, whereas recombination may generate lethals, i.e. individuals with a far below average fitness. These individuals are likely to be scaled to negative fitness values. These exceptions are caught and the affected individuals set to zero fitness. The resulting fitness values are sampled using *stochastic universal sampling* [3, 4] by default, while other sampling methods, such as roulette wheel selection or remainder sampling with or without replacement, are available upon selection.

A variant with the full sifting procedure. As pointed out in [16, 46], the efficiency of evolutionary reordering algorithms as in Fig. 3 can be increased by using "ordinary" sifting (with large maxgrowth factor, say $c = 2$) as an alternative in the recombination phase. As mentioned before, the aim of our paper is to study the gain of the proper genetic operators, and therefore, we do not consider this option here.

4 Alternating Crossover

We suggest a new crossover technique, called alternating crossover, which in combination with sifting as hybridization technique turned out to be very successful. Alternating crossover generates offspring by copying genes alternately from the parents and interleaves them this way. See Figure 4. This creates offspring in which genes that were adjacent in one parent are generally separated by one or more genes from the other parent.

---

[3] Premature convergence e.g. occurs if in the initial population one of the randomly selected individuals represents a fairly good solution already which is far away from the other individuals and if this "superhero" is chosen multiple times for mating and is going to spread its genes throughout the population instantly

---

Alternating Crossover

Input: Parents p_1 and p_2 of length n
Output: Offspring π

$done = \{\}$
$candidate = p_1.\textbf{atPosition}(0)$
$position\_p_1 = 0$
$position\_p_2 = 0$
FOR $(i = 0)$ **TO** $(i = n - 1)$ **DO**
  WHILE $(candidate \in done)$ **DO**
    IF $(i \mod 2 = 0)$ **THEN**
      $candidate = p_1.\textbf{atPosition}(position\_p_1)$
      $position\_p_1 = position\_p_1 + 1$
    ELSE
      $candidate = p_2.\textbf{atPosition}(position\_p_2)$
      $position\_p_2 = position\_p_2 + 1$
    FI
  OD
  $done \cup \{candidate\}$
  $\pi.\textbf{atPosition}(i) = candidate$
OD
return π

Fig. 4. Alternating Crossover

Under normal circumstances this disruption of schemata would be considered harmful, but in conjunction with *sifting* with maxgrowth factor $c = 1$ as hybridization algorithm it bears good results. Sifting performs swaps of neighboring variables and retains the exchange if it was beneficial. This way, every separation of genes introduced during the application of alternate crossover can be revoked if necessary, while on the other hand many genes are tested in the surroundings of their current position. Therefore, alternating crossover in conjunction with sifting exploits the offspring's local neighborhood thoroughly.

Figure 5 depicts an example of an alternating crossover application and highlights the genes in the offspring that were adjacent in a parent and are now in *sifting distance*, i.e. their distance is less than 2. Thus even our minimized sifting procedure is able to restore the original ordering if necessary. (Here, we identify variable z_i with its index i.) We call two genes a and b in sifting distance, when they can be made adjacent by no more than two exchanges of neighboring genes, i.e. when there are at most two genes between a and b. Our minimized sifting procedure moves each gene at least one step in each direction and is therefore able to recover the original ordering should it have been the most beneficial one. In the following example, let the original ordering with adjacent genes a and b be better than the newly generated one:

$$\begin{aligned}
\text{original ordering:} &\quad x\,a\,b\,y \\
\text{newly generated by alternating crossover:} &\quad a\,x\,y\,b \\
\text{exchange neighboring variables } a \text{ and } x: &\quad x\,a\,y\,b \\
\text{exchange neighboring variables } b \text{ and } y: &\quad x\,a\,b\,y
\end{aligned}$$

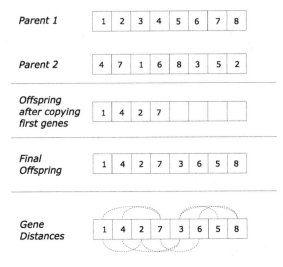

---------- : *former neighbors, now seperated but in sifting distance*

Fig. 5. Example for the operation of *Alternating Crossover*

Since we said the original ordering to be the most beneficial one, sifting would have executed exactly these two variable exchanges.

5   Fitness Calculation via an Optimized Visiting Order

Obtaining the actual fitness value for a variable ordering involves generating the corresponding binary decision diagram via an appropriate sequence of swap-operations. This can be a costly procedure if the ordering differs clearly from the current order. To minimize the number of swaps necessary for fitness calculation we suggest a strategy that attempts to find an efficient visiting order of the individuals of the new population (variable orderings) for which the fitness values (BDD-sizes) are still unknown.

In principle, fitness can be calculated at different times during the run of a genetic algorithm. Calculating fitness for each individual directly after it has been generated has the benefit of being able to decide about the individual's fate at once. If, for example, the offspring generated by a crossover is way worse than its parents it can be discarded in favor of the better parent. On the other hand, this approach does not allow alterations in the order the offspring is tested, which otherwise can be optimized. In the sequel, we explain a strategy to optimize the visiting order of the individuals by providing a formal definition for the distance between variable orderings.

A distance function for variable orderings. In the sequel, we identify any swap-operation with the index of the variable to be swapped with its right neighbor. Thus, for a variable set $\mathcal{Z} = \{z_1, \ldots, z_n\}$ of cardinality n, we denote any swap-operation by an integer $s \in \{1, \ldots, n-1\}$. We write $\pi \triangleright_s \pi'$ to denote that swap-operation s transforms the variable ordering π into the variable ordering π'. By a *swap sequence*, we mean any finite sequence $\sigma = (s_1, s_2, \ldots, s_l)$ of swap-operations. We refer to $|\sigma| = l$ as the length of σ. σ is said to transform π into π', denoted $\pi \triangleright_\sigma \pi'$, if the sequential composition of the swaps s_i transforms π to π', i.e.,

$$\pi \triangleright_\sigma \pi' \text{ if } \pi \triangleright_{s_1} \pi_1 \triangleright_{s_2} \pi_2 \ldots \triangleright_{s_l} \pi_l = \pi'.$$

σ is called a *minimum swap sequence* for (π, π') if σ transforms π to π' and if there is no shorter swap sequence than σ that also transforms π to π'. The distance $\delta(\pi, \pi')$ between two variable orderings π and π' is defined as the length of a minimum swap sequence for (π, π'). That is, $\delta(\pi, \pi') = \min\{|\sigma| : \pi \triangleright_\sigma \pi'\}$.

Proposition 1. δ *is a metric on the the set of variable orderings. That is,*

1. $\delta(\pi, \pi') = 0$ *if* $\pi = \pi'$
2. $\delta(\pi, \pi') = \delta(\pi', \pi)$
3. $\delta(\pi, \pi') \leq \delta(\pi, \hat{\pi}) + \delta(\hat{\pi}, \pi')$

The proof of Proposition 1 is straightforward and omitted here. The orderings with maximum distance between each other are the pairs (π, π^{-1}), were π^{-1} is the inverse ordering of π.

Proposition 2. *If π and π' are variable orderings for a variable set of cardinality n then*

$$\delta(\pi, \pi') \leq \delta(\pi, \pi^{-1}) = \frac{n(n-1)}{2}$$

Proof. The fact that $\delta(\pi, \pi') \leq (n-1) + (n-2) + \ldots + 1 = \frac{n(n-1)}{2}$ is clear as we may consider the swap sequence which first moves the last variable of π' with at most $(n-1)$ swaps at position n, then moves the variable at position $n-1$ in π' with at most $n-2$ swaps at its final position $n-1$, and so on.

It remains to provide the argument why no swap sequence shorter than $\frac{n(n-1)}{2}$ transforms π into π^{-1}. Let π and π' be arbitrary orderings for variables $z_1, \ldots, z_n$ and k_i the number of variables z_j such that $i \neq j$ and (i) z_i occurs in π before z_j and (ii) z_j occurs in π' before π'. That is, $\pi = (\ldots, z_i, \ldots, z_j, \ldots)$ and $\pi' = (\ldots, z_j, \ldots, z_i, \ldots)$. Then, any swap sequence that transforms π into π' has to perform at least k_i swaps that exchange z_i with its right neighbor. Thus, $\delta(\pi, \pi') \geq k_1 + \ldots + k_n$. In the case, $\pi' = \pi^{-1}$, we have $k_i = n - i$, Thus, $\delta(\pi, \pi^{-1}) \geq (n-1) + (n-2) + \ldots + 1 = n(n-1)/2$.   □

Deriving an efficient fitness calculation scheme from the distance graph. The above proposition shows that inversion, a powerful genetic operation, requires a number of swaps quadratic to the length of the inverted segment. This makes an immediate fitness rating of the offspring less desirable in comparison to the opportunity to optimize the order of visiting the individuals. Our strategy for reducing the number of variable swaps, that have to be carried out for computing all fitness values by finding an advantageous visiting order for the individuals, is based on a *distance graph*, a complete graph where the individuals for which the fitness still has to be computed form the vertices, while the edge between two vertices π_1 and π_2 is marked with their distance $\delta(\pi_1, \pi_2)$. (Because of the symmetry of δ the distance graph can be viewed as an undirected graph.) An

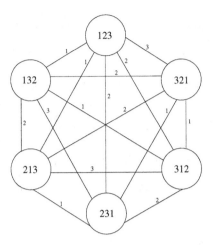

Fig. 6. A distance graph for orderings with three variables

example for a distance graph for three variables[4] is provided in Figure 6. Usually, the distance graph will not contain *all* possible vertices as suggested by the figure, but only those vertices coding for members of the group of offspring whose fitness is still unknown.

Now we could ask for an *optimal visiting strategy* for the individuals, i.e. a visiting order that visits all nodes of the distance graph and which minimizes the sum of all covered distances (the total number of swap operations which have to be carried out). Since we are looking at an instance of the traveling salesperson problem, the question for an optimal visiting order is computationally hard (NP-complete). Instead, we may adapt any heuristic algorithms for the TSP to obtain an efficient, possibly sub-optimal visiting order of the vertices in the distance graph. In our implementation, we employed the *nearest neighbor heuristic* [34] to decide which individual is to be considered next until all fitness values are computed. Our experiments showed that this procedure means a major speed-up towards the regular visiting order, because the calculation of fitness values is one of the most time-consuming but basic and irreplaceable parts of the minimization algorithm.

A variant of the graph-based visiting schema. [16, 46] observed the problem that variable orderings generated by the standard crossover techniques (PMX, OX or CX) might lead to BDDs of unfeasible size. To avoid this problem, we suggest the following variant of our visiting algorithm. If during the execution of a minimum swap sequence from one vertex π to another vertex π' of the distance graph the BDD-size is larger than a certain *threshold* then we may discard π' and, if necessary, generate a new variable ordering π'' via genetic operations (recombination, mutation and sifting as hybridization technique). In this case, of course, the visiting strategy has to be revised dynamically. The threshold can either be a fixed upper bound for the BDD-size or can be determined

---

[4] Again, we identify any variable with its index. E.g., node 123 stands for the ordering $\pi = (z_1, z_2, z_3)$

by a function depending on the fitness values that are already known. Another alternative for the threshold is to use a maxgrowth factor (as it is standard for sifting) for the swap sequences that are executed in the visiting strategy.

In addition, the best intermediate ordering $\hat{\pi}$, obtained by executing the minimum swap sequence from node π to node π' in the distance graph, can be used as an additional candidate for the next generation, provided it is better than π and π'.

Integration of other advanced techniques. Our graph-algorithmic approach for the fitness computation can easily be modified to integrate the three methods suggested by Günther and Drechsler [24] to accelerate evolutionary algorithms for sequencing problems.

(1) For an approach where the BDDs for *several variable orderings* are stored to speed up the fitness calculations (as proposed by [24]) we may also deal with a distance graph, but now equipped with another weight function for the edges. Let $\pi_1, \ldots, \pi_m$ be the variable orderings for which corresponding BDDs are stored. Then, we may use the weight function $\hat{\delta}(\pi, \pi') = \min\{\delta(\pi, \pi'), \delta(\pi_i, \pi') : i = 1, \ldots, m\}$ which captures the possibility to start the computation of the π'-BDD with one of the stored BDDs rather than the π-BDD.
(2) Following [24], we may also use *computed-tables* that store the BDD-sizes for already considered variable orderings. In our setting, this means a simplification of the distance graph which only contains the orderings not considered so far.
(3) The third method suggested in [24] relies on the use of upper and lower bounds for the BDD-sizes that will be obtained through local modifications of the ordering [7]. As shown in [24], this technique in combination with multiple representation as in (1) and computed-tables as in (2) can lead to a speed-up around 80%. This idea can be integrated in our graph-based approach as follows bu choosing a constant d and modifying the visiting strategy as follows: If the current node is π then we use such *approximate fitness values* rather than the precise BDD-sizes for all (possibly, but one) orderings π' with $\delta(\pi, \pi') \leq d$.

6   Experimental Results

To evaluate the performance of the several recombination techniques (crossover, inversion) and the influence of the parameter setting, we implemented the schema sketched in Fig. 3. For all tests we used excerpts of the LGSynth93 benchmark suite (see Fig. 7), obtainable from [31]. We carried out ten runs of our genetic algorithm and present the average BDD size as well as the best result we obtained, in order to visualize the variation in the results. The indicated time shows CPU seconds on a Pentium IV 2.4 GHz PC with 512 MB of RAM running the JJS-BDD package [27] on Linux.

Unless stated otherwise, in all tests the parameters of our genetic algorithm were chosen as follows. The population size is 8, the maxgrowth factor for hybrid sifting is $c = 1$. We carried out experiments with growth factors of 1.1 and 1.2 (not shown here), which resulted in almost identical[5] results, but bearing a longer runtime. For the

---

[5] One benchmark resulted in a BDD two nodes smaller. $size_{avg}$ results were slightly better in most benchmarks

benchmark	original BDD size	inputs	outputs
apex1	6785	45	45
apex2	13418	38	3
apex3	53365	54	50
apex4	1040	9	19
apex5	3944	114	88
apex6	1993	135	99
apex7	1775	49	37
comp	203198	32	3
cps	1869	24	109
dalu	11178	75	16

benchmark	original BDD size	inputs	outputs
des	10771	256	245
duke2	596	22	29
e64	1500	65	65
ex4p	994	84	28
i5	1032	133	66
i6	388	138	67
i7	559	199	67
i8	10366	133	81
vg2	735	25	8

Fig. 7. Benchmarks

Benchmark	order			partially matched			alternating			cycle			inversion		
	size	$size_{avg}$	time	size	$size_{avg}$	time	size	$size_{avg}$	time	size	$size_{avg}$	time	size	$size_{avg}$	time
apex1	1253	1255	40	1246	1258	52	1250	1253	39	1270	1270	33	1246	1253	101
apex2	354	372	32	318	327	54	328	338	22	321	345	23	392	395	22
apex3	841	841	20	839	841	28	839	841	23	841	841	20	840	841	30
apex4	889	889	2	889	889	2	889	889	2	889	889	2	889	889	2
apex5	1044	1044	85	1044	1050	113	1044	1044	72	1073	1082	50	1086	1092	68
apex6	523	532	58	513	527	89	510	531	55	524	531	78	575	587	81
apex7	214	214	6	214	214	6	214	214	6	216	217	5	214	214	6
comp	101	107	33	110	125	28	101	110	36	122	144	37	143	143	20
cps	971	971	18	971	972	12	971	974	13	977	976	10	1010	1010	14
dalu	785	798	157	689	689	248	689	701	138	689	689	205	699	711	192
des	2983	3012	988	2971	2977	723	2958	2974	756	2992	3015	601	2987	2992	953
duke2	336	336	3	336	336	4	336	336	4	336	352	3	336	336	4
e64	129	129	12	129	129	12	129	129	12	129	129	11	129	129	16
ex4p	463	468	16	466	471	26	459	470	16	460	481	17	465	468	21
i5	134	134	17	134	134	16	134	134	18	134	134	18	134	134	35
i6	209	209	14	209	209	15	209	209	15	209	209	14	209	209	14
i7	334	334	39	333	333	59	333	335	50	334	335	52	335	335	38
i8	1277	1280	163	1280	1281	196	1277	1281	149	1285	1344	150	1280	1281	206
vg2	80	80	2	80	80	2	80	80	2	84	84	2	84	84	3

Fig. 8. Comparison between five recombination operators

selection method, we used stochastic universal sampling and realized the concept of elitarism for one individual.

Comparison of the crossover operators. To compare the types of crossover (OX, PMX, CX and AX) and inversion, we restricted our algorithm to the use of a single operator. An inspection of the results for the five operators in Fig. 8 yields that the runtimes all assimilate each other. To compare the quality of the results we take only the best BDD size achieved during the ten runs into account.

order	partially matched	alternating	cycle	inversion
11	14	17	7	7

The above table illustrates for how many benchmark circuits each crossover yielded a best result. (If more than one crossover achieved the best result we awarded a point to each of them.) Thus, alternating crossover bears the best results, followed by partially matched crossover. The combination of different crossover operators is, however, the most promising approach, since the sequential application of different crossovers on the same individual allows more possible outcomes than repetitive application of the same operator. This can also be seen from the results shown in the left column of Fig. 10. For several benchmarks the best result is obtained using a combination of crossovers, in *ex4p* for example, the combination reaches a BDD size of 242 BDD nodes, while the best result of a single operator, in this case alternating crossover, is 459 BDD nodes. Other examples for the superiority of a combination of crossovers to the use of a single operator are *apex1*, *apex3*, *comp* and *des*.

Given our results on the comparison of the recombination techniques (Fig. 8), we argue that the restriction to inversion as the only proper genetic operation in the recombination phase as suggested in [16, 46] shrinks the gain of evolutionary reordering techniques. The motivation given in [16, 46] for omitting crossover techniques was their excessive runtime requirements. However, a comparison of the the time-columns in Fig. 8 shows that – in combination with our graph-based fitness evaluation technique – the crossover techniques are in average no worse than inversion. (Additionally, the generation of too large BDDs can be prevented as described in Section 5.)

Benchmark	regular parameters		alternative parameters		
	size	*time*	*size*	*time*	*pop. size*
apex1	1246	31	1244	828	120
apex2	306	25	302	433	114
apex3	837	24	837	397	120
apex4	889	2	889	5	27
apex5	1044	62	1044	793	120
apex6	498	45	507	601	120
apex7	214	7	214	62	120
comp	95	33	125	221	96
cps	971	11	971	58	72
dalu	689	230	689	1733	120
des	2941	1173	2946	9229	120
duke2	336	4	336	19	66
e64	129	11	129	103	120
ex4p	242	27	460	182	120
i5	134	16	134	204	120
i6	209	15	209	143	120
i7	333	52	333	408	120
i8	1277	187	1277	4366	120
vg2	80	2	80	11	75

Fig. 9. Regular versus alternative parameters

Parameter setting. To illustrate the benefits of our parameter setting and graph-based fitness evaluation technique, we performed tests where we used the parameter setting used in [15]. Here, the population size is set to $\min\{120, 3 \cdot population\, size\}$. The max-

Benchmark	genetic algorithm			sifting		sifting$_{iter}$	
	size	*size$_{avg}$*	*time*	*size*	*time*	*size*	*time*
apex1	1246	1269	31	1381	0.5	1270	3
apex2	306	342	25	589	0.5	502	2
apex3	837	864	24	851	0.2	850	0.8
apex4	889	889	2	889	0.1	889	0.1
apex5	1044	1076	62	1076	0.7	1073	2
apex6	498	569	45	532	0.6	520	3
apex7	214	241	7	297	0.1	248	0.2
comp	95	112	33	95	56	95	68
cps	971	971	11	1010	0.2	1010	0.3
dalu	689	697	230	1552	478	1346	534
des	2941	2968	1173	3242	36	3051	39
duke2	336	340	4	395	0.1	360	0.3
e64	129	129	11	155	0.2	129	0.4
ex4p	242	242	27	512	0.2	507	0.6
i5	134	134	16	134	0.3	134	0.6
i6	209	209	15	215	0.3	209	2
i7	333	334	52	335	0.9	335	2
i8	1277	1280	187	2104	2	2092	5
vg2	80	80	2	157	0.1	152	0.9

Fig. 10. Comparison between our genetic algorithm and sifting

imum growth factor for hybrid sifting is set to $c = 2$. Elitarism is applied to the better half of the population. The results in Figure 9 demonstrate that the alternative choice of parameters rarely achieves a better result than our choice. The best result, obtained for benchmark *apex2*, is only four nodes smaller than our result. On the other hand, the alternative parameters results in a runtime which exceeds ours generally by factor 10 to 20. In summary, as Figure 9 shows, our genetic algorithm with crossover and the graph-based visiting strategy performs very well, already with a small population size.

Comparison of our genetic algorithm with "pure sifting". For a comparison of the schema in Fig. 3 which only uses crossover (but no inversion) against deterministic reordering heuristics, we assigned probability 0.6 to alternating crossover, and 0.2 to both partially matched and cycle crossover. We obtained similar results when cycle crossover is replaced with order crossover or when assigning the same weight to them. As before, the maxgrowth factor for hybrid sifting is 1. On the other hand, we considered sifting and iterated sifting with maxgrowth factor 1.3. Using our genetic algorithm, the resulting BDD in general is considerably smaller than it is after application of sifting. In some examples like *apex2* and *dalu* we even achieve a bisection of the BDD's size. In no case is the best result of ten GA runs worse then the result achieved by sifting. This positive result is obtained at the expense of runtime, which in average is an order of magnitude higher than it is for sifting, on the other hand for benchmarks *comp* and *dalu* the runtimes for sifting even exceed those of our GA. In average, however, runtime for our GA is longer, though it generates a substantially smaller BDD.

In his diploma thesis [29], the first author also reports about experiments with the window permutation algorithm [21]. The obtained results agree with the common observation that window permutation is fast but a rather weak minimization heuristic. Thus, our genetic algorithm yields much better results in terms of quality, in some cases, like *comp*, *des* and *dalu* for instance, the BDD-sizes were even only a fraction ($<$ 1%) of those returned by window permutation, on the price of a longer computation time.

7   Conclusion

The goal of the paper was to study in detail the gain of genetic operations in the context of dynamic reordering algorithms for BDDs. We discussed several crossover variants and suggested a new one, called alternating crossover, which turned out to be very useful in combination with a "minimized version" of sifting as hybridization technique. In addition, we proposed a graph-algorithmic approach to speed up the fitness evaluation which, in case of the variable ordering problem for BDD, is a time-consuming step. In contrast to the observations made by [16, 46] our experiments (see Section 6) show that a random selection between crossover techniques and inversion yields better results than the sparse use of "proper" genetic operations as in [16, 46].

Using the proposed techniques, runtime requirements for genetic reordering algorithms were brought down to a reasonable level, although, concerning the computation time, our techniques are still not competitive to deterministic reordering heuristics such as sifting or window permutation. However, our approach nicely fits in the framework of Drechsler et al. [16, 46] who pointed out that the mixture of genetic techniques with ordinary sifting yields a good balance between speed and quality, as it captures the advantages of both genetic algorithms and comparably fast deterministic reordering algorithms. In addition, we explained that other methods that improve the efficiency, e.g. those suggested in [24], can easily be integrated.

There are various directions in which our algorithm (and its implementation) could be extended. Although we made good experience dealing with sifting and maxgrowth factor 1 as hybridization technique, window permutation is another candidate. Another direction is the consideration of a *group-preserving* variant of our algorithm. In fact, there are several BDD-applications where not all variable permutations should be regarded as potential solutions, but only those that group together certain variables. One example are switching functions with symmetric inputs where typically good orderings put the variables of any symmetry group together. Group-preserving orderings play also a crucial role for symbolic model checking where there are several good reasons (see e.g. [19]) to group any state-variables and its copy (the corresponding next-state variable) together. For such applications where we are given disjoint groups of variables, such that for some application-dependent reasons[6] the variables in either group should be placed together, we suggest to apply the same genetic operations (crossover, mutation, inversion) but with groups of variables rather than single variables. E.g., in case of alternating crossover, we may apply the schema shown in Figure 4 with groups of variables rather than single variables. In a similar way, the other crossover techniques can

---

[6] To treat symmetries, known methods from the literature to derive the symmetry groups automatically from a given BDD can be applied here as well

be modified to treat groups of variables. In the hybridization step, we may apply *group sifting* [40] which relies on the same schema as sifting but moves groups of adjacent variables rather than single variables.

Another future direction is to check whether the concepts of alternating crossover and the graph-algorithmic approach for the fitness calculation are also useful for other permutation-problems.

Acknowledgments

The authors would like to thank the anonymous referees for many helpful comments and Sascha Klüppelholz and Jörn Ossowski for their support with the JJS-BDD-package.

References

1. Akers, S.B., *Binary Decision Diagrams*, IEEE Trans. on Computers, Vol. C-27, 1978
2. Bahar, R.I., *Algebraic Decision Diagrams and their Applications*, Proceedings on the International Conference on Computer Aided Design, 1993
3. Baker, J.E., *Balancing Diversity and Convergence in Genetic Search*, Ph.D. Thesis, 1987
4. Baker, J.E., *Reducing Bias and Inefficiency in the Selection Algorithm*, Proceedings of the Second International Conference on Genetic Algorithms on Genetic algorithms, 1987
5. Bollig, B., Löbbing, M., Wegener, I., *Simulated Annealing to Improve Variable Orderings for OBDDs*, Proceedings of the International Workshop on Logic Synthesis, 1995
6. Bollig, B., Wegener, I., *Improving the Variable Ordering of OBDDs Is NP-Complete*, IEEE Transactions on Computers, Volume 45, 1996
7. Bollig, B., Löbbing, M., Wegener, I., *On the Effect of Local Changes in the Variable Ordering of Ordered Binary Decision Diagrams*, Information Processing Letters, Volume 59, pp 233-239, 1996.
8. Bryant, R.E., *Graph-Based Algorithms for Boolean Function Manipulation*, IEEE Trans. on Computers - Vol. C-35, 1986
9. Bryant, R.E., *On the complexity of VLSI implementations and graph representations of Boolean functions with applications to integer multiplication*, IEEE Transactions on Computers, Vol. 40, pp 205–213, 1991.
10. Clarke, E., Grumberg, O., Peled, D., *Model Checking*, MIT Press, 1999
11. Clarke E., Fujita, M., Zhao, X., *Multi-Terminal Binary Decision Diagrams and Hybrid Decision Diagrams*, Representations of Discrete Functions, Kluwer Academic Publishers, 1996
12. Costa, U., Deharbe, D., Moreira, A., *Advances in BDD reduction using parallel genetic algorithm*, Proceedings of the 10 th International Workshop on Logic Synthesis (IWLS), 2001.
13. Davis, L., *Applying Adaptive Algorithms to Epistatic Domains*, Proceedings of the International Joint Conference on Artificial Intelligence, 1985
14. De Jong, K., *Analysis of the behavior of a class of genetic adaptive systems*, Ph.D. Thesis, University of Michigan, 1975
15. Drechsler, R., Becker, B., Göckel N., *A Genetic Algorithm for Variable Ordering of OBDDs*, IEEE Proceedings, 143(6), pp 363–368, 1996.
16. Drechsler, R., Göckel, N., *Minimization of BDDs by Evolutionary Algorithms*, International Workshop on Logic Synthesis (IWLS), Lake Tahoe, 1997

17. Drechsler, R., Becker, B., Göckel, N., *Learning Heuristics for OBDD Minimization by Evolutionary Algorithms*, Proceedings Parallel Problem Solving from Nature (PPSN), Lecture Notes in Computer Science 1141, pp. 730-739, 1996.

18. Drechsler, R., Becker, B., *Binary Decision Diagrams: Theory and Implementation*, Kluwer Academic Publishers, 1998

19. Enders, R., Filkorn, T., Taubner, D., *Generating BDDs for Symbolic Model checking in CCS*, Distributed Computing, Vol. 6, pp 155-164, 1993.

20. Friedman, S. J., Supowit, K. J., *Finding the Optimal Variable Ordering for Binary Decision Diagrams*, IEEE Transactions on Computers, Volume 39, 1990

21. Fujita, M., Matsunaga, Y.,Kakuda, T., *On Variable Orderings of Binary Decision Diagrams for the Application of Multi-Level Logic Synthesis*, Proceedings of the European Conference on Design Automation, February 1991

22. Goldberg, D.E., Lingle, R., *Alleles, Loci and the Traveling Salesman Problem*, Proceedings of the International Conference on Genetic Algorithms and their Applications, 1985

23. Goldberg, D.E., *Genetic Algorithms in Search, Optimization and Machine Learning*, Addison Wesley, 1989

24. Günther, W., Drechsler, R., *Improving EAs for Sequencing Problems*, Proceedings of the Genetic and Evolutionary Computation Conference, 2000.

25. Hachtel, G., Somenzi, F., *Logic Synthesis and Verification Algorithms*, Kluwer Academic Publishers, 1996

26. Holland, J., *Adaptation in Natural and Artificial Systems*, University of Michigan Press, 1975

27. Klüppelholz, S., Ossowski, J., Tietjen, J., *JJS-BDD an object oriented c++ sobdd library*, University of Bonn, http://jjs-bdd.sourceforge.net/

28. Lee, C.Y., *Representation of switching circuits by binary decision programs*, Bell Systems Technical Journal - Vol. 38, 1959

29. Lenders, W., *Genetic Algorithms for the Variable Ordering Problem of Binary Decision Diagrams*, Diploma Thesis, Institut für Informatik I, Universität Bonn, Germany, 2004.

30. Malik, S., Wang, A.R., Brayton, R.K., Sangiovanni-Vincentelli, A., *Logic verification using binary decision diagrams in logic synthesis environment*, Proceedings ICCAD, pp 6–9, 1988.

31. McElvain, K., *LGSynth93 Benchmark Set: Version 4.0*, obtainable at http://www.cbl.ncsu.edu/CBL_Docs/lgs93.html, 1993

32. McMillan, K. L., *Symbolic Model Checking*, Kluwer Academic Publishers, 1993

33. Meinel, C., Theobald, T., *Algorithms and Data Structures in VLSI Design: OBDD-Foundations and Applications*, Springer-Verlag, 1998

34. Mertens, S., *TSP Algorithms in Action Animated Examples of Heuristic Algorithms*, University of Magdeburg, 1999

35. Minato, S., Ishiura, N., Yajima, S., *Shared Binary Decision Diagram with Attributed Edges for Efficient Boolean Function Manipulation*, In Proceedings of the 27th ACM/IEEE Design Automation Conference (DAC'90), 1990

36. Minato, S., *Binary Decision Diagrams and Applications for VLSI CAD*, Kluwer Academic Publishers, 1996

37. Minato, S., *Zero-Suppressed BDDs and Their Applications*, International Journal on Software Tools for Technology Transfer - Vol. 3, 2001

38. Oliver, I.M., Smith, D.J., Holland, J.R.C., *A Study of Permutation Crossover Operators on the Traveling Salesman Problem*, Proceedings of the Second International Conference on Genetic Algorithms, 1987

39. Ossowski, J., *Symbolic Representation and Manipulation of discrete Functions*, Diploma Thesis, University of Bonn, 2004

40. Panda, S., Somenzi, F., *Who are the Variables in your Neighborhood*, Proceedings of the International Conference on Computer Aided Design, 1995

41. Rudell, R., *Dynamic Variable Ordering for Ordered Binary Decision Diagrams*, Proceedings of the International Conference on Computer-Aided Design, 1993

42. Somenzi, F., "CUDD: CU Decision Diagram Package", University of Colorado at Boulder, 1998

43. Spears, W.M., De Jong, K.A., Bäck, T., Fogel, D.B., de Garis, H., *An Overview of Evolutionary Computation*, Proceedings of the European Conference on Machine Learning, 1993

44. Spears, W.M., *Crossover or Mutation?*, Foundations of Genetic Algorithms 2, Morgan Kaufmann, 1993

45. Tani, S., Hamaguchi, K., *The Complexity of the Optimal Variable Ordering Problems of Shared Binary Decision Diagrams*, Proceedings of the 4th International Symposium on Algorithms and Computation, 1993

46. Thornton, M.A., Williams, J.P., Drechsler, R., Drechsler, N., Wessels, D.M., *SBDD Variable Reordering based on Probabilistic and Evolutionary Algorithms*, Proceedings IEEE Pacific Rim Conference (PACRIM), pp 381–387, 1999.

47. Vose, M.D., Liepins, G.E., *Schema Disruption*, Proceedings of the fourth International Conference on Genetic Algorithms, 1991

48. Wegener, I., *Branching Programs and Binary Decision Diagrams. Theory and Applications*, Monographs on Discrete Mathematics and Applications, SIAM, 2000

Gray, Binary and Real Valued Encodings:
Quad Search and Locality Proofs

Darrell Whitley[1] and Jonathan E. Rowe[2]

[1] Computer Science, Colorado State University, Fort Collins, CO 80523
[2] Computer Science, University of Birmingham, Birmingham B15 2TT, UK

Abstract. This paper looks at proofs concerning Gray codes and the locality of local search; it is shown that in some cases such proofs can be generalized to Binary and real valued encodings, while in other cases such proofs do not hold for Binary encodings. The paper also looks at a modified form of Quad Search that is proven to converge in under 2L evaluations to a global optimum on unimodal 1-D bijective functions and to a local optimum on multimodal 1-D bijective functions. Quad Search directly exploits the locality of the Gray code representation.

1   Introduction

There have been numerous debates about the relative advantages of search algorithms that use Gray code representations compared to Binary and real-valued representations [CS88] [Dav91]. Reflected Gray codes are characterized by a symmetric reflection property, such that a point in one half of a 1-D space folds onto its neighbor in the other half of the search space. This property allow the construction of relatively simple recursive proofs related to the time required to convergence to a local optimum under local search, as well as the convergence of a specialized search strategy called Quad Search [WGW03]. The reflection properties of Gray code are also the foundation for another (new) proof that counts the number of neighbors which in expectation are in the same basin of attraction (or quasi-basin) as a given reference point; the proof assumes one is provided with information about the size of the basin of attraction relative to the size of the search space.

A modified form of Quad Search algorithm is introduced that allows it to provably converge to a local optimum when run on 1-D multimodal functions. An example shows that the original form of Quad Search does not always converge to a local optimum when searching 1-D multimodal functions.

But do these proofs imply that similar proofs do not hold for Binary representations? And what about real valued representations? When possible we constructs similar convergence proofs for Binary and real-valued representations. This paper also compares the assumptions and special purpose operators that are needed for the various convergence proofs.

New proofs are also presented concerning the likelihood of neighbors falling into the "quasi-basin" around the point currently visited under local search. The significance of these results relate to the No Free Lunch theorem [WM95] and the Christensen and Oppacher "Submedian Seeker" algorithm [C01].

A.H. Wright et al. (Eds.): FOGA 2005, LNCS 3469, pp. 21–36, 2005.

2 Gray and Binary Encoding and Quadrants

Gray codes have at least two important properties. First, for any 1-D function (and in each 1-D slice of any multidimensional problem) adjacent neighbors in the real space are also adjacent neighbors in the Gray code hypercube graph [RW97]. Second, the "standard Binary reflected Gray code" folds the search space in each dimension, creating a symmetric reflected neighbor structure [BER76] [RWB04]. For any reference point i, there is exactly one neighbor in the opposite half of the search space. For example, in a 1-D search space of points indexed from 0 to $2^L - 1$, the points i and $(2^L - 1 - i)$ are neighbors. In effect, these neighbors are found by folding or reflecting the 1-D search space about the mid-point between 2^{L-1} and $2^{L-1} + 1$.

There are also reflections between each quartile of the search space. Starting at point i in the first quartile of the search space we can define a set of four integer values where each integer falls in a different quartile, or "quadrant" of the search space. The four points are as follows:

$$
\begin{aligned}
i &= i \\
(2^{L-1} - 1) - i &= i \oplus (2^{L-1} - 1) \\
(2^L - 1) - i &= i \oplus (2^L - 1) \\
(2^{L-1} + i) &= i \oplus (2^L - 1) \oplus (2^{L-1} - 1)
\end{aligned}
\tag{1}
$$

where $\oplus$ denotes exclusive-or on the Binary encoding of the integer. Both $2^L - 1$ and $2^{L-1} - 1$ are Hamming distance 1 from the string of all zeros. Therefore, moving from point i to one of the other 3 points involves flipping the first bit, the second bit, or both. Because exclusive-or just applies a bit-flip mask, this indexing scheme works regardless of the quadrant in which the point i actually appears.

2.1 Quad Search

Consider a unimodal function encoded using an L bit Gray code. Any given point has two neighbors in quadrants outside of its own quadrant; they are the points generated by flipping the two high order bits. The highest order unfixed bit is called the *major* bit. The second highest order unfixed bit is called the *minor* bit. Flipping any other bit must generate a neighbor inside the current quadrant.

Let BB denote the first two bits that map to an arbitrary quadrant. In effect, the pattern $BB**\ldots*$, defines a hyperplane that corresponds to a (hypothetically) continuous quarter, or *quadrant* of the search space. Let N_{BB} refer to a point in quadrant BB. We can flip the major bit to generate the point at $N_{\overline{B}B}$ and the minor bit to generate the point at $N_{B\overline{B}}$. We flip both bits to reach the point at $N_{\overline{BB}}$. Each point is in a different quadrant, each with its respective index.

In Figure 1 the point located at X is a neighbor of point N_{BB} that is located in quadrant BB. X can be left or right of N_{BB}. One of the 4 quadrant neighbors is the current point from which the search is being carried out.

A previously published paper introduces the original *Quad Search* algorithm and proves its convergence time [WGW03]. The new algorithm (see Figure 2) and the proof presented here have been modified to evaluate fewer points and to allow convergence proofs for multimodal 1-D functions. We express the problem in terms of minimization.

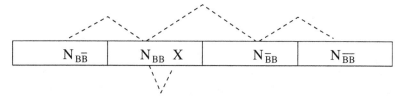

Fig. 1. The neighbor structure across and within quadrants

```
\****************************************************\
\*   BEGIN ONE-DIMENSIONAL QUAD SEARCH ALGORITHM *\
\****************************************************\
```

Step 1: Randomly select point N_{BB}. Evelute point N_{BB} and
   its reflected neighbors $N_{\overline{B}B}, N_{B\overline{B}}$ and $N_{\overline{BB}}$.

Step 2: Without loss of generality, relabel the points so that N_{BB} is the best of these four points.

Step 3: Flip the third unfixed bit in N_{BB} to create point X, and evaluate.
   If $f(X) < f(N_{BB})$ set CASE = 1; else CASE = 2.
      /* Samples a reflected neighbor in quadrant BB */
      /* CASE refers to Case 1 and Case 2 of the proof */

Setp 4: Let **b** represent the third bit in the best-so-far solution.
   If **b** = 0 fix the minor (second) bit.
   If **b** = 1 fix the major (first) bit.
      /* The current best must be in quadrant BB. The third bit determines if */
      /* the local optima is bounded by the leftmost or rightmost quadrant of BB. */

Step 5: Relabel the points so that the new best-so-far is denoted N_{BB} and its reflected
   neighbors $N_{\overline{B}B}, N_{B\overline{B}}$ and $N_{\overline{BB}}$. Three of these points have been evaluated.
      /* This follows from the Quad Search convergence proof. */

Step 6: If CASE = 1, the unevaluated point is at $N_{\overline{B}B}$.   Evaluate $N_{\overline{B}B}$.
      /* If CASE = 2, the additional point is not evaluated. */

Setp 7: If there are more than 2 bits unfixed bits, goto Step 3. Otherwise, STOP.

```
\***********************\
\*   END QUAD SEARCH   *\
\***********************\
```

Fig. 2. The Quad Search algorithm. The proof uses an exclusive-or transform. This makes the proof more direct, but is not needed for the algorithm. However, the algorithm in the untransformed space behaves differently in terms of when the major and minor bits are fixed. Since only the first 3 bits are involved in both algorithm and proof, these can be shown to behave the same by enumeration of the 16 possible conditions (8 bit patterns, and CASE = 1 or CASE = 2)

Theorem 1: *Quad Search converges to the global minimum on bijective unimodal 1-D functions and to a local minimum on multimodal 1-D functions after at most $2L - 1$ evaluations from an arbitrary starting point.*

Proof: The proof is recursive in nature. Assume the search space contains at least 8 points, so that there are at least 2 points in each quadrant. After initializing the search, we show the search space can be cut in half after at most 2 new evaluations.

Sample a point N_{BB} and its reflected neighbors $N_{\overline{B}B}, N_{B\overline{B}}$ and $N_{\overline{BB}}$.

Without loss of generalization, let N_{BB} be the current best solution. Let Z represent a string of zeros of arbitrary length. There exists a bit mask, m, which relabels the points $\{N_{BB}, N_{\overline{B}B}, N_{B\overline{B}}, N_{\overline{BB}}\}$ using exclusive-or so that the minimal neighbor is represented by the string $000Z$. The neighborhood structure is unchanged under exclusive-or. Transforming all the points yields:

$$N_{BB} \oplus m = 000Z, \quad N_{\overline{B}B} \oplus m = 100Z, \quad N_{B\overline{B}} \oplus m = 010Z, \quad N_{\overline{BB}} \oplus m = 110Z$$

where $\oplus$ denotes exclusive-or. We sample one additional point at $001Z$ in quadrant 00.

We first deal with unimodal functions: note the minimum cannot be in quadrant 11.

Case 1. If $f(000Z) > f(001Z) < f(010Z)$, then the global optimum must reside in quadrant 00 or 01. The major bit is fixed to 0 and the search space is reduced. The points $000Z, 001Z, 010Z$, become the points $00Z, 01Z, 10Z$ in the reduced space. Only 2 additional points are needed to continue the search. First we evaluate $11Z$. Then the space is remapped again so that the minimum of these 4 points is at $00Z$. Then we evaluate $001(Z - 1)$.

Case 2. If $f(100Z) > f(000Z) < f(001Z)$, then the global optimum must reside in quadrant 00 or 10. The minor bit is fixed to 0 and the search space is reduced. The points $000Z, 001Z, 100Z$, become the points $00Z, 01Z, 10Z$ in the reduced space. We do not evaluate $11Z$ since we know $f(100Z) > f(000Z) < f(001Z)$ and the minimum of these 3 points is already at $00Z$. We evaluate only 1 additional point at $001(Z - 1)$.

After the first 5 evaluations, the search space contains $2^{(L-1)}$ points. At each iteration, the search space is cut in half after at most 2 new evaluations. After (L-3) iterations the search space is reduced to at most 4 points, since $2^{(L-1)}/2^{(L-3)} = 4$. However, at this point, we have already evaluated $00, 01$ and 10 (remapped so that 00 is the best solution so far); therefore we do not evaluate 11. The total number of evaluations is at most $5 + 2(L - 3) = 2L - 1$. This proves convergence for the unimodal case.

To prove convergence for the multimodal case, it suffices to show that as sections of the search space are eliminated, the remaining space is a continuous interval; if the remaining space is a continuous interval of the original function (where the original 1-D function is defined to wrap around), then cases 1 and 2 above show that the remaining space contains a local optimum of the original function.

The following are specific Subcases of Case 1 and Case 2 above when the best solution so far is an *arbitrary* quadrant. In this case, we don't care where the best solution so far is located, only which quadrants are actually eliminated.

Subcase 1a. Fixing the major bit to 1 eliminates quadrants 00 and 01.
Subcase 1b. Fixing the major bit to 0 eliminates quadrants 10 and 11.
Subcase 2a. Fixing the minor bit to 1 eliminates quadrants 00 and 10.
Subcase 2b. Fixing the minor bit to 0 eliminates quadrants 01 and 11.

Note that under Gray code, the quadrants are (in order) 00, 01, 11, 10. Fixing the major bit (Subcases 1a and 1b) eliminated either the left half of the right half of the space, so that the reduced search space is a continuous interval of the search space at the previous iteration. Fixing the minor bit to 1 (Subcases 2a) eliminates the first and last quadrant, so that again the reduced search space is a continuous interval of the search space at the previous iteration.

Only Subcase 2b "fragments" the search space into two parts, except quadrants 00 (the first) and 10 (the last) are adjacent in the wrapped Gray encoding. Note that in Case 2 above, when a minor bit is set, we automatically eliminate quadrant 11 at the next iteration. Therefore, after Subcase 2b occurs only two possible alternatives exist: Subcase 1b or Subcase 2b must occur.

If Subcase 1b holds, the last half of the search space is eliminated, and if the space was previously fragmented into two parts, one of those parts is discarded, making the search space a continuous interval again.

If Subcase 2b holds again, the space is not remapped (the best known is already in quadrant 00). Therefore, eliminating quadrants 01 and 11 again means that the search space remains in two parts but still adjacent in the wrapped sense.

This means that the search space is never fragmented into more than two parts, and those two parts must always be adjacent in the wrapped sense. Cases 1 and 2 both insure that a local optimum is contained within the interval still to be searched.   □

Obviously, many functions are not bijections. If ties in evaluation are not too common, the algorithm should still display linear time convergence. In general, convergence times that are less than exponential cannot be guaranteed if the space is largely flat.

As bits are fixed, the high order bits used in the computation refer to the two highest order bits *which remain unfixed*. The major bit is always the leftmost bit that has not been fixed. The minor bit is always the second bit from the left that has not been fixed.

2.2 Observations on the Search Behavior of Quad Search

The original Quad Search algorithm provides a form of local search that hybridizes extremely well with genetic algorithms [WGW03]. However, the original version of Quad Search has different behavior in Case 2 which can be summarized as follows.

Previous Case 2. When $f(100Z) > f(000Z) < f(001Z)$, the minor bit is fixed to 0. The points $000Z, 001Z, 100Z$, becomes the points $00Z, 01Z, 10Z$ in the reduced space. Evaluate $11Z$ and $001(Z - 1)$ and continue.

Note that string $11Z$ is evaluated in this version, even though the local optimum is known to be in the interval between $f(10Z) > f(00Z) < f(01Z)$. If $f(11Z) < f(00Z)$ then the search will move to string $11Z$, but there is no longer an interval known to include a local optimum.

This can cause the original Quad Search to converge to a point that is not locally optimal on a multimodal function. The following example shows one can be as many as $2^{L-2} - 2$ positions away from a local optimum.

Assume we have N points numbered from $0...2^L - 1$. The quadrants correspond to the following intervals.

$$0 ... 2^{L-2} - 1; \quad 2^{L-2} ... 2^{L-1} - 1; \quad 2^{L-1} ... 3(2^{L-2}) - 1; \quad 3(2^{L-2}) ... 2^{L-1} - 1$$

Assume we test reflected points located at $0, 2^L - 1, 2^{L-1}, 2^{L-1} - 1$ and find

$$f(2^L - 1) > f(0) < f(2^{L-1} - 1).$$

We next sample at location $2^{L-2} - 1$ and find $f(2^{L-1} - 1) > f(0) < f(2^{L-2} - 1)$. We eliminate half of the space from point 2^{L-2} to $3(2^{L-2}) - 1$.

The original Quad Search next sampled at reflection point $3(2^{L-2})$. Note this is not required, since a local optimum is isolated in the interval $2^{L-1} - 1 \ldots 2^{L-2} - 1$.

Assume $f(3(2^{L-2})) < f(0)$ and the search moves to this new best-so-far. It is possible for the function to be monotonically decreasing from point $3(2^{L-2})$ to $2^{L-1} + 1$; thus, there can man be as many as $2^{L-2} - 1$ improving moves leading to a local optimum located in an eliminated quadrant.

Of course, the original version of Quad Search was only designed to converge on unimodal functions; the modified Quad Search converges to a local optimum on 1-D multimodal bijective functions by only evaluating points in the interval known to contain a local optimum. And the modification to the algorithm is minor.

Another important observation involves running Quad Search on multidimensional bijective functions. Quad Search will *iteratively* converge to a local optimum in each dimension, but the point where Quad Search terminates after exploring each dimension once is only guaranteed to be locally optimal in the last dimension. In this regard, Quad Search is like a line search algorithm. For example, on a 2-dimensional function, dimension 2 is initially fixed to some arbitrary value, while Quad Search locates a local optimum in dimension 1; next the algorithm is run on dimension 2 while dimension 1 is fixed to the point which was found to be locally optimal. Of course, after a local optimum is found in dimension 2, the current best solutions may no longer be locally optimal in dimension 1. However, one can reapply Quad Search (as with line search) until a local optimum has been reached.

It is also important to note that if Quad Search is run on a separable bijective multimodal function, it will converge to a local optimum. And if the separable function is unimodal (for example, as is the sphere function [B96]), then Quad Search will find the global optimum in at most $2L - 1$ evaluations.

2.3   Simple Binary Search for Real Valued Encodings

A similar search can be done using real valued representation that also cuts the search space in half after 2 evaluations. We again assume a 1-D bijective function.

Evaluate points at 0, N/4, N/2, 3N/4, N-1 (rounding if necessary). Note this explicitly divides the space into 4 regions. After the minimum is identified, an optimum must lie in one of two quadrants. For example, let F be the evaluation function. If

$$F(0) > F(N/4) < F(N/2) < F(3N/4) < F(N - 1)$$

then the optimum must be between $F(0)$ and $F(N/2)$. This is both simple and obvious. The process eliminates at least half (+/- 1 with rounding) of the search space from further consideration, and is guaranteed to converge to an optimum in at most $2(\log_2(N) + 1)$ evaluations. This search is very similar to Quad Search, except the resulting search does not search from an arbitrary point.

2.4 Binary Encodings and Quadrants

One of the reasons that Quad Search works using a Gray encoding is because the quadrants themselves have bit prefixes that are also adjacent and wrap. Fixing the first bit to 0 eliminates quadrants 3 and 4. Fixing the first bit to 1 eliminates quadrants 1 and 2. Fixing the second bit to 0 eliminates quadrants 2 and 3. Fixing the second bit to 1 eliminates quadrants 1 and 4. Thus, fixing one of the first two bits will eliminate two adjacent quadrants (where 1 and 4 are adjacent in the wrapped Gray encoding).

This does not happen with Binary neighbors. Consider a unimodal function with an optimum located at a Hamming cliff. One cannot fix bits and limit the search to quadrants 2 and 3 because the points in quadrant 2 start with the prefix 01 under the Binary encoding and the points in quadrant 3 start with the prefix 10. The first bit is either 0 or 1, or the second bit is either 0 or 1. Nothing is resolved. Because of the Hamming cliff, local search under a Binary encoding is not guaranteed to converge to a local optimum. This is already clear from results reported by Whitley et al. [Whi99] with regard to No Free Lunch: a Binary encoding can induce false local optima, even in a 1-D unimodal bijective function.

Of course one can use special operators to force a Binary search. Given a unimodal 1-D bijective function, one can evaluate the two points that make up the dominant Hamming cliff (complementary bit strings located at adjacent points), and determine the gradient at the Hamming Cliff and therefore determine in which half of the space the global optimum in located. The more important point, however, is that local neighborhood search in the standard Binary encoding neighborhoods does not have the same nice convergence properties as a reflected Gray code neighborhood.

3 Quasi-basins, Encodings and Locality

We formally define a **quasi-basin** for a 1-D function, f, with respect to a threshold value, V, where V is a codomain value of f: a quasi-basin is a set of contiguous points in f that are below value V. This in section, we present new proofs that outline sufficient conditions to ensure that the majority of Hamming distance 1 neighbors under Gray and Binary encodings are either in the same basin of attraction, or in the same quasi-basin.

Consider some reference point R in the search space. Whitley, Rowe and Bush [WRB04] show that given a 1-D function, the number of neighbors that are $<= D$ points away from the reference point under Gray code is $\lfloor \log D \rfloor$.

Expressed another way, consider a quasi-basin of size D and a search space of size N where the quasi-basin spans $1/Q$ of the search space (i.e., $D = \frac{1}{Q}N$): under a Gray encoding at most $\lceil \log(Q) \rceil$ bits encode for points that are more than a distance of D points away from R. Note that an increase in precision also increases the size of the search space, so that the quantity N/Q becomes larger and thus $\log(N/Q)$ increases. However Q and $\log(Q)$ remain constant. Thus, at higher precision, the number of neighbors within a distance of N/Q points increases.

Whitley, Rowe and Bush [WRB04] also present the following result without a proof.

Given a quasi-basin that spans $1/Q$ of a 1-D function of size $N = 2^L$ and a reference point R inside the quasi-basin, the expected number of neighbors of R that fall inside the quasi-basin under a reflected Gray code is greater than $\lfloor \log(N/Q) \rfloor - 1$.

In the current paper we provide a proof, but also prove a more general result that allows us to make a similar statement about standard Binary encodings. We first present the proof for the reflected Gray code case, then show that similar bounds hold for the standard Binary encoding.

The significance of this result is that we can outline conditions that would allow a steepest ascent local search bit climber to spend the majority of its time sampling points that are contained in the same quasi-basin and therefore below threshold. This makes it possible to outline sufficient conditions such that steepest ascent local search is provably better than random search. We conjecture that these same conditions allow next-ascent bit climbing as well as Quad Search to out perform random search.

To prove these results, a number of supporting concepts and lemmas are needed.

4   Locality and Neighborhoods

We first define Gray and Binary encoding recursively in order to define other concepts. The Gray encoding is the Standard Binary Reflected Gray code. For strings of length 1, Gray and Binary encodings are the same: the strings 0 and 1 represent integers 0 and 1. We then recursively define either a Binary or Gray code as follows.

Let B_i denotes a Gray encoded string of length L representing integer i, where $0 \leq i \leq 2^L - 1$. Strings of length $L + 1$ are constructed by concatenation and have the form $0B_i$ or $1B_i$ and are defined as follows:

$$0B_0, 0B_1, 0B_2, ... 0B_{2^L-1}, 1B_{2^L-1}, ..., 1B_2, 1B_1, 1B_0$$

Let B_i denotes a Binary encoded string of length L representing integer i, where $0 \leq i \leq 2^L - 1$. Strings of length $L + 1$ are constructed by concatenation and have the form $0B_i$ or $1B_i$ and are defined as follows:

$$0B_0, 0B_1, 0B_2, ... 0B_{2^L-1}, 1B_0, 1B_1, 1B_2, ..., 1B_{2^L-1}$$

While these definitions are well known and obvious, we note that the Gray code folds the search space around a $reflection$ located between $0B_{2^L-1}, 1B_{2^L-1}$. In the Binary case, there is an analogous mid-point $transition$ between $0B_{2^L-1}, 1B_0$ in the recursive construction. We will refer to both of these as $transition$ $points$; these points are important when documenting the locality of Binary and Gray encodings.

The placement of any transition point automatically implies the location of other transition points under the recursive definitions of both Binary and Gray code. We can define an arbitrary key-$transition$ around which neighborhoods can be defined.

We will define **core neighborhoods** as 2^k adjacent points with k connections that are fully contained within the core neighborhood. Core neighborhoods need not have the same connectivity, but all members have the same $number$ of core neighbors. Binary and Gray codes have the same "core neighborhoods" for any set of 2^k points that are within the quasi-basin and adjacent to a transition point.

We will assume that a key-$transition$ $point$ could occur at any position within the quasi-basin. We will count over all possible placements of transition points with the quasi-basin, then characterize what neighborhood structures occur for each possible transition point placement.

For example, let | denote the placement of a *key-transition* point in a quasi-basin made up of 7 points. The following represents the number of core neighbors under both Gray and Binary and the key transition denoted by | shifted into each possible position.

```
0 1 1 2 2 2 2 |
  1 1 2 2 2 2 | 0
    1 2 2 2 2 | 1 1
      2 2 2 2 | 1 1 0
        0 1 1 | 2 2 2 2
          1 1 | 2 2 2 2 0
            0 | 2 2 2 2 1 1
              | 2 2 2 2 1 1 0
```

All of this is obvious given the recursive definitions of Binary and Gray encodings. However we are interested in how Binary and Gray encoding differ in the construction of the non-core neighborhood connections.

4.1 The Matrix $\mathbf{M}^x$

We define a lower triangle matrix $\mathbf{M}^x$ using a recursive definition such that $\mathbf{M}^1 = [1]$. For $x > 1$ the lower triangle matrix $\mathbf{M}^x$ can be decomposed into a 2^{x-1} by 2^{x-1} square matrix whose elements are all the integer x, plus 2 identical copies of lower triangle matrix $\mathbf{M}^{x-1}$. The square matrix occupies the first 2^{x-1} columns of the last 2^{x-1} rows of $\mathbf{M}^x$. The first $2^{x-1} - 1$ rows of $\mathbf{M}^x$ correspond to the recursively defined matrix $\mathbf{M}^{x-1}$. Finally, another copy of $\mathbf{M}^{x-1}$ is appended to the last $2^{x-1} - 1$ rows of the square matrix.

The elements of every matrix $\mathbf{M}^x$ can also be reorganized into a 2^{x-1} by $2^x - 1$ rectangular matrix where all of the rows are identical, such that there are 2^{x-1} copies of x, followed by 2^{x-2} copies of $x - 1$, ..., ending with 2^0 copies of 1. This directly follows from the fact that each of the 2 $\mathbf{M}^{x-1}$ recursive submatrices is 1/2 the size of the square matrix associated with $\mathbf{M}^x$. Thus the quantity

$$\frac{\sum_{i=1}^{x} i(2^{i-1})}{2^x - 1} = \frac{2^x(x - 1) + 1}{2^x - 1} = \frac{2^x x}{2^x - 1} - 1$$

is both the average over the last row of $\mathbf{M}^x$, as well as the average over all of $\mathbf{M}^x$.

The following represents the lower triangle matrix $\mathbf{M}^3$, as well as the corresponding 2^{x-1} by $2^x - 1$ (i.e., 4 by 7) rectangular matrix.

```
1
2 2
2 2 1
3 3 3 3            3 3 3 3 2 2 1
3 3 3 3 1          3 3 3 3 2 2 1
3 3 3 3 2 2        3 3 3 3 2 2 1
3 3 3 3 2 2 1      3 3 3 3 2 2 1
```

We also note that the lower triangle matrix $\mathbf{M}^x$ has elements that are the same as the core neighbor counts for a quasi-basin of $2^x - 1$ points under both Gray and Binary that occur to the right of all possible placements of a key-transition, *except* the counts are incremented by 1.

Lemma 1: *Let $F(x)$ compute the average value over the elements of matrix $\mathbf{M}^x$:*

$$x - 1 < F(x) = \frac{2^x x}{2^x - 1} - 1 \le x$$

Proof: From the recursive definition of $\mathbf{M}^x$ and simple induction

$$F(x) = \frac{(2^{x-1})^2 * x + 2(F(x-1)(2^{x-2})(2^{x-1} - 1)}{(2^{x-1})(2^x - 1)} = \frac{2^x x}{2^x - 1} - 1.$$

To show that the bounds hold note:

$$F(x) = \frac{2^x x}{2^x - 1} - 1 = \frac{2^x (x - 1) + 1}{2^x - 1} > \frac{2^x (x - 1)}{2^x} = x - 1$$

When $x > 1$ this implies $x < 2^x - 1$ and $F(x) = \frac{x2^x - (2^x - 1)}{2^x - 1} < \frac{x2^x - x}{2^x - 1} = x$.
When $x = 1$ this implies $x = 2^x - 1$ and $F(x) = x = 1$.

4.2   The Matrix $\mathcal{M}^x$

We next define a new lower triangle matrix $\mathcal{M}^x$ using a constructive definition where $\mathcal{M}^1 = [0]$. Informally, the lower triangle matrix $\mathcal{M}^x$ is the same as $\mathbf{M}^x$ except the square portion of $\mathcal{M}^x$ is assigned the value $x - 1$ instead of x. Formally, the lower triangle matrix $\mathcal{M}^x$ can be decomposed into a 2^{x-1} by 2^{x-1} square matrix whose elements are all the integer $x - 1$, plus 2 identical lower triangle matrices corresponding to $\mathbf{M}^{x-1}$. The square matrix occupies the first 2^{x-1} columns of the last 2^{x-1} rows of $\mathcal{M}^x$. The first $2^{x-1} - 1$ rows of $\mathcal{M}^x$ correspond to the lower triangle matrix $\mathbf{M}^{x-1}$; Finally, another copy of $\mathbf{M}^{x-1}$ is appended to the last $2^{x-1} - 1$ rows of the square matrix of $\mathcal{M}$. The following is an example of $\mathcal{M}^4$.

```
        1
        2 2
        2 2 1
        3 3 3 3
        3 3 3 3 1
        3 3 3 3 2 2
        3 3 3 3 2 2 1
        3 3 3 3 3 3 3 3
        3 3 3 3 3 3 3 3 1
        3 3 3 3 3 3 3 3 2 2
   ->   3 3 3 3 3 3 3 3 2 2 1
        3 3 3 3 3 3 3 3 3 3 3 3
        3 3 3 3 3 3 3 3 3 3 3 3 1
        3 3 3 3 3 3 3 3 3 3 3 3 2 2
        3 3 3 3 3 3 3 3 3 3 3 3 2 2 1
```

The arrow points to row $2^{x-1} + 2^{x-2} - 1$ which has a special importance. The first 2^{x-2} elements have value $x - 1$, and the remaining elements are identical to the last row of matrix $\mathbf{M}^{x-1}$. Recall that the last row of matrix $\mathbf{M}^{x-1}$ has the same average value as the entire $\mathbf{M}^{x-1}$ lower triangle matrix. Thus, we know the average value of the elements of this row.

The average value over all of $\mathcal{M}^x$ is greater than $F(x-1)$ but less than $x - 1$, since all of the square portion of $\mathcal{M}^x$ has value $x - 1$ while the submatrices of $\mathcal{M}^x$ correspond to $\mathbf{M}^{x-1}$ and therefore have average value $x - 2 < F(x-1) < x - 1$. We next show that the average value of the elements of every row of $\mathcal{M}^x$ numbered from 2^{x-1} to $2^x - 1$ is greater or equal to $F(x-1)$.

Lemma 2: Given a lower triangle matrix $\mathcal{M}^x$, $x \geq 2$ each row indexed from $2x - 1$ to $2^x - 1$ has an average value greater than $x - 2$.

Proof: The proof is constructive. By construction row $2^{x-1} + 2^{x-2}$ is composed entirely of element $x - 1$. The rows from $2^{x-1} + 1$ to $2^{x-1} + 2^{x-2} - 2$ of matrix $\mathcal{M}^x$ can be constructed from row $2^{x-1} + 2^{x-2} - 1$ by repeatedly deleting "blocks" entirely composed of elements with value y, where y does not occur in the row under construction, and y is less than or equal to $x - 2$.

The average value of row $2^{x-1} + 2^{x-2} - 1$ can be bounded as follows

$$x - 1 > \frac{(x-1)2^{x-2} + F(x-1)(2^{x-1} - 1)}{2^{x-1} + 2^{x-2} - 1} > x - 2$$

because we can also regroup the elements and characterize row $2^{x-1} + 2^{x-2} - 1$ as having the value $x - 1$ in the first 2^{x-2} positions with the last $2^{x-1} - 1$ elements corresponding to the last row of matrix $\mathbf{M}^{x-1}$.

In general, rows $2^{x-1} + 1$ to $2^{x-1} + 2^{x-2} - 1$ have $x - 1$ as the first 2^{x-1} elements. The remaining elements corresponding to some row of the lower triangle matrix $\mathbf{M}^{x-2}$. This follows from the definition of $\mathcal{M}$ and the recursive construction $\mathbf{M}$. Row $2^{x-1} + 2^{x-2} - 1$ contains the last row of matrix $\mathbf{M}^{x-2}$, and therefore contains all of the blocks needed to construct all the other rows, since the last row is a slice of all the square matrices used in the recursive construction of $\mathbf{M}$. The largest value of $\mathbf{M}^{x-2}$ is $x - 2$.

When we delete an element from row $2^{x-1} + 2^{x-2} - 1$ to create any row from $2^{x-1} + 1$ to $2^{x-1} + 2^{x-2} - 2$ the average value of the elements in the new row must be greater than

$$\frac{(x-1)2^{x-2} + F(x-1)(2^{x-1} - 1)}{2^{x-1} + 2^{x-2} - 1} = \frac{3(2^{x-2})(x-1) - 2^{x-1} + 1}{3(2^{x-2}) - 1} > (x-1) - \frac{2}{3}$$

This follows from the fact that deleting a below average element from a set of numbers increases the average value of the set.

Finally, the last 2^{x-2} rows $\mathcal{M}^x$ are identical to rows 2^{x-1} to $2^{x-1} + 2^{x-2} - 1$ except the elements are shifted right and $2^x - 2$ additional copies of of $x - 1$ are added. Therefore, the last 2^{x-2} rows of $\mathcal{M}^x$ also have an average value greater than $x - 2$. □

4.3 Gray Codes and Quasi-basins

Consider a contiguous set of K points that form a quasi-basin in a 1-D function. Assume these K points are intersected by a barrier. We consider all possible placements of the barrier relative to the set of points. We then compute a tight bound on the average number of neighbors that exist when the barrier is a reflection point in a Gray encoding.

The following illustration represents a quasi-basin over 7 points. Each row represents the number of neighbors for each point, such that those neighbors fall inside the quasi-basin under a Gray code, where the main reflection point of the Gray code is at the location marked by the symbol $|$.

```
2 3 2 3 3 3 2 |
  2 2 3 3 2 3 | 1
    1 3 2 3 3 | 2 2
      2 3 3 3 | 2 3 2
        2 3 2 | 3 3 3 2
          2 2 | 3 3 2 3 1
            1 | 3 2 3 3 2 2
              | 2 3 3 3 2 3 2
```

Theorem 2: Given a quasi-basin of size S in a 1-D function, the expected number of neighbors that fall in the quasi-basin under Gray code is greater than $\lfloor \log_2(S) \rfloor - 1$.

Proof: For $S = 2^{x-1}$ to $2^x - 1$ we note that $\lfloor \log_2(S) \rfloor - 1 = x - 2$.

We consider all possible placements of the Gray code reflection key-transition, and average over all possible neighborhood arrangements. This means that all probabilities in the expected value calculation are uniform. The neighborhood structure is symmetric around the key-transition; we can therefore consider only the lower triangle matrix which lies to the right of the key-transition.

When $S = 2^x - 1$ we can use the matrix $\mathcal{M}^x$ to bound the expected number of neighbors for every point in the quasi-basin on one side of the main barrier/reflection for all possible placements of the reflection points.

Let S vary from 2^{x-1} to $2^x - 1$. A group of 2^k points adjacent to a reflection must have at least k core neighbors under Gray code; the square submatrix of $\mathcal{M}^x$ counts only these core neighbors. The lower triangle submatrices corresponding to $\mathbf{M}^{x-1}$ counts core neighbors, plus 1 more in each position. Under Gray code, there is at least one additional neighbor if there is another point in the space in a position symmetric around a reflection. This is the case for all elements in the positions that correspond to the two $\mathbf{M}^{x-1}$ lower triangle matrices. There are at least 2^{x-1} points in the quasi-basin. The square submatrix of $\mathbf{M}^{x-1}$ has 2^{x-2} core neighbors, so there must be an additional 2^{x-2} neighbors that fall either to the left or right of a reflection to either side of the 2^{x-2} core neighbors; thus each point in the square submatrix has one additional neighbor. The lower triangle submatrices of $\mathbf{M}^{x-1}$ also have one additional neighbor across the reflection that occurs immediately to the left of these lower triangle submatrices.

When $2^{x-1} \leq S \leq 2^x - 1$ we compute a bound on the average over the first S rows of matrix $\mathcal{M}^x$. The submatrix of $\mathbf{M}^{x-1}$ makes up the first $2^{x-2} - 1$ rows of matrix $\mathcal{M}^x$; by lemma 1, the average value of the elements in $\mathbf{M}^{x-1}$ is greater than $x - 2$. We select the next $S - (2^{x-2} - 1)$ rows that are needed from matrix $\mathcal{M}^x$. From lemma 2, the average value over each of these rows is greater than $x - 2$.

Thus, for any point in a quasi-basin of size S, the expected number of neighbors that also fall in the quasi-basin is greater than $\lfloor \log_2(S) \rfloor - 1$. $\square$

Corollary: *Given a quasi-basin that spans $1/Q$ of the search space and a reference point R that falls in the quasi-basin, the majority of the neighbors of R under a reflected Gray code representation of a search space of size $N = 2^L$ will be subthreshold in expectation when $\lfloor \log(N/Q) \rfloor - 1 > \log(Q) + 1$.*

This result has strong implications for local search. It means that under the conditions just outlined, we can expect a majority of the neighbors that are sampled under local search using a Hamming distance-1 Gray encoded bit representation neighborhood to also be below threshold when searching from a subthreshold point. It also follows from these observations that the percentage of below threshold neighbors increases at higher precision.

5   Binary Codes and Quasi-basins

We have already noted that Binary encodings have the same core neighborhood as Gray. But what about additional, non-core neighbors?

We start with the special case where there are exactly $2^x - 1$ elements in the quasi-basin. The number of neighbors when $x = 3$ and $2^x - 1 = 7$ is illustrated by the following example.

```
1 2 3 2 3 3 3 |
  2 2 2 2 3 3 | 0
    1 2 2 2 3 | 1 1
      2 2 2 2 | 2 1 1
        1 1 2 | 2 2 2 2
          1 1 | 3 2 2 2 1
            0 | 3 3 2 2 2 2
              | 3 3 3 2 3 2 1
```

We will again work with the lower triangle form of the matrix. This lower triangle matrix can still be recursively decomposed but the elements in the topmost recursive matrix and the rightmost recursive matrix differ by 1 in all positions. Furthermore, the square matrix has the value $x - 1$ in all positions except in its own lower triangle: the value x appears in the lower triangle of the square matrix. The reflected neighbors are gone, which is why the topmost recursive matrix is 1 less in every position. Each element corresponding to a position in the rightmost recursive lower triangle matrix has a neighbor at distance 2^{x-1} in the lower minor triangle of the square matrix.

Lemma 4: For a quasi-basin of exactly $2^x - 1$ elements, the number of neighbors that are in the quasi-basin is exactly $x - 1$ under a Binary encoding.

Proof: The proof is by induction. This is true by inspection for $x = 1, 2$ and 3.

We will again use a recursive lower triangle representation that decomposed into a square matrix and two recursively defined lower triangle submatrices. These count the core neighbors, plus those additional neighbors that occur due to the Binary encoding.

Assume the lemma is true for case $x - 1$. Then the topmost recursive matrix represents the lower triangle matrix associated with $x - 1$ and has an average value per element of exactly $x - 2$ by the inductive hypothesis; the topmost recursive matrix has $2^{x-2}(2^{x-1} - 1)$ elements. The rightmost recursively defined lower triangle submatrix has an average value per element of exactly $x - 1$. This includes the core neighbors, plus one additional neighbor, since any point more than 2^{x-1} positions to the right of the key transition must have a (non-core) neighbor in the square submatrix.

The square submatrix has $2^{x-1}2^{x-1}$ elements and $2^{x-2}(2^{x-1} - 1)$ in its own lower minor triangle, which connect to an element in the rightmost recursively defined lower triangle submatrix. Adding these together yields.

$$2^{x-2}(2^{x-1} - 1)(x - 2) + 2^{x-2}(2^{x-1} - 1)(x - 1) + 2^{x-1}2^{x-1}(x - 1) + 2^{x-2}(2^{x-1} - 1)$$

$$= [(2^{x-1})^2 + 2^{x-1}(2^{x-1} - 1)](x - 1)$$

Therefore, since the number of elements in the recursively defined matrix is $(2^{x-1})^2 + 2^{x-1}(2^{x-1} - 1)$ the average value per element is exactly $x - 1$. $\square$

Theorem 3: Given a quasi-basin of size S in a 1-D function, the expected number of neighbors that fall in the quasi-basin under Binary code is greater than $\lfloor \log_2(S) \rfloor - 1$.

Proof: For $S = 2^{x-1}$ to $2^x - 1$ we note that $\lfloor log_2(S) \rfloor - 1 = x - 2$.

We again consider all possible positions in the quasi-basin for which the number of neighbors is being computed, and all possible placements of the key-transition barrier. Thus, all probabilities in the expected value calculation are uniform.

If $S = 2^{x-1} - 1$ (this is one element smaller than is allowed for S), the number of neighbors is precisely $x - 2$ on average (lemma 4).

We will let S vary from 2^{x-1} to $2^x - 1$. A group of 2^i points adjacent to a key-transition must have at least i core neighbors; the square submatrix of $\mathcal{M}^x$ counts only these core neighbors. For rows 2^{x-1} to $2^x - 1$ the lower triangle submatrix corresponding to $\mathcal{M}^{x-1}$ which is right of the square submatrix is such that each element must have one neighbor at a distance of 2^{x-1} to the right, since the square submatrix of $\mathcal{M}^x$ is of size 2^{x-1}.

When $2^{x-1} \leq S \leq 2^x - 1$ we select the $S - 2^{x-2} - 1$ rows that are needed from matrix $\mathcal{M}^x$; from lemmas 1 and 2, all of these rows have average value greater than $x - 2$. $\square$

Corollary: *Given a quasi-basin that spans $1/Q$ of the search space and a reference point R that falls in the quasi-basin, the majority of the neighbors of R under a Binary representation of a search space of size $N = 2^L$ will be subthreshold in expectation when $\lfloor \log(N/Q) \rfloor - 1 > \log(Q) + 1$.*

6   Discussion

A steepest ascent local search algorithm currently at a subthreshold point can only move to an equal or better point which must also be subthreshold. And as precision increases, the number of subthreshold neighbors also increases, since $\lfloor \log(N/Q) \rfloor - 1$ increases while Q remains constant. This assumes the quasi-basin is not divided by increasing the precision.

The above analysis would need to hold for each dimension of a multidimensional search space, but these results suggest there are very general conditions where a steepest ascent bit-climbing algorithm using either a Binary or a Gray code representation can display subthreshold-seeking behavior. This also assumes the search algorithm can absorb the start-up costs of locating a subthreshold starting point.

Whitley, Rowe and Bush [WRB04] proposed a subthreshold bit climber called LS-SubT that uses a Gray Code representation. LS-SubT first samples 1000 random points, and then climbs from the 100 best of these points. In this way, LS-SubT estimates a threshold value and attempts to spend a majority of its time in the best 10 percent of the search space.

Experimental results on a number of benchmarks indicate that LS-SubT sometimes produces statistically significant better solution quality compared to steepest ascent bit climbing local search. LS-SubT never produces statistically significant worse performance than steepest ascent bit climbing local search. The data suggests two additional observations about subthreshold-seeking behavior [C01] [WRB04]. First, the sampling used by LS-SubT results in a higher proportion of subthreshold points compared to steepest ascent bit climbing local search. Second, a larger proportion of subthreshold neighbors are sampled for searches using higher precision. At 10 bits of precision per parameter, LS-SubT sampled subthreshold points more than 55 percent of the time. At 10 bits of precision, LS-SubT also used statistically significanbt fewer evaluations than steepest ascent local search. At 20 bits of precision per parameter, and at least 80 percent of the points samples by LS-SubT were subthreshold.

The locality results presented in this paper can perhaps be generalized to cover both next ascent bit climbers such as Random Bit Climbing (RBC) as well as Quad Search. While Quad Search looks different, it progressively looks closer and closer to the current best solution. Consider that RBC looks at all L bit flips before testing any bit again. Under RBC, checking the highest order bit samples a point in the other "half" of Hamming space (note this point is not necessarily far away under Gray code). Assuming that RBC does not find an improving move, the bits could be re-sorted so that high-order bits are sampled first, in effect reducing the search space by half after each bit flip. Under Quad Search, two points are checked in the other half of the space before the search space is reduced by half. Again assuming that Quad Search does not find an improving move, it checks at most $2L$ points while reducing the search space by half after every 2 bit flips. From this perspective, RBC and Quad Search have similar locality: the percentage of points that are close the the current best solutions should be nearly identical in expectation (except the modified Quad Search will potentially skip some evaluations.) This suggests that the only major adjustment needed to generalize the locality results for steepest ascent is to account for the fact that the current solution is changing location in both next ascent and Quad Search. We conjecture this makes no difference in the expected number of neighbors that fall in the quasi-basin.

7   Conclusions

Experiments have shown that a hybridization of Quad Search with a steady state genetic algorithm finds much better solutions on difficult benchmark problems than hybrid genetic algorithms that are combined with steepest ascent local search or RBC (Random

Bit Climbing). The apparent reason for this is that Quad Search very quickly converges to good solutions.

The modified Quad Search iteratively optimizes in each dimension so that it is guaranteed to converge to a locally optimal solution *in that dimension* before it moves to the next dimension. As with line search, one pass of the Quad Search is not guaranteed to converge to a locally optimal solution. On the other hand, this may make Quad Search more stochastic in its exploration of the search space compared to steepest ascent and next ascent.

Quad Search exploits unique properties of Gray codes and there would appear to be no direct analogy to Quad Search using Binary representations that provides the same generality and flexibility.

This paper also looks at locality, showing that most neighbors samples under both Gray and Binary representations are very close to the current reference point around which the neighborhood is defined. If that reference point R is in a quasi-basin that spans $1/Q$ of the search space then the majority of the neighbors of R under a reflected Gray code representation or a Binary representation of a search space of size $N = 2^L$ will be subthreshold in expectation when $\lfloor \log(N/Q) \rfloor - 1 > \log(Q) + 1$. This outlines conditions that will allow a steepest ascent bit climber to spend a majority of its time below threshold, and thus out perform random search. We conjecture the same results also hold for Quad Search.

References

[B96]   T. Bäck. *Evolutionary Algorithms in Theory and Practice*. Oxford University Press, 1996.

[BER76]   James R. Bitner, Gideon Ehrlich, and Edward M. Reingold. Efficient Generation of the Binary Reflected Gray Code and Its Applications. *Communications of the ACM*, 19(9):517–521, 1976.

[C01]   S. Christensen and F. Oppacher (2001). What can we learn from No Free Lunch? In *GECCO-01*, pages 1219–1226. Morgan Kaufmann, 2001.

[CS88]   R. Caruana and J. Schaffer. Representation and Hidden Bias: Gray vs. Binary Coding for Genetic Algorithms. In *Proc. of the 5th Int'l. Conf. on Machine Learning*. Morgan Kaufmann, 1988.

[Dav91]   Lawrence Davis. *Handbook of Genetic Algorithms*. Van Nostrand Reinhold, New York, 1991.

[RW97]   S. Rana and D. Whitley. Representations, Search and Local Optima. In *Proceedings of the 14th National Conference on Artificial Intelligence AAAI-97*, pages 497–502. MIT Press, 1997.

[RWB04]   J. Rowe, D. Whitley, L. Barbulescu, and J.P. Watson. Properties of Gray and Binary Representations. *Evolutionary Computation*, 12:in press, 2004.

[WGW03]   D. Whitley, D. Garrett, and J. Watson. Genetic Quad Search. In *GECCO 2003*, pages 1469–1480. Springer, 2003.

[Whi99]   D. Whitley. A Free Lunch Proof for Gray versus Binary Encodings. In *GECCO-99*, pages 726–733. Morgan Kaufmann, 1999.

[WM95]   David H. Wolpert and William G. Macready. No free lunch theorems for search. Technical Report SFI-TR-95-02-010, Santa Fe Institute, July 1995.

[WRB04]   D. Whitley, J. Rowe, and K. Bush. Subthreshold Seeking Behavior and Robust Local Search. In *GECCO 2004*, pages v2:282–293. Springer, 2004.

A Comparison of Simulated Annealing with a Simple Evolutionary Algorithm

Thomas Jansen

FB 4, LS 2, Universität Dortmund, 44221 Dortmund, Germany
Thomas.Jansen@udo.edu

Abstract. Evolutionary algorithms belong to the class of general randomized search heuristics. Theoretical investigations often concentrate on simple instances like the well-known (1+1) EA. This EA is very similar to simulated annealing, another general randomized search heuristic. These two algorithms are systematically compared under the perspective of the expected optimization time when optimizing pseudo-boolean functions. It is investigated how well the algorithmic similarities can be exploited to transfer analytical results from one algorithm to the other. Limitations of such an approach are illustrated by the presentation of example functions where the performance of the two algorithms differs in an extreme way. Furthermore, an attempt is made to characterize classes of functions where such a transfer of results is more successful.

1 Introduction

General randomized search heuristics are a popular tool for optimization when the problem or objective function is not well-understood, no good problem-specific algorithm is known, and there are not sufficient resources to develop such an algorithm. A prominent member of this broad class of heuristics are evolutionary algorithms. Of course, also other kinds of such algorithms are known: simulated annealing, tabu search, and randomized hill-climbing are some examples. For evolutionary algorithms, there is not only an overwhelming amount of reports on successful applications. There is also a growing body of serious theoretical work.

The different randomized search heuristics have different origins. Yet they share a lot of common properties. For evolutionary algorithms like genetic algorithms, evolution strategies, and evolutionary programming, this is well-known [1]. But this holds for other search heuristics, too.

Here, we consider two specific randomized search heuristics that have very different origins and motivations and yet are very similar. They are both general, not problem-specific randomized search heuristics, both can be applied for problems that can be coded as maximization of a pseudo-boolean function, i. e. a function $f: \{0,1\}^n \to \mathbb{R}$. Adopting the perspective of the evolutionary algorithm community, both can be described as using a population of size just 1 and utilizing an offspring population that also consists of one individual, only. They apply different random operators, called mutation in the context of evolutionary

A.H. Wright et al. (Eds.): FOGA 2005, LNCS 3469, pp. 37–57, 2005.

algorithms, in order to generate such an offspring. Then, based on different selection mechanism, it is decided whether this offspring replaces its parent. For the analysis we ignore the problem of choosing some stopping criterion and consider the number of mutation-selection-rounds (called generations) until some optimal solution to the problem is found for the very first time. We call the expectation of this random variable the expected optimization time.

The evolutionary algorithm uses standard bit-flip-mutations where each bit is flipped independently of the other bits with some mutation probability p_m. We use $p_m = 1/n$ which is the most recommended static choice and do neither discuss the use of other fixed mutation probabilities that may be beneficial [9] nor the application of any dynamic or adaptive mechanism for changing the mutation probability during a run. The selection mechanism applied is known as plus-selection from evolution strategies: in the case of $(1+1)$ selection, the offspring y replaces its parents if and only if its fitness is not inferior to its parent's fitness, i.e., $f(y) \geq f(x)$. One may argue that $(1+1)$ ES would be an appropriate name for this evolutionary algorithm. However, due to its application in the binary search space whereas evolution strategies traditionally work more often in $\mathbb{R}^n$, the name $(1+1)$ EA is more common. This extremely simple EA may be the most simple EA possible and it has been subject to numerous studies [5, 6, 14, 16, 20]. Using an even simpler mutation like single bit-flips would reduce it to a random local search, using a less strict selection pushes it towards pure random search, both search methods being essentially different from evolutionary algorithms.

The other search heuristics is very well known as simulated annealing [12]. The "mutation operator" applied utilizes single bit-flips. Of course, in general simulated annealing works on a neighborhood that may be defined in almost arbitrary ways. However, unless more is known about the objective function, using single bit-flips is a reasonable and common choice. In order to escape from local optima a less strict selection mechanism is needed. Here, the offspring replaces its parent not only when its fitness is not inferior but also in other cases with some probability that depends on the difference in fitness between parent and offspring and a time dependent parameter called temperature. We give a precise definition of the selection mechanism in the following section.

It is noteworthy that both algorithms find a global optimum for any objective function with probability 1. This makes a crucial difference to simple local search methods. For simulated annealing, however, this global convergence requires that the temperature is set in an appropriate way. One may want to argue that the freedom of choice with respect to temperature makes us compare a single evolutionary algorithm with a whole class of simulated annealing algorithms. We believe that the comparison made is nonetheless fair since in practice both algorithms are sometimes applied in this way: the $(1+1)$ EA as algorithm that needs no tuning of parameters whereas simulated annealing requires appropriate control of the temperature. Obviously, this makes simulated annealing more powerful in some sense by the cost of increasing the effort of users needed.

Clearly, the $(1+1)$ EA bears a very strong resemblance to simulated annealing. Therefore, it is natural to speculate that it might be possible to transfer analytical results concerning the (1+1) EA to simulated annealing and vice versa. Clearly, this would be beneficial for the theoretical knowledge on both algorithms. We investigate this possibility and try to understand what conditions allow for a direct transfer of results. One direction of theoretical research is the rigorous analysis of the performance of simple evolutionary algorithms on well-structured or at least well-understood objective functions. We employ this kind of approach to tackle the question of transfer of analytical results. We are able to demonstrate by example that there are strong limitations to such an approach. It is an important subject to future research to carry over such analyses to more realistic optimization problems. For simulated annealing some work in this direction has been done [2, 11]. In particular, Wegener [19] has presented a comparison of simulated annealing with its fixed-temperature counterpart, called Metropolis algorithm, on the minimum spanning tree problem.

This is by no means the first attempt to compare evolutionary algorithms with simulated annealing. However, such comparisons tend to be either purely empiric [8, 18] or very general [7]. Here, we try to find a balance between the attempt to be as general as possible and the wish to see concrete and meaningful results that have consequences in practical settings. We do this by considering concrete functions and function classes but presenting theoretical, not merely empirical analyses.

In the next section we give precise definitions of both algorithms and explain our analytical framework. In Section 3 we investigate the behavior of both algorithms in some exemplifying situations. We demonstrate that the performance of the two algorithms can differ in an extreme way and that each algorithm has the potential to outperform the other by far. We use these results as motivation for the search for classes of functions where the performance is similar in Section 4. Finally, in Section 5, we conclude with a short summary and remarks on possible future research.

2   Algorithms and Analytical Framework

We want to compare the (1+1) EA and simulated annealing. Both algorithms operate on some objective function $f\colon S \to \mathbb{R}$. In the context of evolutionary algorithms, this function f is called a fitness function and the EA tries to find points in the search space S with maximal f-value. In the context of simulated annealing the function is described as an energy function and the algorithm tries to find points that minimize f. We concentrate on pseudo-boolean functions, i. e. functions $f\colon \{0,1\}^n \to \mathbb{R}$. For the sake of notational simplicity we describe both algorithms in such a way that they both maximize f. This is nothing more than a notational change to simulated annealing. Both algorithms operate in rounds which remain mostly unchanged. They both operate based on one current point in the search space $x \in \{0,1\}^n$, which is called population in the case of the (1+1) EA. In each round both algorithms generate one new point $y \in \{0,1\}^n$ which is a variation of x. Simulated annealing chooses one point uniformly at random from a given neighborhood of x. We define this neighborhood to consist of all points with Hamming distance exactly 1 to x – this is a very common choice. The $(1+1)$ EA copies x but replaces each bit with probability $1/n$ by its complement in this process, independently for each bit. Thus, on average x and y differ by exactly one bit, too – but the Hamming distance may be arbitrarily large, although the probability quickly decreases with increasing Hamming distance. After the new point y is created, both algorithms decide whether the new point y replaces the old point x. If $f(y) \geq f(x)$, both algorithms replace x by y. If $f(y) < f(x)$, only simulated annealing still replaces x by y with some positive probability. This probability for accepting a decreasing step depends on $f(y) - f(x)$ and a time-dependent parameter which is called temperature. This temperature is changed with time according to a fixed scheme which is called a cooling schedule. Lower temperatures imply a lower probability to accept worsenings such that for temperature 0 the probability decreases to 0. In our notation, we use a function $\alpha\colon \mathbb{N} \to [1; \infty[$ where $1/(\alpha(t) - 1)$ is the temperature in the usual sense. Thus increasing values of α imply a decreasing temperature. Note that, since simulated annealing strictly searches within a small neighborhood of the current search point, this accepting of worsenings is crucial for a proof of global convergence.

In practice, both algorithms are terminated when some stopping criterion is fulfilled. Here, we describe the algorithms as infinite random processes without stopping criterion. We are interested in the first point of time when a point with maximal function value is found.

(1+1) EA	**Simulated Annealing**
0. $t := 1$	0. $t := 1$
1. Choose $x_t \in \{0, 1\}^n$ u. a. r.	1. Choose $x_t \in \{0, 1\}^n$ u. a. r.
2. $y := x_t$; Independently for each bit y_i, with probability $1/n$, set $y_i := 1 - y_i$.	2. $y := x_t$; Choose $i \in \{1, \ldots, n\}$ uniformly at random and set $y_i := 1 - y_i$.
3. If $f(y) \geq f(x_t)$, set $x_{t+1} := y$, else $x_{t+1} := x_t$.	3. With prob. $\min\{1, \alpha(t)^{f(y) - f(x_t)}\}$, set $x_{t+1} := y$, else $x_{t+1} := x_t$.
4. $t := t + 1$	4. $t := t + 1$
5. Continue at line 2.	5. Continue at line 2.

Both algorithms find a global optimum of any pseudo-boolean function f with probability 1 – given an unlimited number of steps and if, for simulated annealing, α grows sufficiently slow. Thus the most interesting question to ask is how long this takes. For both algorithms, we define a random variable that we call T_{EA} for the $(1+1)$ EA and T_{SA} for simulated annealing. Both variables are defined by $\min\{t \geq 1 \mid x_t = \max\{f(x') \mid x' \in \{0, 1\}^n\}\}$. We call T_{EA} and T_{SA} the optimization time and are mostly interested in $\mathrm{E}\,(T_{\mathrm{EA}})$ and $\mathrm{E}\,(T_{\mathrm{SA}})$, the expected optimization times.

For a comparison of the $(1+1)$ EA and simulated annealing we consider both algorithms on the same function f and compare the expected optimization times $\mathrm{E}\,(T_{\mathrm{EA}})$ and $\mathrm{E}\,(T_{\mathrm{SA}})$. We use the well-known notations for the growth of functions and describe both expected optimization times as functions of the dimension of the search space n where n grows to infinity [3]. Thus, we obtain our results in the form of upper and lower bounds that are not exact but only asymptotic. In the case that asymptotic upper and lower bounds are matching we call this bound asymptotically tight.

Definition 1. *Let* $f, g \colon \mathbb{N} \to \mathbb{R}^+$ *be two functions. We say* $f = O(g)$ *iff* $\exists n_0 \in \mathbb{N}, c \in \mathbb{R}^+ \colon \forall n \geq n_0 \colon f(n) \leq c \cdot g(n)$. *We say* $f = \Omega(g)$ *iff* $g = O(f)$. *We say* $f = \Theta(g)$ *iff* $f = O(g)$ *and* $f = \Omega(g)$. *We say* $f = o(g)$ *iff* $\lim_{n \to \infty} f(n)/g(n) = 0$. *We say* $f = \omega(g)$ *iff* $g = o(f)$.

3   Limitations for Direct Transfers of Results

It is very difficult if not simply impossible to come up with a universal, fair, and meaningful comparison of two general randomized search heuristics. In order to come to any results a number of decisions have to be made. We compare the (1+1) EA with simulated annealing. While the (1+1) EA can reach any point in the search space in only one step simulated annealing has to make many steps and may have to accept a lot of decreasing steps in order to do so. This requires the choice of an appropriate cooling schedule. Sticking to the idea of simulated annealing we only consider cooling schedules where the temperature does not increase, i. e., non-decreasing functions $\alpha \colon \mathbb{N} \to [1; \infty[$. It is known that in some circumstances increasing the temperature can be beneficial [4]. However, we consider such "cooling" schedules to be degenerated and inappropriate for simulated annealing.

Still, the choice of a cooling schedule introduces a great degree of freedom that is not present when using the (1+1) EA. We accept this difference here without trying to find any balancing mechanisms for the (1+1) EA. For a concrete objective function $f \colon \{0,1\}^n \to \mathbb{R}$, we consider simulated annealing to be superior if we can find a cooling schedule such that simulated annealing clearly outperforms the (1+1) EA by means of expected optimization time. On the other hand, we consider the (1+1) EA to be superior if it outperforms simulated annealing for any choice of a cooling schedule. Being more restrictive here would not be appropriate: it is very easy to hinder simulated annealing to have acceptable performance by choosing a bad cooling schedule.

If the cooling schedule α is a constant function, i. e., $\alpha(t) = \alpha \in [1; \infty[$ for all t, the resulting algorithm is known as Metropolis algorithm [13]. Obviously, the analysis becomes much simpler in this case. Even though it is known that non-static choices of α can be crucial for the success of simulated annealing [4, 17], in many cases the Metropolis algorithm already shows good performance. Furthermore, results for static α often yield upper or lower bounds for the non-static case [4].

Often, functions of unitation (functions where the function value does only depend on the number of ones in the input) are considered as example functions since they are particularly simple to describe and understand. For such functions, it is useful to introduce some additional notations. Assume that we consider an objective function $f \colon \{0,1\}^n \to \mathbb{R}$ that is a function of unitation and that reaches its unique global optimum for $x = 1^n$. Let T be the random variable describing the optimization time. Let $T^{(i)}$ denote the random variable that describes the optimization time when the initial bit string contains exactly i ones. Applying the total probability theorem we see that

$$\mathrm{E}\,(T) = \sum_{i=0}^{n} \binom{n}{i} \cdot 2^{-n} \cdot \mathrm{E}\left(T^{(i)}\right)$$

holds. Let $T^{(i)+}$ denote the random variable describing the number of steps until the number of ones in the current string x is increased for the first time, given that the current string contains exactly i ones. Using these notions we cite the following results from [4]. Note, that the analysis of simulated annealing as presented here depends heavily on the particular choice of the neighborhood and the specific properties of f.

Theorem 1. *Consider simulated annealing with static α on an objective function $f\colon \{0,1\}^n \to \mathrm{I\!R}$ that is a function of unitation and has its unique global optimum at 1^n. Let p_i^+ denote the probability to increase the number of ones in the current string from i to $i+1$. Let p_i^- denote the probability to decrease the number of ones in the current string from i to $i-1$. Using the notation from above, the following holds for all $i \in \{1, \ldots, n-1\}$.*

$$E\left(T_{SA}^{(i)+}\right) = \frac{1}{p_i^+} + \frac{p_i^-}{p_i^+} \cdot E\left(T_{SA}^{(i-1)+}\right)$$

$$E\left(T_{SA}^{(i)+}\right) = \sum_{k=0}^{i} \left(\frac{1}{p_k^+} \cdot \prod_{l=k+1}^{i} \frac{p_l^-}{p_l^+}\right)$$

Probably the best known example function in the context of evolutionary algorithms is ONEMAX, given by $\mathrm{ONEMAX}(x) = \sum_{i=1}^{n} x_i$. It is known that $\mathrm{E}\,(T_{\mathrm{EA}}) = \Theta(n \log n)$ holds for ONEMAX [5]. It will turn out to be helpful to have results on $\mathrm{E}\,(T_{\mathrm{SA}})$ for ONEMAX and some static choices of α.

Theorem 2. *Consider simulated annealing with constant α on ONEMAX. For any constant $\varepsilon > 0$ and any $\alpha \geq ((1+\varepsilon)/2) \cdot (n-2)$, $E(T_{SA}) = O(n \log n)$ holds.*

Proof. For ONEMAX, $p_i^+ = (n-i)/n$ and $p_i^- = i/(\alpha n)$ hold. This yields

$$\mathrm{E}\left(T_{\mathrm{SA}}^{(i)+}\right) = \frac{n}{n-i} + \frac{i}{\alpha(n-i)} \cdot \mathrm{E}\left(T_{\mathrm{SA}}^{(i-1)+}\right)$$

according to Theorem 1. If we had $\alpha \geq \varepsilon \cdot (i/n) \cdot \mathrm{E}\left(T_{\mathrm{SA}}^{(i-1)+}\right)$ for all i, we had

$$\mathrm{E}\left(T_{\mathrm{SA}}^{(i)+}\right) \leq \frac{n}{n-i} + \frac{1}{\varepsilon} \cdot \frac{n}{n-i} = \left(1 + \frac{1}{\varepsilon}\right) \frac{n}{n-i}$$

for all i. If we substitute this for $\mathrm{E}\left(T_{\mathrm{SA}}^{(i-1)+}\right)$ in our condition on α, we see that $\alpha \geq (1+\varepsilon) \cdot i/(n-i)$ suffices. Since $i/(n-i)$ with $i \in \{1, \ldots, n-2\}$ is maximal for $i = n-2$, $\alpha \geq (1+\varepsilon) \cdot (n-2)/2$ is sufficient. We have

$$\mathrm{E}\left(T_{\mathrm{SA}}\right) \leq \sum_{i=0}^{n-1} \mathrm{E}\left(T_{\mathrm{SA}}^{(i)+}\right)$$

and use $\mathrm{E}\left(T_{\mathrm{SA}}^{(i)+}\right) \leq (1+1/\varepsilon) \cdot n/(n-i)$. This yields

$$\mathrm{E}\left(T_{\mathrm{SA}}\right) \leq \sum_{i=0}^{n-1} \left(1+\frac{1}{\varepsilon}\right) \frac{n}{n-i} < \left(1+\frac{1}{\varepsilon}\right) n(1+\ln n)$$

and we have $\mathrm{E}\left(T_{\mathrm{SA}}\right) = O(n \log n)$ since $\varepsilon > 0$ is a constant. $\qquad\square$

Theorem 3. *Consider simulated annealing with constant α on* ONEMAX*. For any $\alpha \leq (\sqrt{n}-1)/2$, $E(T_{SA}) = \Omega(2^{\sqrt{n}} \cdot \sqrt{n})$ holds. For any α, $E(T_{SA}) = \Omega(n \log n)$ holds.*

Proof. On ONEMAX, we have

$$\mathrm{E}\left(T_{\mathrm{SA}}^{(i)+}\right) = \frac{n}{n-i} + \frac{i}{\alpha(n-i)} \cdot \mathrm{E}\left(T_{\mathrm{SA}}^{(i-1)+}\right) \tag{1}$$

which yields

$$\mathrm{E}\left(T_{\mathrm{SA}}^{(i)+}\right) \geq \frac{n}{n-i} + \frac{2i}{(\sqrt{n}-1) \cdot (n-i)} \cdot \mathrm{E}\left(T_{\mathrm{SA}}^{(i-1)+}\right)$$

due to the upper bound on α. For $i \geq n-\sqrt{n}$ we have

$$\mathrm{E}\left(T_{\mathrm{SA}}^{(i)+}\right) \geq \frac{n}{n-i} + \frac{2(n-\sqrt{n})}{(\sqrt{n}-1) \cdot \sqrt{n}} \cdot \mathrm{E}\left(T_{\mathrm{SA}}^{(i-1)+}\right) = \frac{n}{n-i} + 2\mathrm{E}\left(T_{\mathrm{SA}}^{(i-1)+}\right)$$

Using $\mathrm{E}\left(T_{\mathrm{SA}}^{(n-\sqrt{n})+}\right) \geq \sqrt{n}$ we get $\mathrm{E}\left(T_{\mathrm{SA}}^{(n-1)+}\right) \geq 2^{\sqrt{n}-2} \cdot \sqrt{n}$. With probability $1 - 2^{-n}$ the initial population differs from 1^n. Since simulated annealing can increase the number of ones in its population by at most 1 in each step, we know that 1^n can only be reached via some population with exactly $n-1$ ones. This yields $\mathrm{E}\left(T_{\mathrm{SA}}\right) \geq (1-2^{-n}) \cdot \mathrm{E}\left(T_{\mathrm{SA}}^{(n-1)+}\right)$. From this the desired result follows.

From Equation (1) we see that $\mathrm{E}\left(T_{\mathrm{SA}}^{(i)+}\right)$ decreases with increasing values of α. For $\alpha \to \infty$ we get $\mathrm{E}\left(T_{\mathrm{SA}}^{(i)+}\right) = n/(n-i)$. Since we have $\mathrm{E}\left(T_{\mathrm{SA}}\right) \geq (1/2) \cdot \sum_{i=\lceil n/2 \rceil}^{n} \mathrm{E}\left(T_{\mathrm{SA}}^{(i)+}\right)$, we get $\mathrm{E}\left(T_{\mathrm{SA}}\right) = \Omega(n \log n)$ as an immediate consequence.

$\qquad\square$

The results on ONEMAX seem to contradict our claim that a direct transfer of results is not possible. The (1+1) EA and simulated annealing both optimize ONEMAX on average in $O(n \log n)$ steps. But even this simple functions has the potential to demonstrate the differences between the two optimization heuristics. We define our first example function $f_1 \colon \{0, 1\}^n \to \mathbb{R}$ by

$$f_1(x) := \begin{cases} \text{ONEMAX}(x) & \text{if } \text{ONEMAX}(x) \neq n - 1 \\ 0 & \text{otherwise} \end{cases}$$

for all $x \in \{0, 1\}^n$. The only difference to ONEMAX, a sharp decrease in function values for all strings with exactly $n - 1$ ones, is no difficult obstacle to the (1+1) EA but it is for simulated annealing.

Theorem 4. On f_1, $E(T_{EA}) = \Theta(n^2)$ holds.

Proof. For the lower bound, it suffices to observe that the last step that leads to the unique global optimum is a mutation of at least 2 specific bits, given that the initial string contains at least two zero bits. Thus, we have $(1 - (n+1)2^{-n}) \cdot n^2$ as a lower bound on the expected optimization time. For the upper bound we can employ the method of fitness based partitions [5]. The probability to increase the current function value by at least 1 is given by

$$\binom{n - \text{ONEMAX}(x)}{1} \frac{1}{n} \left(1 - \frac{1}{n}\right)^{n-1} \geq \frac{n - \text{ONEMAX}(x)}{en},$$

$$\text{if } \text{ONEMAX}(x) < n - 2,$$

$$\frac{1}{n^2} \left(1 - \frac{1}{n}\right)^{n-2} \geq \frac{1}{en^2}, \quad \text{if } \text{ONEMAX}(x) = n - 2, \text{ and}$$

$$\frac{1}{n} \left(1 - \frac{1}{n}\right)^{n-1} \geq \frac{1}{en}, \quad \text{if } \text{ONEMAX}(x) = n - 1,$$

for all $x \in \{0, 1\}^n$ different from the unique global optimum. By adding up the corresponding expected waiting times we get

$$\left(\sum_{i=0}^{n-3} \frac{en}{n - i}\right) + en^2 + en = O\left(n^2\right)$$

as upper bound on the expected optimization time.   □

Theorem 5. On f_1, $E(T_{SA}) = \Omega\left(2^{\sqrt{n}} \cdot \sqrt{n}\right)$ holds for any non-increasing cooling schedule, i. e., any non-decreasing function $\alpha \colon \mathbb{N} \to [1; \infty[$.

Proof. We know from Theorem 3, that as long as $\alpha(t) \leq (\sqrt{n} - 1)/2$ holds, the expected time needed to reach a current string with $n - 2$ ones is bounded below by $\Omega(2^{\sqrt{n}}\sqrt{n})$. In order to be faster, we need $\alpha(t) > (\sqrt{n} - 1)/2$. Then, the expected time until a step from a current string with $n - 2$ ones to a string with $n - 1$ ones is accepted is bounded below by $\alpha(t)^{n-2}$. Together, this yields the claimed lower bound.   □

The difficulties of simulated annealing are easy to understand: In order to reach the unique global optimum a temporary decrease in function value of $n - 2$ has to be accepted. In order to do this, $\alpha(t)$ needs to be small; $\alpha(t) = 1 + O(\log(n)/n)$ is needed for a polynomial expected waiting time. But any $\alpha(t) \leq (\sqrt{n} - 1)/2$ is too small to even come so far in expected polynomial time. We recognize a clear advantage for the (1+1) EA due to is insensitivity towards changes in function values and its ability to "jump" across small regions of low function values.

Our next example function f_2 will demonstrate a similarly drastic advantage for simulated annealing over the (1+1) EA. The idea is simple: The (1+1) EA can "jump" across regions of low function values in polynomial time only if the jump needed is not too large. Simulated annealing can walk through such valleys given that the decrease in function value is not too large – almost independently of the size of the valley.

We define this example function $f_2 \colon \{0,1\}^n \to \mathbb{R}$ by

$$f_2(x) := \begin{cases} 2^n \cdot n + 1 & \text{if } x = 0^n \\ 2^n \cdot n - (n - i) & \text{if } x = 1^i 0^{n-i}, \ i \in \{1, \dots, n\} \\ 2^n \cdot \text{ONEMAX}(x) & \text{otherwise} \end{cases}$$

for all $x \in \{0,1\}^n$.

Theorem 6. *On f_2, $E(T_{EA}) = \Theta(n^n)$ holds.*

Proof. The upper bound is trivial, since for any string y the probability to reach y from the current string x is bounded below by n^{-n}. This yields n^n as upper bound on $E(T_{EA})$ for any objective function. For the proof of the lower bound, we note that the (1+1) EA reaches 1^n on average in $O(n \log n)$ steps given that it never encounters any string $1^i 0^{n-i}$. Then, a mutation of all n bits simultaneously is needed in order to reach the unique global optimum 0^n. This yields $E(T_{EA}) \geq n^n \cdot p$, if p is a lower bound on the probability of reaching 1^n as current string at some point of time. It is easy to see that $p = \Omega(1)$ holds: Consider levels of strings $L_i = \{y \in \{0,1\}^n \mid \text{ONEMAX}(y) = i\}$. With probability exponentially close to 1, the (1+1) EA does not reach a level L_i with $i < n/3$ before reaching 1^n. For all i with $n/3 \leq i \leq n - 3$, the level L_i contains $\Omega(n^3)$ strings. On average, the (1+1) EA spends $O(n)$ generations on each level L_i. For symmetry reasons, we get that the (1+1) EA reaches a string $1^j 0^{n-j}$ on such a level with probability $O(1/n^2)$, only. For L_{n-2}, we get $O(1/n)$ as bound for this probability. For L_{n-1} we take a closer look. The probability for a mutation that changes the current population but stays within the level is $\Theta(1/n)$. Thus, on average only $O(1)$ strings are visited within this level before L_n is entered. Thus, we get $p = 1 - O(1/n)$ altogether. □

Theorem 7. *On f_2, $E(T_{SA}) = O(n^3)$ holds.*

Proof. We choose $\alpha(t) := n^{4/2^n}$ fixed and independent of t. We consider a run of simulated annealing on f_2 and partition it into two phases. The first phase starts with random initialization and ends when some string $x = 1^i 0^{n-i}$ with $i \in \{1, 2, \ldots, n\}$ becomes current string. The second phase starts after the first phase and ends when the unique global optimum is reached. Obviously, we get an upper bound on $\mathrm{E}(T_{\mathrm{SA}})$ if we add up upper bounds for the expected lengths of the two phases.

On average, the first phase ends within $O(n \log n)$ steps. This follows from Theorem 2: If we change the differences in function values of neighboring points in the search space from d to $s \cdot d$ and change α to $\alpha^{1/s}$ at the same time, nothing changes for simulated annealing. Here, we increase the differences in function values from 1 to 2^n and have $\alpha = n^{4/2^n}$. Thus, this is equivalent to having $\alpha = n^4$ in the scenario of Theorem 2.

Now we consider the second phase. We start with a string $x = 1^i 0^{n-i}$ for some $i \in \{1, \ldots, n\}$. The probability to accept some string that is neither of this form nor the global optimum is in each step bounded above by $\alpha^{-2^n} = 1/n^4$. If this happens, we consider the second phase to be a failure and reconsider the run, again with two phases. If we succeed to prove $O(n^3)$ as upper bound on the average length of the second phase, then the probability of such a failure is bounded above by $O(1/n)$ and the expected number of such repetitions does not harm our upper bound. Thus, we now have to deal with the situation that we only have current strings of the form $1^i 0^{n-i}$.

We denote the probability to decrease the number of leading ones in x from i to $i-1$ by p_i^-. Accordingly, p_i^+ denotes the probability to increase this number from i to $i+1$. For i with $0 < i < n$, p_i^- and p_i^+ are independent of i. Obviously, $p_i^+ = 1/n$ and $p_i^- = 1/(\alpha n) = 1/n^{1+4/2^n}$ holds in this case. We consider cn^3 steps where c is some constant sufficiently large. Application of Chernoff bounds yields that with probability very close to 1 we have $c'n^2$ steps where the value of i changes within these cn^3 steps (or the global optimum is reached). By choosing c large we can get c' as large as we want. We consider these $T := c'n^2$ steps. Let the conditional probability to increase or decrease i in such a step be denoted by q_i^+ or q_i^- respectively. By definition, $q_i^+ = p_i^+/(p_i^+ + p_i^-)$ and $q_i^- = p_i^-/(p_i^+ + p_i^-)$ holds. If we have at least $(T + n)/2$ decreasing events within these steps, the global optimum must be reached. In order to estimate the probability for this we note that $q_i^- = \frac{1}{1+n^{4/2^n}}$ holds. Using $e^x \leq 1 + 2x$ for not too large positive values of x we get

$$q_i^- = \frac{1}{1 + n^{4/2^n}} \geq \frac{1}{2 + (8 \ln n)/2^n} = \frac{1}{2} - \frac{(8 \ln n)/2^n}{4 + (18 \ln n)/2^n} = \frac{1}{2} - \frac{8 \ln n}{2^{n+2} + 18 \ln n}$$

and see that $q_i^- \approx 1/2$ holds. For $q_i^- = 1/2$, the desired result would be clear. Instead of T random experiments where we have success probability q_i^- we consider T pairs of experiments, where each pair consists of one random experiment with success probability $1/2$ and another, independent random experiment with success probability $(16 \ln n)/(2^{n+2} + 18 \ln n)$. We replace one success in the original experiment with a pair of (success/no success) in the corresponding pair. The probability for such an outcome in the pair equals

$$\frac{1}{2} \cdot \left(1 - \frac{16 \ln n}{2^{n+2} + 18 \ln n}\right) = \frac{1}{2} - \frac{8 \ln n}{2^{n+2} + 18 \ln n} = \bar{q_i},$$

so the probability for at least $(T+n)/2$ successes is unchanged. Since we consider only $T = O(n^2)$ random experiments, the probability to have at least one pair with a success in the second part is exponentially small. Thus, we can work under the assumption that this does not happen and have $1/2$ as remaining probability for a success. Let S denote the random number of total successes in T experiments. We have $\text{Prob}(S = \lfloor T/2 \rfloor) \geq \text{Prob}(S = j)$ for any j and know that $\text{Prob}(S = \lfloor T/2 \rfloor) \leq 2/\sqrt{T}$ holds. Thus, $\text{Prob}(S \geq (T + n)/2) \geq 1/2 - (n/2) \cdot 2/\sqrt{T} = 1/2 - 1/\sqrt{c'}$ holds. Choosing c such that $c' > 4$ holds yields the following result: With probability $\Omega(1)$ within $O(n^3)$ steps the global optimum is reached. Therefore, the expected number of such "runs" of length $O(n^3)$ needed to find the global optimum is $O(1)$ and we get $O(n^3)$ in total as upper bound on the expected optimization time of simulated annealing on f_2. □

Sometimes the expected optimization time can be a very misleading measure. There are example functions known where the expected optimization time is exponential but with probability very close to 1 a global optimum is found within a polynomial number of steps. In such cases independent restarts can decrease the expected optimization time drastically. We strengthen the differences in performance between the (1+1) EA and simulated annealing we have seen so far by discussing why restarts are not helpful here.

In the case of f_1, the exponential lower bound for simulated annealing is based on the fact that at some point of time a string with less than $n - 1$ ones is current string which holds with a probability exponentially close to 1. Then, a worsening of $n - 2$ has to be accepted causing an exponential waiting time with probability exponentially close to 1. Thus, restarts cannot change any of this.

In the case of f_2, it seems to be the case that the expected optimization time of the (1+1) EA decreases if we increase the chance of finding some string $1^i 0^{n-i}$ with $i < n$ as first point of this kind. This can in fact be achieved by restarts as the probability to reach some point of this kind with $n - i = O(1)$ is not very small. But it is easy to see that this does not really help: Increasing the number of leading ones (and thereby increasing the Hamming distance to the global optimum) increases the function value. Such a step has probability $\Omega(1/n)$. Thus, with probability exponentially close to 1 the (1+1) EA reaches 1^n as current string before reaching 0^n – even if it reaches some arbitrary $x = 1^i 0^{n-i}$ with $i = \Omega(n)$ before.

We have learned that there can be enormous differences in performance between the (1+1) EA and simulated annealing – even for very simply structured functions. Such performance differences can occur with any of the two algorithms being the superior one. This shows that a simple and direct transfer of results is not possible. Both example functions are based on extreme changes in function values of neighboring strings. We will use this observation in the next section as motivation for a class of functions where we hope that a transfer of results may be possible.

4   Looking for Classes of Functions that Allow for a Transfer of Results

When looking for classes of functions where the (1+1) EA and simulated annealing have similar performance, we have to rule out functions like the two example functions from Section 3. Both functions, f_1 and f_2, have extreme differences in function values for neighboring points in the search space. We can exclude such functions by considering classes of functions where the fitness landscapes are smooth in some way. But we have seen that it is not the absolute difference in function values that matters: the interplay between this difference and the current value of $\alpha(t)$ determines the behavior of simulated annealing. Arbitrary scalings of the fitness differences can be compensated by an appropriate change of the cooling schedule. The following definition takes this into account.

Definition 2. A *pseudo-boolean function* $f: \{0,1\}^n \to \mathbb{R}$ is called smooth integer (s. i.), if $f(x) \in \mathbb{Z}$ for all $x \in \{0,1\}^n$ and if $|f(x) - f(y)| \leq 1$ for all $x, y \in \{0,1\}^n$ with $H(x,y) \leq 1$, where $H(x,y)$ denotes the Hamming distance between x and y.

It is clear that smooth integer functions cannot separate simulated annealing and the (1+1) EA in the way f_1 or f_2 do. What remains to be shown is that they do not only hinder such constructs but do indeed imply similarities in the way the (1+1) EA and simulated annealing optimize a function.

There are different known example functions that have the property to be smooth integer: ONEMAX, the needle-in-the-haystack functions, and Prügel-Bennett's F_n [15] are such functions with very different properties otherwise. Other known example functions can be changed so that they become s. i. without losing their main properties. The function JUMP$_k$ [10] is an example.

We have seen that the (1+1) EA and simulated annealing perform very similarly on ONEMAX. The same is known for needle-in-the-haystack functions [6]. Both functions have in common that there are no "valleys": there is no need for a mutation of several bits simultaneously for the (1+1) EA and there is no need for simulated annealing to accept any worsenings. We will consider a very simple class of functions, inspired by JUMP$_k$ [10], and show that this is not necessary for similar performance.

Definition 3. For $k \in \{1, 2, \ldots, \lfloor (n-1)/2 \rfloor\}$ the function $f_k: \{0,1\}^n \to \mathbb{R}$ is defined by

$$f_k(x) := \begin{cases} 2n - 4k - 2 - \text{ONEMAX}(x) & \text{if } n - 2k \leq \text{ONEMAX}(x) \leq n - k - 1 \\ \text{ONEMAX}(x) - 2k & \text{if } n - k \leq \text{ONEMAX}(x) \\ \text{ONEMAX}(x) & \text{otherwise} \end{cases}$$

for all $x \in \{0,1\}^n$.

A function f_k partitions the search space into three disjoint regions. In the first region, where $0 \leq \text{ONEMAX}(x) < n - 2k$ holds, the function value is given by $\text{ONEMAX}(x)$. Then follows the second region, where $n - 2k \leq \text{ONEMAX}(x) < n - k$ holds; there the function values decrease with the number of ones in x. Finally, in the last region, where $\text{ONEMAX}(x) \geq n - k$ holds, the function values increase again until they reach the global maximum $n - 2k$ for $x = 1^n$. A typical example is $f_4 \colon \{0, 1\}^{20} \to \mathbb{R}$ shown in Figure 1.

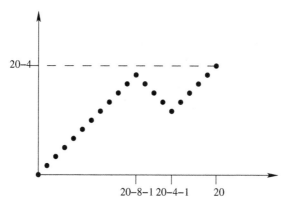

Fig. 1. The function $f_4 \colon \{0, 1\}^{20} \to \mathbb{R}$

It is interesting to note that f_k is not only smooth integer but also a function of unitation, just like ONEMAX is and like a needle-in-the-haystack function can be. We will only deal with such functions here. It is clear that different and interesting functions exist which are smooth integer but no functions of unitation. However, we consider these functions to be out of the scope of this paper.

Before investigating the behavior of the $(1+1)$ EA and simulated annealing on f_k, we want to make a general remark on functions of unitation which are smooth integer. As far as the performance of the $(1+1)$ EA and simulated annealing are concerned, we can partition all these functions into at most 3^n disjoint sets, such that both algorithms have identical behavior for all functions of one such set – and this holds for each set. The reason is that the $(1+1)$ EA is only sensitive towards the ordering of the function values and simulated annealing only to the difference between neighboring function values. Thus, we can choose $f(0^n) = z$ with any value $z \in \mathbb{Z}$ and define the function f by a vector $d = (d_1, d_2, \ldots, d_n) \in \{-1, 0, 1\}^n$ with the following interpretation. We have

$$f(x) = z + \sum_{i=1}^{\text{ONEMAX}(x)} d_i$$

for all $x \in \{0, 1\}^n$. Since the value of z does not matter, we can use $z = 0$ for such functions. Using this notation, ONEMAX is defined by $d_i = 1$ for all $i \in \{1, \ldots, n\}$, a needle-in-the-haystack function by $d_i = 0$ for all $i \in \{1, \ldots, n-1\}$ and $d_n = 1$. We get f_k by $d_i = 1$ for all $i \in \{1, \ldots, n - 2k - 1\}$, $d_i = -1$ for all $i \in \{n - 2k, \ldots, n - k - 1\}$ and $d_i = 1$ for all other i.

Our idea of the function f_k is that of Figure 1: a ONEMAX-like function with a rather small valley just in front of the unique global optimum. This idea is not realized by f_k if k becomes so large that the valley occupies about half of the search space or even more like it is the case for $k \geq n/4$. Even though it is not technically difficult to extend our formal analysis to f_k for such values of k we will not do so since it distracts us from our line of thought. Since the most interesting cases are those where the algorithm used has polynomial expected optimization time, we restrict ourselves to $k = \Theta(1)$ here.

Theorem 8. *For f_k with $k = \Theta(1)$, $E(T_{EA}) = \Theta(n^{2k+1})$ holds.*

Proof. Let p denote the probability that $\text{ONEMAX}(x) = n - 2k - 1$ holds for the current string x at some point of time. In order to reach the unique global optimum from there, a direct mutation to some y with $\text{ONEMAX}(y) \geq n - 1$ is needed. We distinguish two cases. It may be the case that the unique global optimum is reached via a direct mutation. Such a mutation has probability $(1/n)^{2k+1}(1 - 1/n)^{n-2k-1} = \Theta(1/n^{2k+1})$. Otherwise, a string y with $\text{ONEMAX}(y) = n - 1$ is reached. Such a mutation has probability

$$\binom{2k+1}{2k}\left(\frac{1}{n}\right)^{2k}\left(1 - \frac{1}{n}\right)^{n-2k} + \binom{n-2k-1}{1}\left(\frac{1}{n}\right)^{2k+2}\left(1 - \frac{1}{n}\right)^{n-2k-2}$$

$$= \Theta\left(\frac{1}{n^{2k}}\right).$$

Given such a string y as current string x, the number of ones in the current string may either increase, decrease, or remain unchanged. The probability to increase it to n equals $(1/n)(1 - 1/n)^{n-1} = \Theta(1/n)$. The probability to decrease it is bounded below by

$$\binom{n-1}{2k}\left(\frac{1}{n}\right)^{2k}\left(1 - \frac{1}{n}\right)^{n-2k} = \Theta(1).$$

We see that the probability to increase the number of ones from $n - 2k - 1$ to $n-1$ and increase it to n before decreasing it to $n-2k-1$ or even smaller values is $\Theta(1/n^{2k+1})$. Thus, $E(T_{EA}) = \Omega(p \cdot n^{2k+1})$ follows.

It is well known that the (1+1) EA optimizes OneMax on average in time $O(n \log n)$ [14]. Thus, we can conclude that on average it finds a current string x with $\text{ONEMAX}(x) = n - 2k - 1$ within $O(n \log n)$ steps. This implies $E(T_{EA}) = O(n^{2k+1})$.

Now we need a lower bound on p. With probability exponentially close to 1, we have $\text{ONEMAX}(x_0) < (3/5)n$ for the initial current $x = x_0$. Thus, we are "left of the valley" initially. The probability to have a mutation leading one step closer to "the top of the hill", i.e., a local maximum immediately to the left of the valley, is always clearly larger than the probability for a mutation leading over the top of that hill into the valley. Mutations leading to points left of a local minimum in the bottom of the valley will, with high probability, lead back to the local maximum just in front of the valley. In particular, as long as we are in the valley (not in a local maximum), the probability to reach the local maximum in front of the valley before moving closer to the global maximum (including reaching it) is always $1 - O(1/n)$. This is due to the facts that $k = \Theta(1)$, the probability to move to the left is always $\Theta(1)$, and the probability to move to the right is always $O(1/n)$. We conclude that $p = 1 - O(1/n)$ and thus $\mathrm{E}\,(T_{\mathrm{EA}}) = \Theta(n^{2k+1})$ holds. $\qquad\square$

The performance of the $(1+1)$ EA is governed by the time needed to jump over the valley just before reaching the unique global optimum. Simulated annealing will have to walk through this valley accepting k decreases in function value consecutively. It is not surprising that this dominates the expected optimization time of simulated annealing.

Theorem 9. *For f_k with $k = \Theta(1)$, $E(T_{SA}) = O(n^{2k+1})$ holds. For $\alpha \geq (1 + \varepsilon)n/2$, where $\varepsilon > 0$ is a constant and $\alpha = O(n)$ holds, $E(T_{SA}) = \Theta(n^{2k+1})$ holds for f_k with $k = \Theta(1)$.*

Proof. The proof works for any cooling schedule which guarantees (1) $\alpha(t) \geq (1 + \varepsilon)n/2$ for all $t \geq t_0$, where $\varepsilon > 0$ is a constant and $t_0 = O(n^{2k+1})$, and (2) $\alpha(t) = O(n)$ for all t. We choose $\alpha := n$ fixed for the sake of simplicity of notation.

We begin with considering f_1 and use

$$\mathrm{E}\,(T_i^+) = \frac{1}{p_i^+} + \frac{p_i^-}{p_i^+} \cdot \mathrm{E}\,(T_{i-1}^+)$$

from Theorem 1 in order to derive an upper bound on $\mathrm{E}\,(T_{\mathrm{SA}}) \leq \sum_{i=0}^{n-1} \mathrm{E}\,(T_i^+)$.

For $i \leq n - 4$, nothing is different from ONEMAX, so that $\mathrm{E}\,(T_{n-4}^+) = \Theta(n)$ holds. For $i = n-3$, we are in a situation where each mutation leads to a decrease in function value. We get

$$\mathrm{E}\,(T_{n-3}^+) = \frac{n^2}{3} + \frac{n^2}{3} \cdot \frac{n-3}{n^2} \mathrm{E}\,(T_{n-4}^+) = \Theta\,(n^2)$$

as a consequence. For $i = n - 2$, we are in a situation where each mutation leads to an increase in function value. We get

$$\mathrm{E}\,(T_{n-2}^+) = \frac{n}{2} + \frac{n}{2} \cdot \frac{n-2}{n} \mathrm{E}\,(T_{n-3}^+) = \Theta\,(n^3)$$

as a consequence. Finally, for $i = n - 1$, we are again in the same situation as for ONEMAX which implies

$$\mathrm{E}\,(T_{n-1}^+) = n + n \cdot \frac{n-1}{n^2} \mathrm{E}\,(T_{n-2}^+) = \Theta\,(n^3).$$

This implies $\mathrm{E}\,(T_{\mathrm{SA}}) = \Theta(n^3)$ on f_1, which is in accordance with the desired result. We repeat the proof for f_2 and look for differences and similarities. Since the size of the "valley" changes with k, the absolute number of ones in the strings cannot be compared directly.

On f_2, for $i \leq n - 6$, everything is the same as for ONEMAX. We call $i \in \{0, 1, \ldots, n - 6\}$ type I levels here. This corresponds to $i \in \{0, 1, \ldots, n - 4\}$ on f_1. We get $E\left(T_{n-6}^+\right) = \Theta(n)$ as a direct consequence.

For $i = n - 5$, we are in the situation where each mutation leads to a worse offspring. This corresponds to $i = n - 3$ on f_1. We call this level type II. We get $E\left(T_{n-5}^+\right) = \Theta(n^2)$.

For $i = n - 4$, we are in a situation which is inverse to ONEMAX: increasing the number of ones decreases the function value and vice versa. We call such levels type III levels. Note that there is no type III level for f_1. We get

$$E\left(T_{n-4}^+\right) = \frac{n^2}{4} + \frac{n^2}{4} \cdot \frac{n-4}{n} E\left(T_{n-5}^+\right) = \Theta\left(n^4\right)$$

here.

For $i = n - 3$ we are on f_2 in the same situation as for $i = n - 2$ on f_1: each mutation implies an increase in function value. We get

$$E\left(T_{n-3}^+\right) = \frac{n}{3} + \frac{n}{3} \cdot \frac{n-3}{n} E\left(T_{n-4}^+\right) = \Theta\left(n^5\right)$$

as a consequence. We call this level type IV.

For $i = n - 2$ and $i = n - 1$ we are in a ONEMAX-like situation, again. We conclude that $E\left(T_{n-2}^+\right) = O(n^5)$ and $E\left(T_{n-1}^+\right) = O(n^5)$ hold. These levels are called type V levels.

We see that only type III levels are new for f_2 in comparison to f_1. For larger values of k, no new situations occur. We summarize what we have for general, yet constant values of k. Levels $i \in \{0, 1, \ldots, n - 2k - 2\}$ are type I levels where $E\left(T_i^+\right) = \Theta(n/(n - i))$ holds. The only type II level is $i = n - 2k - 1$. Here $E\left(T_{n-2k-1}^+\right) = \Theta(n^2)$ holds. After that we have the type III levels for $i \in \{n - 2k, n - 2k, \ldots, n - k - 2\}$, where $E\left(T_i^+\right) = \Theta(n^{2(i-n+2k)+4})$ holds. Note that for $k = 1$, $n - 2k = n - 2 > n - 3 = n - k - 2$ holds, so that there is no type III level for f_1. We see that the optimization time sharply increases here with the required decrease in function values as could be expected. For $i = n - k - 1$ we have the only type IV level, where $E\left(T_{n-k-1}^+\right) = \Theta(n^{2k+1})$ holds. The following levels $i \in \{n - k, n - k + 1, \ldots, n - 1\}$ are of type V and we have $E\left(T_i^+\right) = \Theta(n^{2k+1})$ there. Since we have $k = \Theta(1)$, the results on $E\left(T_i^+\right)$ imply the desired result on $E\left(T_{SA}\right)$ for f_k. □

Theorem 9 contains a general upper bound, only. The possibility remains that simulated annealing outperforms the $(1+1)$ EA on f_k using an appropriate cooling schedule. We do not have a formal proof that this is not the case. But we can offer some arguments that make this seem to be unlikely. First, we note that switching from a fixed value of α to a non-increasing cooling schedule is not helpful. In the beginning f_k looks like ONEMAX. In this situation large values of α are helpful. Only later, in the valley, smaller values for α can lead to a speed-up. This is the opposite of a non-decreasing cooling schedule, thus a constant value seems to be an appropriate choice.

We can prove that for $\alpha \leq (\sqrt{n}-1)/2$ simulated annealing has exponential expected optimization time. We can prove a polynomial upper bound only for $\alpha \geq (1+\varepsilon)n/2$. If we had a proof that for $\alpha \leq cn$ with some constant $0 < c < 1/2$ the expected optimization time of simulated annealing on ONEMAX is not polynomial, then we could actually prove that non-decreasing cooling schedules cannot be helpful and that no other constant value for α can cause a substantial speed-up.

The example functions f_k show that "valleys" alone are not sufficient to separate the (1+1) EA from simulated annealing on s.i. functions. Neither regions with increasing function values nor regions with constant function values (so called plateaus) can cause a separation since both algorithms work in the same way there. These arguments are far from a formal proof that the (1+1) EA and simulated annealing necessarily show similar performance on s.i. functions. But the results here seem to indicate that such a proof may exist.

However, a note of caution should be added. We seem to have reason to believe that the performance of the (1+1) EA and simulated annealing on s.i. functions which are functions of unitation are necessarily very similar. We present another family of example functions which demonstrate that large (yet polynomial) performance differences can occur. Due to the missing lower bound for a range of possible fixed values of α on ONEMAX we cannot prove that simulated annealing does perform worse for all cooling schedules. But for all cooling schedules which enable us to prove polynomial upper bounds at all, the upper bound for simulated annealing will be considerably worse than the one for the (1+1) EA.

The idea of the example function g_k (again $k \in \mathbb{N}$, $k = \Theta(1)$) is to concatenate the valleys of f_i with $i \leq k$ in ascending order in such a way that the peaks at the end of the valleys grow. We have $g_1 = f_1$ and get different functions for $k > 1$. One example, $g_3 \colon \{0,1\}^{30} \to \mathbb{R}$, is visualized in Figure 2.

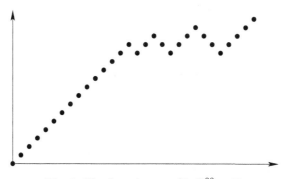

Fig. 2. The function $g_3 \colon \{0,1\}^{30} \to \mathbb{R}$

A formal definition is somewhat involved. We choose to define the function by a vector of differences between function values $d = (d_1, \ldots, d_n) \in \{-1,0,1\}^n$. For the definition of this vector d it is convenient to define points of change p_j, first, where $j \in \{1, 2, \ldots, 2k+1\}$. Given these points of change, we define

$d_1 = d_2 = \cdots = d_{p_1} = 1$, $d_{p_1+1} = \cdots = d_{p_2} = -1$, $d_{p_2+1} = \cdots = d_{p_3} = 1$, and so on. For the points of change, we define $p_1 = n - k(k+2)$, $p_{2j} = p_{2j-1} + j$, and $p_{2j+1} = p_{2j} + j + 1$ for $j \in \{1, 2, \ldots, k\}$. Using $g_k(0^n) = 0$ for all k and n this completes the formal definition of g_k. As an example, consider g_3 for $n = 30$. We get $p_1 = 15$, $p_2 = 16$, $p_3 = 18$, $p_4 = 20$, $p_5 = 23$, $p_6 = 26$, and $p_7 = 30$, leading to g_3 as it can be seen in Figure 2.

It is not difficult to prove an acceptable upper bound on the optimization time on g_k for the (1+1) EA. The algorithm can "jump from hill top to hill top." This way, the expected optimization time is dominated by the final valley.

Theorem 10. *For g_k with $k = \Theta(1)$, $E(T_{EA}) = \Theta(n^{2k+1})$ holds.*

Proof. For the lower bound it suffices to note that the final valley has to be crossed which is identical to the situation for f_k. For the upper bound, we note that for each valley a mutation of $2i + 1$ specific bits is enough to cross it. This yields

$$E(T_{\mathrm{EA}}) = O\left(\sum_{i=1}^{k} n^{2i+1}\right) = O\left(n^{2k+1}\right)$$

as upper bound.   □

For simulated annealing, the situation is similar to the situation for f_k. We can only prove an upper bound that is based on a special fixed value for α. However, we have the same reasons as mentioned above to believe that the upper bound given may be tight.

Theorem 11. *For g_k with $k = \Theta(1)$, $E(T_{SA}) = O(n^{k^2+k+1})$ holds. For $\alpha \geq (1+\varepsilon)n/2$, where $\varepsilon > 0$ is a constant and $\alpha = O(n)$ holds, $E(T_{SA}) = \Theta(n^{k^2+k+1})$ holds for f_k with $k = \Theta(1)$.*

Proof. Like in the proof of Theorem 9, this proof works for any cooling schedule which guarantees (1) $\alpha(t) \geq (1+\varepsilon)n/2$ for all $t \geq t_0$, where $\varepsilon > 0$ is a constant and $t_0 = O(n^{k^2+k+1})$, and (2) $\alpha(t) = O(n)$ for all t. Again, we choose $\alpha := n$ fixed for the sake of simplicity of notation.

We partition the different levels of ones according to the types in the proof of Theorem 9. We make the following observations. Type I levels only add a term $\Theta(n)$ to the expected optimization time from the previous level. Type II levels add a term $\Theta(n^2)$ to the n-fold of the expected optimization time from the previous level. Type III levels add a term $\Theta(n^2)$ to the n^2-fold of the expected optimization time from the previous level. Type IV levels add a term $\Theta(n)$ to the n-fold of the expected optimization time from the previous level. Finally, type V levels just add a term $\Theta(n)$ to the expected optimization time of the previous level.

We have k levels of type II and type IV each. This adds $2k$ to the exponent of the expected optimization time. We have k valleys, the first without type III level, all others with one type III level more than the previous valley. This adds up to $k(k-1)/2$ type III levels, adding $k(k-1)$ to the exponent of the expected optimization time. We add 1 for the first $n - k(k + 2)$ levels and get $\Theta(n^{k(k-1)+2k+1}) = \Theta(n^{k^2+k+1})$ as bound as claimed. The linear and quadratic terms added do not lead to an increase in the asymptotic notation since we have $k = O(1)$. $\qquad\qquad\qquad\qquad\qquad\qquad\qquad\qquad\qquad\qquad\qquad\qquad\qquad\qquad\qquad\quad$ □

One may believe that increasing α in a way that $\alpha = \omega(n)$ holds may decrease the expected optimization time since it seems to decrease the influence of the smaller valleys. However, this is not the case. At the type II level at the beginning of a valley with depth j we have $\mathrm{E}\left(T_i^+\right) = \Omega(\alpha n + n^2)$ when we use an arbitrarily fixed α. The choice of α has influence on the expected time until the bottom of the valley is reached. We get $\mathrm{E}\left(T_{i+j}^+\right) = \Omega(\alpha^j n^{j+1} + \alpha^{j-1} n^{j+2})$ there. Indeed, when climbing up the valley we get a factor $\Theta(n/\alpha)$ that may diminish the influence of the time spent before. This adds up to $\mathrm{E}\left(T_{i+2j}^+\right) = \Omega(n^{2j+1} + n^{2j+2}/\alpha)$ just in front of the next hill top. We see that the α-independent term n^{2j+1} cannot be avoided. And exactly this term dominates the expected optimization time in the proof of Theorem 11.

When we consider the quotient of the upper bound for simulated annealing and the expected optimization time for the (1+1) EA, we see that it is always polynomially bounded for $k = \Theta(1)$. But simultaneously it can become arbitrarily large. It is easy to extend the proofs to the case $k = O(\log n)$, which would yield a super-polynomially growing quotient. However, neither the (1+1) EA nor simulated annealing have polynomial expected optimization time in this case. Therefore, we do not consider such a result to be interesting.

5   Conclusions

We have presented a systematic comparison of simulated annealing and the (1+1) EA under the aspect of expected optimization time. The comparison is inspired by the extreme similarities of these two randomized search heuristic. The main hope is to find mechanisms to transfer analytical results from one of these two algorithms to the other. This concrete study may be seen as an example of an approach that tries to find a kind of taxonomy for general randomized search heuristics. Having such a general goal in mind the steps taken here may seem appear to be too timid. Despite its obvious desirability, we consider such a general approach to be too ambitious given the current knowledge on the analysis of randomized search heuristics.

By means of two example functions we learned that in spite of their similarities the performances of simulated annealing and the (1+1) EA can be drastically different. This leads to a clearer formulation of our goal. We are looking for classes of functions where both algorithms have similar expected optimization times. For the two example functions which demonstrated extreme performance differences extreme differences between neighboring points in the search space are crucial. Therefore, the definition of smooth integer functions is motivated. The absolute difference in function values between neighboring points is limited to 1, scaling tricks are prohibited since the function values are restricted to be integers. The class of smooth integer functions, especially the sub-class of functions which are smooth integer and functions of unitation, has been investigated more closely. One family of example functions showed that a single valley in the search space that needs to be crossed causes an obstacle that is of equal difficulty to the (1+1) EA and simulated annealing. A second family of example functions showed that simple concatenation of valleys is already sufficient to demonstrate arbitrarily large, yet polynomial differences in performance between the two randomized search heuristics.

It is still unclear whether the performance of simulated annealing and the (1+1) EA on smooth integer functions or at least on smooth integer functions which are functions of unitation is necessarily similar. We believe that this question deserves some attention in the future. Learning how to identify classes of functions where similar yet different randomized search heuristics may become a valuable tool in order to classify objective functions as well as randomized search heuristics.

There are two open problems that deserve to be stated explicitly. First, the performance of simulated annealing with constant α, i.e., the Metropolis algorithm, on OneMax needs to be determined asymptotically tight for all possible values of α. The results here leave a gap for $\sqrt{n}/2 \leq \alpha \leq n/2$. Second, either an explicit definition of an example function that is smooth integer and causes $E(T_A)$ to be exponential but causes $E(T_B)$ to be polynomial (with A = EA and B = SA or the other way round) or a formal proof that such a function does not exist is needed.

References

1. T. Bäck (1996): *Evolutionary Algorithms in Theory and Practice*. Oxford University Press.
2. T. Carson, R. Impagliazzo (2001): Hill-climbing finds random planted bisections. *Proceedings of the ACM-SIAM Symposium on Discrete Algorithms (SODA 2001)*. 903–909.
3. T. Cormen, C. Leiserson, R. Rivest, C. Stein (2001): *Introduction to Algorithms*. MIT Press.
4. S. Droste, T. Jansen, I. Wegener (2001): Dynamic parameter control in simple evolutionary algorithms. In W. N. Martin, W. M. Spears (Eds.): *Foundations of Genetic Algorithms 6 (FOGA 2000)*. Morgan Kaufmann. 275–294.
5. S. Droste, T. Jansen, I. Wegener (2002): On the analysis of the (1+1) evolutionary algorithm. *Theoretical Computer Science* 276:51–81.
6. J. Garnier, L. Kallel, M. Schoenauer (1999): Rigorous hitting times for binary mutations. *Evolutionary Computation* 7(2):173–203.
7. W. E. Hart (1995): A theoretical comparison of evolutionary algorithms and simulated annealing. In *Proceedings of the Fifth Annual Conference on Evolutionary Programming* 147–154.
8. L. Ingber, B. Rosen (1992): Genetic algorithms and very fast reannealing: a comparison. *Mathematical and Computer Modeling* 16:87–100.

9. T. Jansen, I. Wegener (2000): On the choice of the mutation probability for the (1+1) EA. In M. Schoenauer, K. Deb, G. Rudolph, X. Yao, E. Lutton, J.J. Merelo, H.-P. Schwefel (Eds.): *Sixth International Conference on Parallel Problem Solving From Nature (PPSN VI)*. Springer, LNCS 1917, Springer. 89–98.

10. T. Jansen, I. Wegener (2002): On the analysis of evolutionary algorithms - a proof that crossover really can help. *Algorithmica* 34(1):47–66.

11. M. R. Jerrum, G. Sorkin (1993): Simulated annealing for graph bisection. *Proceedings of the IEEE Symposium on Foundations of Computer Science (FOCS '93)*. IEEE Press. 94–103.

12. S. Kirkpatrick, C. D. Gelatt, M. P. Vecchi (1983): Optimization by simulated annealing. *Science* 220, 671–680.

13. N. Metropolis, A. W. Rosenbluth, M. N. Rosenbluth, A. H. Teller, E. Teller (1953): Equation of state calculation by fast computing machines. *Journal of Chemical Physics* 21, 1078–1092.

14. H. Mühlenbein (1992): How genetic algorithms really work. Mutation and Hill-climbing. In Männer, Manderick (Eds): *Proceedings of the 2nd Parallel Problem Solving from Nature (PPSN II)*. North-Holland. 15–25.

15. A. Prügel-Bennett (2004): When a genetic algorithm outperforms hill-climbing. *Theoretical Computer Science* 320(1):135–153.

16. G. Rudolph (1997): *Convergence Properties of Evolutionary Algorithms*. Dr. Kovač.

17. G. B. Sorkin (1991): Efficient simulated annealing on fractal energy landscapes. *Algorithmica* 6, 346–418.

18. N. L. J. Ulder, E. H. L. Aarts, H.-J. Bandelt, P. J. M. van Laarhoven, E. Pesch. (1994): Genetic local search algorithms for the traveling salesman problem. In *First International Conference on Parallel Problem Solving from Nature (PPSN I)*, 109–116.

19. I. Wegener (2004): Simulated annealing beats Metropolis in combinatorial optimization. Technical Report SFB 531, CI 181/04. University of Dortmund, Germany.

20. I. Wegener, C. Witt (2003): On the optimization of monotone polynomials by the (1+1) EA and randomized local search. In *Genetic and Evolutionary Computation Conference (GECCO 2003)*, Springer. 622–633.

NP-Completeness
of Deciding Binary Genetic Encodability

Andreas Blass[*] and Boris Mitavskiy

[1] Department of Mathematics, University of Michigan, Ann Arbor, MI, 48109, United States
ablass@umich.edu
[2] School of Computer Science, University of Birmingham, Birmingham B15 2TT, Great Britain
B.S.Mitavskiy@cs.bham.ac.uk

Abstract. In previous work of the second author a rigorous mathematical foundation for re-encoding one evolutionary search algorithm by another has been developed. A natural issue to consider then is the complexity of deciding whether or not a given evolutionary algorithm can be re-encoded by one of the standard classical evolutionary algorithms such as a binary genetic algorithm. In the current paper we prove that, in general, this decision problem is NP-complete.

1 Introduction

In recent years evolutionary algorithms have been widely exploited to solve various complex optimization problems. In order to apply an evolutionary algorithm to attack a specific optimization problem, one needs to model the algorithm in a suitable manner. The importance of finding appropriate models is emphasized in much of the research literature: see, for instance, the introduction to chapter 17 of [13], [14], [12] and [11]. The general methodology for how to construct the search space and the appropriate recombination operators with the aim of applying the classical genetic algorithm first appeared in [9]. However, there is a variety of different types of EAs which people use. This might be, for example, nonlinear GP with homologous crossover introduced by Poli (see [8] for a detailed description of how this algorithm works), or, even more so, it might be a special type of an algorithm used to attack a specific problem. It is in general interesting to know if it is possible to re-encode a given algorithm by a binary genetic algorithm. In [6] and in [7] a rigorous mathematical framework was introduced, allowing one to re-encode one evolutionary algorithm by another.

In particular, necessary and sufficient conditions for a given evolutionary search algorithm to be embeddable into a binary genetic (or semi-genetic) algorithm have been established. The aim of the current paper is to investigate the computational complexity of deciding if a given evolutionary search algorithm can be re-encoded by another (probably a more commonly used) evolutionary algorithm. The main results of the current paper demonstrate that deciding if a given evolutionary algorithm can be embedded into a binary semi-genetic algorithm can be done in polynomial time, while the more useful, analogous decision problem pertaining to the classical genetic algorithm is, unfortunately, NP-complete.

[*] Partially supported by NSF grant DMS–0070723.

A.H. Wright et al. (Eds.): FOGA 2005, LNCS 3469, pp. 58–74, 2005.
© Springer-Verlag Berlin Heidelberg 2005

The paper is organized as follows: In the next two sections we introduce the basic notation and framework used throughout the paper. Next, in Section 4 we demonstrate how the classical types of algorithms fit into this framework. Section 5 is devoted to a summary of the previous work which sets the foundation for the results of the current paper. The main results are then presented in the final Section 6.

2 Notation

Ω is a finite set, called a *search space*.

$f : \Omega \rightarrow (0, \infty)$ is a function, called a *fitness function*. The goal is to find a maximum of the function f.

$\mathcal{F}$ is a collection of binary operations on Ω. Intuitively $\mathcal{F}$ can be thought of as the collection of *reproduction transformations*: two parents produce one offspring. The family of asexual reproductions or *mutations* (these are unary operations on Ω, i. e. functions from Ω into itself) will be denoted by $\mathcal{M}$. By a *search system* we mean a search space Ω together with families $\mathcal{F}$ and $\mathcal{M}$ of reproduction transformations and mutations. We shall denote a search system either by $(\Omega, \mathcal{F}, \mathcal{M})$ or simply by Ω; the latter notation follows the convention, common in many parts of mathematics, of using the same symbol, in this case Ω, for a mathematical structure, in this case a search system, and for its underlying set.

Remark: In general there is no reason to assume that a child has exactly two parents. All of the results in this paper are valid for families of q-ary operations on Ω. The only reason $\mathcal{F}$ is assumed to be a family of binary transformations is to alleviate the complexity of notation. In the general case, the definition of search system should, of course, include all the reproduction transformations, regardless of the number of arguments. Search systems were called "heuristic tuples" in earlier work of the second author. They are almost the same as what are called "algebras" in universal algebra, but the morphisms of search systems, defined in Section 5, are different from homomorphisms of algebras.

3 How Does an Evolutionary Algorithm Work?

A typical evolutionary algorithm works as follows: A population $P = \begin{pmatrix} x_1 \\ x_2 \\ \vdots \\ x_{2m} \end{pmatrix}$ with

$x_i \in \Omega$ is selected randomly. The algorithm cycles through the following stages:

Evaluation

Individuals of P are evaluated:

$$\begin{pmatrix} x_1 \\ x_2 \\ \vdots \\ x_{2m} \end{pmatrix} \begin{matrix} \rightarrow & f(x_1) \\ \rightarrow & f(x_2) \\ \vdots & \vdots \\ \rightarrow & f(x_{2m}) \end{matrix}$$

Selection

A new population

$$P' = \begin{pmatrix} y_1 \\ y_2 \\ \vdots \\ y_{2m} \end{pmatrix}$$

is obtained by choosing each y_i independently by the following random process. Choose at random j in the range $1 \le j \le 2m$, the probability of any j being $\dfrac{f(x_j)}{\sum_{l=1}^{2m} f(x_l)}$. Then set $y_i = x_j$.

Thus, all of the individuals of P' are among those of P, and the expectation of the number of occurrences of any individual of P in P' is proportional to the number of occurrences of that individual in P times the individual's fitness value. In particular, the fitter the individual is, the more copies of that individual are likely to be present in P'. On the other hand, the individuals having relatively small fitness value are not likely to enter into P' at all. This is designed to imitate the natural "survival of the fittest" principle.

Partition

The individuals of P' are partitioned into m pairwise disjoint couples for mating according to some probabilistic rule: For instance the couples could be

$$Q_1 = \begin{pmatrix} y_{i_1^1} \\ y_{i_2^1} \end{pmatrix} \quad Q_2 = \begin{pmatrix} y_{i_1^2} \\ y_{i_2^2} \end{pmatrix} \quad \cdots \quad Q_j = \begin{pmatrix} y_{i_1^j} \\ y_{i_2^j} \end{pmatrix} \quad \cdots \quad Q_m = \begin{pmatrix} y_{i_1^m} \\ y_{i_2^m} \end{pmatrix}$$

Reproduction

Replace every one of the selected couples $Q_j = \begin{pmatrix} y_{i_1^j} \\ y_{i_2^j} \end{pmatrix}$ with the couple

$$Q' = \begin{pmatrix} T_1(y_{i_1^j}, y_{i_2^j}) \\ T_2(y_{i_1^j}, y_{i_2^j}) \end{pmatrix}$$

for some couple of transformations $(T_1, T_2) \in \mathcal{F}^2$. The couple (T_1, T_2) is selected according to a fixed probability distribution on $\mathcal{F}^2$. This gives us a new population

$$P'' = \begin{pmatrix} z_1 \\ z_2 \\ \vdots \\ z_{2m} \end{pmatrix}$$

Mutation

Finally, with small probability we replace z_i with $F(z_i)$ for some randomly chosen $F \in \mathcal{M}$. The choices for different i's are independent. This, once again, gives us a new population $P''' = \begin{pmatrix} w_1 \\ w_2 \\ \vdots \\ w_{2m} \end{pmatrix}$

Upon completion of mutation start all over with P''' as the initial population. The cycle is repeated a certain number of times depending on the problem. A more general and extensive description is given in [13]. The importance of choosing a reasonable representation for a specific problem is emphasized in some of the modern research. See, for instance, [10]. A few special types of evolutionary algorithms are introduced in the next section.

4   Special Evolutionary Algorithms

Classical Genetic Algorithm with Masked Crossover: Let $\Omega = \prod_{i=1}^{n} A_i$. For every subset $M \subseteq \{1, 2, \ldots, n\}$, define a binary operation L_M on Ω as follows:

$$L_M(\mathbf{a}, \mathbf{b}) = (x_1, x_2, \ldots, x_i, \ldots, x_n)$$

where $\mathbf{a} = (a_1, a_2, \ldots, a_n)$ and $\mathbf{b} = (b_1, \ldots, b_n) \in \Omega$ and $x_i = \begin{cases} a_i & \text{if } i \in M \\ b_i & \text{otherwise.} \end{cases}$

This L_M is a masked crossover operator with mask M. Let $\mathcal{F} = \{L_M \mid M \subseteq \{1, 2, \ldots, n\}\}$. That is, $\mathcal{F}$ in this example is simply the family of masked crossover transformations. The probability distribution on the set $\mathcal{F}^2$ is concentrated on the pairs of the form $(L_M, L_{\bar{M}})$ where $\bar{M}$ denotes the complement of the set M in $\{1, 2, \ldots, n\}$.

Example: Let $n = 5$ and $A_i = \{0, 1, \ldots, i+1\}$. Suppose a given population P consists of 6 individuals which are the rows of the matrix

$$\begin{pmatrix} 2\ 3\ 4\ 5\ 6 \\ 0\ 1\ 2\ 3\ 4 \\ 1\ 2\ 3\ 4\ 5 \\ 0\ 0\ 1\ 2\ 3 \\ 1\ 1\ 0\ 1\ 2 \\ 1\ 2\ 1\ 5\ 4 \end{pmatrix}$$

Say, after the selection stage is complete one obtains the following population

$$P' = \begin{pmatrix} 2\ 3\ 4\ 5\ 6 \\ 2\ 3\ 4\ 5\ 6 \\ 1\ 2\ 3\ 4\ 5 \\ 0\ 0\ 1\ 2\ 3 \\ 0\ 1\ 2\ 3\ 4 \\ 1\ 2\ 3\ 4\ 5 \end{pmatrix}$$

Now the following individuals are paired for mating (masked crossover in this case):

$$Q_1 = \begin{pmatrix} 2\ 3\ 4\ 5\ 6 \\ 0\ 0\ 1\ 2\ 3 \end{pmatrix}, \quad Q_2 = \begin{pmatrix} 2\ 3\ 4\ 5\ 6 \\ 1\ 2\ 3\ 4\ 5 \end{pmatrix}, \quad \text{and } Q_3 = \begin{pmatrix} 0\ 1\ 2\ 3\ 4 \\ 1\ 2\ 3\ 4\ 5 \end{pmatrix}$$

Suppose we have chosen the masks $M_1 = \{1, 4, 5\}$, $M_2 = \{1, 2\}$ and $M_3 = \{3, 4\}$ for the crossover of pairs Q_1, Q_2 and Q_3 respectively. In the language of this paper it means we have chosen the pairs of transformations $(L_{M_1}, L_{\bar{M}_1})$ for Q_1, $(L_{M_2}, L_{\bar{M}_2})$ for Q_2 and $(L_{M_3}, L_{\bar{M}_3})$ for Q_3 respectively. Upon applying these we obtain

$$Q_1 \to \begin{pmatrix} L_{M_1}((2,\,3,\,4,\,5,\,6),\,(0,\,0,\,1,\,2,\,3)) \\ L_{\bar{M}_1}((2,\,3,\,4,\,5,\,6),\,(0,\,0,\,1,\,2,\,3)) \end{pmatrix} = \begin{pmatrix} 2\,0\,1\,5\,6 \\ 0\,3\,4\,2\,3 \end{pmatrix},$$

likewise

$$Q_2 \to \begin{pmatrix} L_{M_2}((2,\,3,\,4,\,5,\,6),\,(1,\,2,\,3,\,4,\,5)) \\ L_{\bar{M}_2}((2,\,3,\,4,\,5,\,6),\,(1,\,2,\,3,\,4,\,5)) \end{pmatrix} = \begin{pmatrix} 2\,3\,3\,4\,5 \\ 1\,2\,4\,5\,6 \end{pmatrix},$$

and, finally,

$$Q_3 \to \begin{pmatrix} L_{M_3}((0,\,1,\,2,\,3,\,4),\,(1,\,2,\,3,\,4,\,5)) \\ L_{\bar{M}_2}((0,\,1,\,2,\,3,\,4),\,(1,\,2,\,3,\,4,\,5)) \end{pmatrix} = \begin{pmatrix} 1\,2\,2\,3\,5 \\ 0\,1\,3\,4\,4 \end{pmatrix}.$$

The family $\mathcal{M}$ of mutation transformations in this example (and in all of the following ones) consists of the transformations $M_a : \Omega \to \Omega$, where $a \in \prod_{i \in S} A_i$ and $S \subseteq \{1, 2, \ldots, n\}$. The transformation M_a sends any $\mathbf{x} = (x_1, x_2, \ldots, x_n) \in \Omega$ to the $\mathbf{y} = (y_1, y_2 \ldots, y_n) \in \Omega$ whose components are $y_q = \begin{cases} a_q & \text{if } q \in S \\ x_q & \text{otherwise.} \end{cases}$ In other words, M_a simply replaces the qth coordinate of its argument with $a_q \in A_q$ whenever $q \in S$.

Binary Genetic Algorithm with Masked Crossover
When every $A_i = \{0, 1\}$ (which means that $\Omega = \{0, 1\}^n$) in the example above, one obtains the classical binary genetic algorithm.

Random Respectful Recombination
Random Respectful Recombination first appeared in [9]. Here the search space Ω and the family of mutation transformations, $\mathcal{M}$, are exactly the same as in the example of classical genetic algorithm, and the family of mating transformations is described below. As in [5], we call these mating transformations Holland transformations because their corresponding family of fixed subsets is precisely the collection of subsets of Ω determined by the classical Holland schemata together with the empty set. (See examples following Corollary 8 in the next section.) For every given point $\mathbf{u} = (u_1, u_2, \ldots, u_n) \in \Omega$ define a Holland transformation $T_{\mathbf{u}} : \Omega^2 \to \Omega$ as follows: for every $\mathbf{a} = (a_1, a_2, \ldots, a_n)$ and $\mathbf{b} = (b_1, b_2, \ldots, b_n) \in \Omega$

$$T_{\mathbf{u}}(\mathbf{a}, \mathbf{b}) = (x_1, x_2, \ldots, x_n)$$

where

$$x_i = \begin{cases} a_i \text{ if } a_i = b_i \\ u_i \text{ otherwise} \end{cases}$$

In other words, if the ith coordinates of $\mathbf{a}$ and $\mathbf{b}$ coincide, then the ith coordinate of $T_{\mathbf{u}}(\mathbf{a}, \mathbf{b})$ also coincides with them. If, on the other hand, the ith coordinates of $\mathbf{a}$ and $\mathbf{b}$ differ, then the ith coordinate of $T_{\mathbf{u}}(\mathbf{a}, \mathbf{b})$ is that of $\mathbf{u}$, namely, u_i. Let $\mathcal{F} = \{T_{\mathbf{u}} \mid \mathbf{u} \in$

$\Omega\}$ be the family of all Holland transformations. The probability distribution on $\mathcal{F}$ may be chosen in different ways depending on the circumstances, but this is not relevant to the objective of the current paper.

Every transformation in the pair $(T_{\mathbf{u}}, T_{\mathbf{v}})$ is chosen independently.

Binary Random Respectful Recombination

The search space Ω and the family of mating transformations $\mathcal{F}$ and the family of mutations $\mathcal{M}$ are exactly the same as these for the binary genetic algorithm with masked crossover described above. The only difference is that the probability distribution on $\mathcal{F}^2$ is now completely uniform (rather than being concentrated on the diagonal-like subset described in the classical genetic algorithm example). For instance, if $n = 5$, $M_1 = \{2, 3, 4\}$, $M_2 = \{1, 3, 5\}$ and the pair (T_{M_1}, T_{M_2}) is selected for mating, we have, for instance,

$$\begin{pmatrix} 1\,0\,0\,1\,1 \\ 1\,1\,0\,0\,1 \end{pmatrix} \longmapsto \begin{pmatrix} T_{M_1}((1, 0, 0, 1, 1),\ (1, 1, 0, 0, 1)) \\ T_{M_2}((1, 0, 0, 1, 1),\ (1, 1, 0, 0, 1)) \end{pmatrix} = \begin{pmatrix} 1\,0\,0\,1\,1 \\ 1\,1\,0\,0\,1 \end{pmatrix}$$

The same definition could be applied in the non-binary case, but then it would not agree with the random respectful recombination of [9] (as described above). The difference is that, when the two parents have different alleles of a certain gene, then random respectful recombination allows the offspring to have any allele of that gene, while the present definition only allows the offspring to have either of the two alleles present in the parents. The two notions are equivalent just when there are only two possible alleles.

The following type of algorithm may seem useless at first. Its importance will become clear in the next section when we present the binary embedding theorem which shows that the binary semi-genetic algorithm possesses an interesting universal property.

Binary Semi-genetic Algorithm

The search space $\Omega = \{0, 1\}^n$, just as in the case of the binary genetic algorithm. The family of mating transformations $\mathcal{F}$ is defined as follows: Fix an individual $\mathbf{u} = (u_1, u_2, \ldots, u_n) \in \Omega$. Define a semi-crossover transformation $F_{\mathbf{u}} : \Omega^2 \to \Omega$ as follows: For any given pair $(\mathbf{x}, \mathbf{y}) \in \Omega^2$ with $\mathbf{x} = (x_1, x_2, \ldots, x_n)$ and $\mathbf{y} = (y_1, y_2, \ldots, y_n)$ we have $F_{\mathbf{u}}(\mathbf{x}, \mathbf{y}) = \mathbf{z} = (z_1, z_2, \ldots z_n) \in \Omega$ where

$$z_i = \begin{cases} 1 & \text{if } x_i = y_i = 1 \\ u_i & \text{otherwise} \end{cases}$$

In other words, $F_{\mathbf{u}}$ preserves the ith gene if it is equal to 1 in both of the parents and replaces it with u_i otherwise. Let $\mathcal{F} = \{F_{\mathbf{u}} \mid \mathbf{u} \in \Omega\}$ be the family of all semi-crossover transformations. The family of mutation transformations $\mathcal{M}$ is exactly the same as in the examples above.

Example: With $n = 5$ and $\mathbf{u}_1 = (0, 1, 1, 0, 1)$, $\mathbf{u}_2 = (0, 1, 0, 0, 1)$ we have

$$\begin{pmatrix} 1\,0\,0\,1\,1 \\ 1\,1\,0\,0\,1 \end{pmatrix} \longmapsto \begin{pmatrix} F_{\mathbf{u}_1}((1, 0, 0, 1, 1),\ (1, 1, 0, 0, 1)) \\ F_{\mathbf{u}_2}((1, 0, 0, 1, 1),\ (1, 1, 0, 0, 1)) \end{pmatrix} = \begin{pmatrix} 1\,1\,1\,0\,1 \\ 1\,1\,0\,0\,1 \end{pmatrix}$$

Notice that, if 1 is present in the ith position of both parents, then it remains in the ith position of both offspring. There are absolutely no other restrictions, though.

In practice the choice of the search space Ω is primarily determined by the specific problem and related circumstances. The general methodology for the construction of the search spaces first appeared in the work of Radcliffe (see, for instance, [9]). Radcliffe introduced the notion of a forma which captures the essential properties of the Holland schemata in a representation independent setting. A forma is simply a partition of the search space into equivalence classes. A given collection of forma with suitable properties (see [9]) is, in a sense, no different from the collection of the classical Holland schemata provided that one encodes the search space using the "genetic representation function" which is also introduced in [9]. The connection between all of the possible families of mating transformations on a given search space Ω and the corresponding families of invariant subsets established in [5] has been exploited in [6] and in [7] to extend Radcliffe's notion of the genetic representation function to compare various evolutionary algorithms via possible encodings of their search spaces (see corollary 12 of [6]). In particular, necessary and sufficient conditions, stated in terms of the internal structure of the search space, for re-encoding a given algorithm by a search system corresponding to a binary genetic algorithm have been established (see Theorem 14 of [6] or, more generally, Theorem 3.7 of [7]). These ideas will be summarized in the next section.

5   Summary of Previous Work

As we have seen in the previous sections, a given evolutionary search algorithm is determined primarily by the ordered 4-tuple $(\Omega, \mathcal{F}, \mathcal{M}, f)$. In the current paper we shall be primarily concerned with the search space Ω and the family of mating transformations $\mathcal{F}$. The family of mutations $\mathcal{M}$ is of less importance because it is ergodic, meaning that the only invariant subsets under $\mathcal{M}$ are $\emptyset$ and the entire search space Ω. The notion of an invariant subset is defined below. The reason why invariant subsets play a significant role in the current paper is Theorem 8. We give the definition for a family Γ of operations of any number of arguments, but we shall use it in this paper only for the family $\mathcal{F}$ of binary operations of a search system.

Definition 1 For a given family of m-ary operations Γ on a set Ω (that is, functions from Ω^m into Ω) a subset $S \subseteq \Omega$ is *invariant* under Γ if and only if for all $T \in \Gamma$ we have $T(S^m) \subseteq S$. We shall denote by Λ_Γ the family of all invariant subsets of Ω under Γ. In other words, $\Lambda_\Gamma = \{S \mid S \subseteq \Omega, \ T(S^m) \subseteq S \ \forall \ T \in \Gamma\}$.

Below we list the families of invariant subsets for each of the examples of Section 4:

Classical Genetic Algorithm. In this case, the family of invariant subsets $\Lambda_\mathcal{F}$ is $\{\prod_{i=1}^{n} T_i \mid T_i \subseteq A_i\}$. This is precisely the family of subsets determined by Antonisse's schemata (see corollary 2.4 of [5]).

Random Respectful Recombination. $\Lambda_\mathcal{F} = \{\prod_{i=1}^{n} T_i \mid T_i = \{a\}$ for some $a \in A_i$ or $T_i = A_i\} \cup \{\emptyset\}$. This is precisely the family of subsets determined by the Holland schemata together with the empty set (see corollary 3.5 of [5]).

Binary Semi-genetic Algorithm. It is not hard to verify that $\Lambda_{\mathcal{F}} = \{\prod_{i=1}^{n} T_i \mid T_i = \{1\}$ or $T_i = \{0, 1\}\} \cup \{\emptyset\}$. This is precisely the family of subsets determined by Holland schemata whose fixed positions can only equal 1 (can't equal 0).

The mathematical properties of the family of invariant subsets, Λ_{Γ} have been described in detail in [5]. In the current presentation we just mention a few facts and notions which will be of particular importance here.

It is easy to verify (see Proposition A1 of [5]) that the family Λ_{Γ} is closed under arbitrary intersections and contains Ω. It then follows that for every element $x \in \Omega$ there is a unique smallest element of Λ_{Γ} containing x (namely the intersection of all the members of Λ_{Γ} containing x).

Definition 2 Given a search system $\Omega = (\Omega, \mathcal{F}, \mathcal{M})$, denote by S_x^{Ω} the smallest element of $\Lambda_{\mathcal{F}}$ containing x. When the search system Ω is clear from the context we shall just write S_x instead of S_x^{Ω}.

The following definition is a natural extension of the notion of a genetic representation function introduced in [9].

Definition 3 Given two search systems $\Omega_1 = (\Omega_1, \mathcal{F}_1, \mathcal{M}_1)$ and $\Omega_2 = (\Omega_2, \mathcal{F}_2, \mathcal{M}_2)$, a *morphism* $\delta : \Omega_1 \to \Omega_2$ is just a function $\delta : \Omega_1 \to \Omega_2$ which respects the reproduction transformations in the following sense: for each $T \in \mathcal{F}_1$ and each $\mathbf{x} = (x, y) \in \Omega_1^2$ there exists $F_{(x, y)} \in \mathcal{F}_2$ such that $\delta(T(x, y)) = F_{(x, y)}(\delta(x), \delta(y))$ (see Figure 1). Analogously, we must have, for each $M \in \mathcal{M}_1$ and each $x \in \Omega$ some $H_x \in \mathcal{M}_2$ such that $\delta(M(x)) = H_x(\delta(x))$.

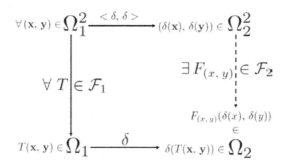

Fig. 1. The morphism $\delta : \Omega_1 \to \Omega_2$

A morphism $\delta : \Omega_1 \to \Omega_2$ provides the means for encoding the search system Ω_1 by the search system Ω_2. Unless the underlying function δ is one to one, there is some nontrivial coarse graining involved. We therefore give a special name to those morphisms whose underlying functions are injective.

Definition 4 We say that a morphism $\delta : \Omega_1 \hookrightarrow \Omega_2$ is an *embedding* if the underlying function $\delta : \Omega_1 \to \Omega_2$ is one-to-one.

search systems and the morphisms between them form a well-defined category (see [3] for details about the notion of category). Some universal constructions on this category have been studied in [7]. The central result of [6] is a connection between the family of invariant subsets and the family of all possible re-encoding morphisms between two given search systems. This connection is analogous to the corresponding connection between the family of open subsets and continuous maps in topology and between sigma-algebras and measurable functions in analysis:

Proposition 5 *Let $\delta : \Omega_1 \to \Omega_2$ be a morphism of search systems. Then $S \in \Lambda_{\mathcal{F}_2} \Longrightarrow \delta^{-1}(S) \in \Lambda_{\mathcal{F}_1}$. In words, a preimage of an invariant set under a morphism is invariant.*

The converse of Proposition 5 holds under the following technical requirement:

Definition 6 We say that a given family of m-ary operations Γ on a set Ω (that is a family of functions from Ω^m to Ω) is *composition closed* if the following two conditions hold:
1. For all $T_0, T_1, T_2, \ldots, T_m \in \Gamma$, the composite operation $T : \Omega^m \to \Omega$ defined by $T(\mathbf{x}) = T_0(T_1(\mathbf{x}), T_2(\mathbf{x}), \ldots, T_m(\mathbf{x}))$ is also a member of Γ.
2. For all $S \subseteq \Omega$, we have $\bigcup_{T \in \Gamma} T(S^m) \supseteq S$.

It is fairly straightforward to verify that every one of the families of mating transformations involved in the examples of Section 4 is composition closed. In fact, it was already shown in [5] that the families of masked crossover transformations and Holland transformations (those which are used for modelling random respectful recombination) are composition closed (see Proposition 2.1 and Theorem 3.6 of [5]). It was also shown (see proposition 9 of [6]) that the family of binary semi-crossover transformations is composition closed.

As noted before, for any family of m-ary operations on Ω the corresponding family of invariant subsets Λ_Γ is closed under arbitrary intersections. Moreover, for any function $\delta : \Omega_1 \to \Omega_2$, the inverse image of the intersection of two subsets of Ω_2 is the intersection of the inverse images of these subsets: $\delta^{-1}(U \cap V) = \delta^{-1}(U) \cap \delta^{-1}(V)$. This motivates the following definition:

Definition 7 Given a family of m-ary operations Γ on Ω, we say that a family of subsets $\widetilde{\Lambda_\Gamma} \subseteq \Lambda_\Gamma$ is a *base* of Λ_Γ if every set $K \in \Lambda_\Gamma$ is the intersection of some sets in $\widetilde{\Lambda_\Gamma}$. (Equivalently, $K = \bigcap_{S \in \widetilde{\Lambda_\Gamma}, \, S \supseteq K} S$).

We now continue with the examples following Definition 1 and list bases for each of them:

Classical Genetic Algorithm. In this case a base for $\Lambda_{\mathcal{F}}$ is the family $\widetilde{\Lambda_{\mathcal{F}}} = \{\prod_{i=1}^n T_i \mid T_i = A_i \text{ for all but one } i\}$. The reader can see that $|\widetilde{\Lambda_{\mathcal{F}}}| = (\sum_{i=1}^n 2^{|A_i|}) - n + 1$. Every element of $\widetilde{\Lambda_{\mathcal{F}}}$ can be thought of as a union of subsets determined by Holland schemata having exactly one fixed position at the same gene.

Random Respectful Recombination. In this case a base for $\Lambda_{\mathcal{F}}$ is the family $\widetilde{\Lambda_{\mathcal{F}}}$ consisting of all products $\prod_{i=1}^n T_i$ where T_i is a one-element set for one value of i, and $T_i = A_i$ for all other values of i. This is precisely the family of subsets determined by Holland schemata having exactly one fixed position.

Binary Semi-genetic Algorithm. It is not hard to verify that in this case a base for $\Lambda_{\mathcal{F}}$ is the family $\widetilde{\Lambda_{\mathcal{F}}}$ consisting of all products $\prod_{i=1}^{n} T_i$ where $T_i = \{1\}$ for one value of i, and $T_i = A_i$ for all other values of i. Notice that $\widetilde{\Lambda_{\mathcal{F}}}$ is precisely the family of subsets determined by Holland schemata having exactly one fixed position, and that fixed position is equal to 1.

The following fact is the central result of [6]:

Theorem 8 *Let $\Omega_1 = (\Omega_1, \mathcal{F}_1, \mathcal{M}_1)$ and $\Omega_2 = (\Omega_2, \mathcal{F}_2, \mathcal{M}_2)$ denote search systems with $\mathcal{F}_2$ and $\mathcal{M}_2$ being composition closed, and let $\delta : \Omega_1 \to \Omega_2$ be a function. Then the following are equivalent:*

    1. $S \in \widetilde{\Lambda_{\mathcal{F}_2}} \implies \delta^{-1}(S) \in \Lambda_{\mathcal{F}_1}$.
    2. $S \in \Lambda_{\mathcal{F}_2} \implies \delta^{-1}(S) \in \Lambda_{\mathcal{F}_1}$.
    3. $\delta : \Omega_1 \to \Omega_2$ is a morphism of search systems.

In the current paper we shall be primarily concerned with the issue of whether or not a given search system can be embedded into a search system representing the binary semi-genetic or the binary genetic algorithm in the sense of Definition 4. We therefore introduce the following definition.

Definition 9 We say that a given search system $\Omega = (\Omega, \mathcal{F}, \mathcal{M}_1)$ is *semi-genetic (genetic)* if it can be embedded, in the sense of Definition 4, into the search system $(\{0, 1\}^n, \mathcal{F}_M, \mathcal{M})$ for some n where $\mathcal{F}_M$ is the family of all semi-crossover transformations (the family of all masked crossover transformations) and $\mathcal{M}$ is the family of mutations (the same in all of the examples) as introduced in Section 4.

In [6] necessary and sufficient conditions for a given search system to be semi-genetic have been established[1]:

Theorem 10 *Given a search system $\Omega = (\Omega, \mathcal{F}, \mathcal{M})$, the following are equivalent:*

    1. Ω can be embedded into an n-dimensional semi-genetic search system for some n.
    2. $\forall x, y \in \Omega$ with $x \neq y$ we have either $x \notin S_y^{\Omega}$ (see Definition 2) or vice versa: $y \notin S_x^{\Omega}$.
    3. $\forall x, y \in \Omega$ with $x \neq y$ we have $S_x^{\Omega} \neq S_y^{\Omega}$. (Another way to say this is that the map sending x to S_x^{Ω} is one-to-one.)

Moreover, if an embedding exists for some n, then there exists one for $n = |\Omega|$. We also must have $n \geq \lceil \log_2 |\Omega| \rceil$.

Once we are equipped with Theorem 8, it is not hard to establish a criterion analogous to Theorem 10 for a given search system to be genetic:

Theorem 11 *A given search system $\Omega_1 = (\Omega_1, \mathcal{F}_1, \mathcal{M}_1)$ is genetic if and only if for each $x \neq y$ in Ω_1 there exists a complementary pair of invariant subsets A and B ($A \cap B = \emptyset$ and $A \cup B = \Omega_1$, and $A, B \in \Lambda_{\mathcal{F}}$) with $x \in A$ and $y \in B$.*

---

[1] In fact, a lot more has been accomplished in [6] and in [7]. A one-to-one correspondence between all possible morphisms (re-encodings) of the search space by a binary semi-genetic (genetic) algorithm and certain corresponding collections of ordered tuples of invariant subsets of the search space has been established (see Theorems 14 and 20 of [6])

Proof. Suppose first that Ω_1 is genetic. Let $\delta : \Omega_1 \rightarrow \Omega$ be an embedding where $\Omega = (\{0, 1\}^n, \mathcal{F}_M, \mathcal{M})$ is the search system describing the binary genetic algorithm with masked crossover as in Section 4. Fix $x \neq y \in \Omega_1$. Then, since δ is an embedding, $\delta(x) = (x_1, x_2, \ldots, x_n) \neq \delta(y) = (y_1, y_2, \ldots, y_n) \in \{0, 1\}^n$. But this means that $x_i \neq y_i$ for some i. Consider the Holland schemata $H_1 = \{0, 1\}^{i-1} \times \{x_i\} \times \{0, 1\}^{n-i}$ and $H_2 = \{0, 1\}^{i-1} \times \{y_i\} \times \{0, 1\}^{n-i}$. Notice that $\delta(x) \in H_1$ and $\delta(y) \in H_2$, $H_1 \cap H_2 = \emptyset$ and $H_1 \cup H_2 = \{0, 1\}^n$. Now simply let $A = \delta^{-1}(H_1) \in \Lambda_{\mathcal{F}}$ and $B = \delta^{-1}(H_2) \in \Lambda_{\mathcal{F}}$. Since δ is a morphism, A and B are as required.

Now suppose that for all $x \neq y \in \Omega_1$ there exists a complementary pair of invariant subsets A and B with $x \in A$ and $y \in B$. Then we can choose a sequence of invariant sets $A_1, A_2, \ldots, A_n$ with invariant complements A_i^c such that each pair $x \neq y \in \Omega_1$ is separated by the chosen sets, i.e., there is i with $x \in A_i$ and $y \in A_i^c$ or vice versa. Now consider the map $\delta : \Omega_1 \rightarrow \{0, 1\}^n$ defined as follows for all $x \in \Omega_1$:

$$\delta(x) = (x_1, x_2, \ldots, x_n) \text{ where } x_i = \begin{cases} 1 & \text{if } x \in A_i \\ 0 & \text{if } x \in A_i^c \end{cases}. \text{ We observe that } \delta \text{ is a mor-}$$

phism: Indeed, according to examples following Theorem 8, Holland schemata with one fixed position (subsets of the form $H_i^1 = \{0, 1\}^{i-1} \times \{1\} \times \{0, 1\}^{n-i}$ and $H_i^0 = \{0, 1\}^{i-1} \times \{0\} \times \{0, 1\}^{n-i}$) form a base of $\Lambda_{\mathcal{F}_M}$ and we have $\delta^{-1}(H_i^1) = A_i \in \Lambda_{\mathcal{F}_1}$ and $\delta^{-1}(H_i^0) = A_i^c \in \Lambda_{\mathcal{F}_1}$ so that $\delta : \Omega_1 \rightarrow \Omega$ is a morphism of search systems thanks to Theorem 8. It remains to show that δ is one-to-one: Fix $x \neq y \in \Omega_1$. Then for at least one i we have $x \in A_i$ and $y \in A_i^c$ or vice versa. But then we have $x_i \neq y_i$ and so, according to the definition of δ, $\delta(x) \neq \delta(y)$, and the desired conclusion follows. We deduce now that δ is an embedding so that Ω_1 is, indeed, genetic.   □

6   Complexity of Deciding if a Given Search System Is Genetic

In the previous section we have summarized the results which establish some conditions to tell us when a given algorithm can be re-encoded by a binary semi-genetic or by a binary genetic algorithm. In the current section we shall investigate the complexity of deciding whether or not condition 2 of Theorem 10 and the condition of Theorem 11 are satisfied. We shall see below that deciding whether or not a given algorithm is semi-genetic (in the sense of Definition 9) can be done in polynomial time, while deciding if a given algorithm is genetic (also in the sense of Definition 9) is an NP-complete problem. Here both "polynomial time" and "NP" are with respect to the size of the representation, i.e., $|\Omega| + |\mathcal{F}|$.

Theorem 12 *The following problem can be solved in polynomial time with respect to the size of the input provided that there exists a constant q such that for every $F \in \mathcal{F}$ the computation of $F(x, y)$ is done in $O(n^q)$ steps[2].*

Instance of the problem: A search system $\Omega = (\Omega, \mathcal{F}, \mathcal{M})$ and individuals $x, y \in \Omega$.
Question: Is it true that $y \notin S_x$?

---

[2] A very reasonable assumption since there would be little point in running such an evolutionary search algorithm otherwise

Proof. We construct a polynomial time procedure to answer this question: According to proposition A.4 of [5] we have a nested chain of inclusions $\{x\} \subseteq \mathcal{F}(\{x\}) \subsetneq \mathcal{F}(\mathcal{F}(\{x\})) \subsetneq \ldots \subsetneq \mathcal{F}^k(\{x\}) = \mathcal{F}^{k+1}(\{x\}) = S_x$. Moreover, it is clear that $k \leq |\Omega|$. The following algorithm will therefore answer the question: "Is it true that $y \notin S_x$?"

Step 1: Set $K := \{x\}$ and $l = 1$

Step 2: For each $(u, v) \in K^2$ and each $T \in \mathcal{F}$ compute $T(u, v)$. If $T(u, v) = y$ then stop and return "no". If not, let $K := K \cup \{T(u, v) \,|\, (u, v) \in K$ and $T \in \mathcal{F}\}$ and $l := l + 1$. If $l = |\Omega| + 1$ then stop and return "yes". Otherwise repeat step 2.

It remains to show that the algorithm above solves the problem in $O(n^m)$ steps where m is a fixed integer and $n = |\Omega| + |\mathcal{F}|$. But notice that the computational part of step 2 takes no longer than $|\Omega|^2 \cdot |\mathcal{F}| \cdot O(n^q)$ steps for the integer q such that for every $F \in \mathcal{F}$ the computation of $F(x, y)$ is done in $O(n^q)$ steps (see the assumption). Moreover, step 2 is repeated at most $|\Omega| \leq n$ times so that the total amount of time it takes the algorithm to run is $|\Omega| \cdot |\Omega|^2 \cdot |\mathcal{F}| \cdot O(n^q) \leq O(n^{q+4})$ steps and the argument is complete. $\qquad\square$

Since the number of pairs of elements in a search space is quadratic (we only need that it is bounded by a polynomial) with respect to the size of the search space itself, Theorem 12 together with Theorem 10 immediately implies:

Corollary 13 *Given a search system $\Omega = (\Omega, \mathcal{F}, \mathcal{M})$ the decision whether or not Ω is semi-genetic can be made in polynomial time with respect to the size of the input provided that computation of $F(x, y)$ for all x, y can be done in polynomial time uniformly (meaning that the time bound is independent of x and y).*

The situation turns out to be less pleasant in the case of deciding whether or not a given search system is genetic, i. e. whether the condition of Theorem 11 is satisfied, as the following theorem shows:

Theorem 14 *The following problem is NP-complete.*

    Instance of the problem: A search system $\Omega = (\Omega, \mathcal{F}, \mathcal{M})$ and individuals $x, y \in \Omega$. We assume given a polynomial time algorithm for evaluating the functions in $\mathcal{F}$ on arguments in Ω.

    Question: Does there exist a subset $A \subseteq \Omega$ such that $x \in A$ while $y \in A^c$ and both A and $A^c \in \Lambda_{\mathcal{F}}$?

Proof. It is easy to see that the problem is in NP; just guess an appropriate A and check that it works. The challenging part is to build an appropriate polynomial time reduction. We reduce from the "Not All Equal 3-SAT" problem (see section A9.1, page 259 of [2]) In order to state the "Not All Equal 3-SAT" problem, we need the notion of a literal and of a clause:

Definition 15 Given a set U of boolean variables, a literal over U is either a variable, say a, from U or the negation of the variable $a \in U$, denoted by $\bar{a}$. We sometimes use a bar over a literal that is not a variable; then $\bar{\bar{a}}$ means simply a.

Definition 16 A 3-clause over a set U of boolean variables is a disjunction of some three literals over U. If the literals involved in C are a, b and c then we shall write $C = a \vee b \vee c$.

The "Not All Equal 3-SAT" problem is given by:

Instance: A collection $\mathcal{C}$ of 3-clauses over a set U of variables.

Question: Does there exist a truth assignment such that every clause in $\mathcal{C}$ contains a literal whose value is true and a literal whose value is false?

The "Not All Equal 3-SAT" problem is known to be NP-complete (see [2]). We now proceed building the reduction. So fix an instance of the "Not All Equal 3-SAT", i. e. a collection $\mathcal{C}$ of 3-clauses over a set of variables U. Let the search space Ω consist of all the literals involved in the clauses of $\mathcal{C}$. We still have to define the family of reproduction transformations $\mathcal{F}$ and the distinguished pair of points x and y. ($\mathcal{M}$ does not play any role here since it is always assumed to be ergodic.) We define one such transformation for every clause $C \in \mathcal{C}$, say $C = a \vee b \vee c$, where a, b and c denote arbitrary literals. We define the transformation $T_C : \Omega^2 \to \Omega$ as follows:

$$T_C(a, b) = T_C(b, a) = \bar{c}$$
$$T_C(a, c) = T_C(c, a) = \bar{b}$$
$$T_C(b, c) = T_C(c, b) = \bar{a}.$$

For all other pairs $(u, v) \in \Omega^2$ we define $T_C(u, v) = u$. Now, for each $w \in \Omega$ define a transformation $F_w : \Omega^2 \to \Omega$ as follows: For pairs of the form $(a, \bar{a})$ let $F_w(a, \bar{a}) = w$. For all other pairs let $F_w(u, v) = u$. Finally let $\mathcal{F} = \{T_C \mid C \in \mathcal{C}\} \cup \{F_w \mid w \in \Omega\}$. To define the distinguished pair of points in Ω, fix a literal $d \in \Omega$ and let $x = d$ and $y = \bar{d}$. We have now constructed an instance of the problem we are interested in. Clearly the construction above is done in polynomial time with respect to the input size. It remains to show that there exists a truth assignment such that every clause contains a literal whose value is true and a literal whose value is false $\iff$ there exists a subset $A \subseteq \Omega$ such that $x \in A$ while $y \in A^c$ and both A and $A^c \in \Lambda_{\mathcal{F}}$.

Proof (of the $\implies$ direction). Fix a truth assignment f as in the assumption. Let $A = \{u \mid f(u) = T\}$. Then $A^c = \{u \mid f(u) = F\}$. Since $x = d$ and $y = \bar{d}$, it must be the case that either $x \in A$ and $y \in A^c$ or vice versa. It only remains to show that both A and A^c are invariant under $\mathcal{F}$. So, fix individuals (literals) u and $v \in A$ (or u and $v \in A^c$). Choose any transformation $G \in \mathcal{F}$. Then $G = F_w$ for some literal $w \in \Omega$ or $G = T_C$ for some clause $C = a \vee b \vee c$. Now observe that we can't have $(u, v) = (a, \bar{a})$ for any $a \in \Omega$ since both u and v have the same truth value. Therefore, every transformation of the form F_w maps (u, v) into u and, hence, leaves both A and A^c invariant. Every T_C does the same thing unless both u and v appear in some clause C. Let z denote the remaining literal in the clause C. Since $f(u) = f(v)$, we must have $f(z) \neq f(u)$ (this was the assumption about the truth assignment). But then $f(T_C(u, v)) = f(\bar{z}) = f(u) = f(v)$ and so T_C leaves both A and A^c invariant. This shows that both A and A^c are invariant under $\mathcal{F}$. $\qquad\square$

Proof (of the $\impliedby$ direction). Now suppose there exist a complementary pair of subsets of Ω invariant under $\mathcal{F}$, say A and A^c, with $x \in A$ and $y \in A^c$. We have to produce a truth assignment f on Ω such that, for each clause $C \in \mathcal{C}$, not all the literals have the same truth value. Now for all literals $u \in A$ let $f(u) = T$ and for $u \in A^c$ let $f(u) = F$. We still have to show that f is a well-defined truth assignment, meaning that literals u and $\bar{u}$ are never assigned the same truth value, or, equivalently, it is not the case that both

u and $\bar{u} \in A$ or both u and $\bar{u} \in A^c$. This is precisely the purpose of the transformations F_w: if, say, u and $\bar{u} \in A$, then for all $z \in \Omega$ we have $F_z(u, \bar{u}) = z \in A$ and therefore $A = \Omega$, contradicting the fact that $A^c \neq \emptyset$ since $y \in A^c$. Analogously, the assumption that u and $\bar{u} \notin A$ leads to a contradiction. We conclude now that f is, indeed, a well-defined truth assignment. It remains to show that, for each clause $C \in \mathcal{C}$, not all the literals have the same truth value. So fix a clause $C \in \mathcal{C}$. Say $C = a \vee b \vee c$. Since there are exactly three literals in C, some two of them must have the same truth value. Without loss of generality assume these are a and b. Whichever of A and A^c contains a and b must, since it is invariant, also contain $T_C(a, b) = \bar{c}$. By what we have already proved, while checking that f is well-defined, this set cannot contain c. Thus, c does not have the same truth value as a and b. The desired conclusion is now established.   □

We have shown now that "Not All Equal 3-SAT" problem can be reduced in polynomial time to the problem in the statement of the theorem, and "Not All Equal 3-SAT" is known to be NP-complete. The desired conclusion now follows.   □

Theorem 14 gives us the NP-completeness of testing whether, for a given Ω and $\mathcal{F}$, a given pair $x \neq y$ can be separated by complementary, invariant sets. To determine whether $(\Omega, \mathcal{F}, \mathcal{M})$ is genetic, we would have to test whether *every* pair $x \neq y$ in Ω can be separated. This problem is clearly also in NP, since it amounts to checking $|\Omega|^2 - |\Omega|$ instances of the question in Theorem 14. (In fact, by symmetry, it suffices to check only $\dfrac{|\Omega|(|\Omega| - 1)}{2}$ instances.) But we cannot immediately conclude that it is NP-complete. It seems harder than the problem in Theorem 14, since it involves many instances of that problem, but it is conceivable that there could be a way to determine whether all pairs are separable without checking each (or any) one individually. The following theorem settles this question. For brevity, we say that two elements can be separated (in a given search system) if there is a complementary pair of invariant sets, each containing one of the two elements.

Theorem 17 *The following problem is NP-complete.*
Instance of the problem: A search system $\Omega = (\Omega, \mathcal{F}, \mathcal{M})$ with a polynomial time algorithm for evaluating the functions in $\mathcal{F}$ on arguments in Ω.
Question: Can every pair of distinct elements of Ω be separated?

Proof. We have already observed that the problem in the theorem is in NP. To prove completeness, we reduce the problem from Theorem 14 to the problem in the present theorem. This will suffice, since the former problem is already known to be NP-complete.

So let Ω, $\mathcal{F}$, and x, y constitute an instance of the problem from Theorem 14. We must convert it, by a polynomial time computation, into Ω' and $\mathcal{F}'$ such that x and y can be separated in Ω if and only if all pairs of distinct elements can be separated in Ω'. Here we have simplified notation by ignoring the $\mathcal{M}$ component of search systems, since it is irrelevant to the problem.

We may assume, when constructing Ω' and $\mathcal{F}'$, that $x \neq y$. Indeed, if $x = y$ then they obviously cannot be separated, so we need only produce some Ω' and $\mathcal{F}'$ in which not all pairs of distinct elements can be separated, and this task is trivial. We may further assume that $\mathcal{F}$ is nonempty. Indeed, if $\mathcal{F} = \emptyset$, then all subsets of Ω are invariant, so x and y (being distinct) can be separated, and we need only produce Ω' and $\mathcal{F}'$ in which all pairs can be separated; this too is trivial.

Henceforth, we therefore assume that $\mathcal{F}$ is nonempty and that x and y are distinct. We define Ω' to be the following set, consisting of four copies of Ω plus two additional elements.

$$\Omega' = (\Omega \times \{1, 2, 3, 4\}) \cup \{5, 6\}.$$

We abbreviate the ordered pairs $\langle a, 1 \rangle$ in Ω' as $a1$, and similarly with 2, 3, or 4 in place of 1. The family $\mathcal{F}'$ contains, for each $f \in \mathcal{F}$, an associated function $f' : (\Omega')^2 \to \Omega'$ defined as follows.

$$f'(a1, b2) = f(a, b)3$$
$$f'(i, aj) = a4 \text{ for } i \in \{5, 6\} \text{ and } j \in \{1, 2, 3\}$$
$$f'(x4, y4) = 5$$
$$f'(y4, x4) = 6$$
$$f'(u, v) = u \text{ for all } u, v \text{ not covered by the previous lines.}$$

We have used here the assumption that $x \neq y$ because otherwise the third and fourth lines of the definition of f' would contradict each other. It is clear that Ω' and $\mathcal{F}'$ can be computed from Ω and $\mathcal{F}$ in polynomial time. We must verify that they have the required separation properties, and we break this verification into two lemmas.

Lemma 18 *In $(\Omega', \mathcal{F}')$, every two distinct elements, except possibly 5 and 6, can be separated.*

Proof. By inspecting the definition of f', we find that the following sets and their complements (in Ω') are invariant under $\mathcal{F}'$.

1. $(\Omega \times \{4\}) \cup \{5, 6\}$
2. $(\Omega \times \{1, 4\}) \cup \{5, 6\}$
3. $(\Omega \times \{2, 4\}) \cup \{5, 6\}$
4. $((\Omega - \{a\}) \times \{4\}) \cup \{5, 6\}$ for any $a \in \Omega$.

These sets suffice to separate any two elements of Ω' that come from different sets in the following list, which partitions Ω':

$$\Omega \times \{1\}, \quad \Omega \times \{2\}, \quad \Omega \times \{3\}, \quad \Omega \times \{4\}, \quad \{5, 6\}$$

It remains to separate any two distinct elements from the same block of this partition except for 5 and 6.

If the two points come from $\Omega \times \{4\}$, then they can be separated by a set as in item 4 of the list above. If they come from $\Omega \times \{1\}$, then, calling them $a1$ and $b1$, we can separate them because the singleton $\{a1\}$ and its complement are invariant. The same argument applies if they come from $\Omega \times \{2\}$. Finally, if they are $a3$ and $b3$, then they are separated by

$$(\Omega \times \{1, 4\}) \cup \{a3, 5, 6\}$$

and its complement, both of which are invariant.   □

Lemma 19 *5 and 6 can be separated in* $(\Omega', \mathcal{F}')$ *if and only if x and y can be separated in* $(\Omega, \mathcal{F})$.

Proof. Suppose first that A and A^c are invariant and separate x and y in Ω. Then $(A \times \{1,2,3,4\}) \cup \{5\}$ and $(A^c \times \{1,2,3,4\}) \cup \{6\}$ are invariant and separate 5 and 6 in Ω'.

Conversely, suppose B and B^c are invariant in Ω' and separate 5 and 6; say $5 \in B$ and $6 \in B^c$. By virtue of the second equation in the definition of f' (applied with $i = 5$) and the invariance of B, we know that if B contains $a1$, $a2$, or $a3$, then it must also contain $a4$ (for the same $a \in \Omega$). (We have used here the assumption that $\mathcal{F} \neq \emptyset$, because we need an f' to use in this invariance argument.) The same argument applies to B^c if we use $i = 6$ instead of 5. As a result, for each $a \in \Omega$, all four of $a1$, $a2$, $a3$, and $a4$ lie in the same one of B and B^c. That is, there is an $A \subseteq \Omega$ such that

$$B = (A \times \{1,2,3,4\}) \cup \{5\} \quad \text{and} \quad B^c = (A^c \times \{1,2,3,4\}) \cup \{6\}.$$

By virtue of the first equation in the definition of f', the invariance of B and B^c under f' implies the invariance of A and A^c under f. Furthermore, if x and y were in the same one of A and A^c, then the last two equations in the definition of f' would force 5 and 6 into the same one of B and B^c, contrary to our assumption. Thus, A and A^c are invariant subsets of Ω separating x and y, as required.   □

The two lemmas together tell us that we have a reduction of the problem in the present theorem to the one in Theorem 14.   □

7 Conclusions

In the current paper it has been rigorously established that deciding whether or not a given search algorithm can be re-encoded by a binary genetic algorithm is, in the very general case, a complicated (NP-complete) problem. It should be pointed out though, that many well-known types of algorithms, such as the non-linear genetic programming with homologous crossover, can be easily embedded into a binary genetic algorithm. The situation is somewhat analogous to that in analysis: according to the Brownian motion model, the path of a particle is continuous and nowhere differentiable with probability 1, while most continuous functions used in calculus (various combinations of elementary functions, their integrals...) are at least piecewise differentiable.

References

1. Davis, L. (1985) Applying Adoptive Algorithms to Epistatic Domains, *Proceedings of the International Joint Conference on Artificial Intelligence*, pages 162-164.
2. Garey, M. and Johnson, D. (1979) Computers and Intractability. A Guide to the Theory of NP-Completeness. W. H. Freeman and Company.
3. Mac Lane, S. (1971) Categories for the Working Mathematician. *Graduate Texts in Mathematics 5*, Springer-Verlag.
4. Michalewicz, Z. (1996). Genetic algorithms + data structures = evolution programs. Springer-Verlag.

5. Mitavskiy B. (2004). Crossover invariant subsets of the search space for evolutionary algorithms. *Evolutionary Computation*, 12(1): 19-46.
   http://www.math.lsa.umich.edu/~bmitavsk/

6. Mitavskiy B. (2003). Comparing evolutionary computation techniques via their representation. In E. Cantú-Paz *et al.* editors, *Proceedings of the Genetic and Evolutionary Computation Conference (GECCO)*, Vol. 1, pages 1196-1209, Springer-Verlag.

7. Mitavskiy, B. (2003). A category theoretic method for comparing evolutionary computation techniques via their representation. In B. ten-Cate, editor, *Proceedings of the Eighth ESSLLI Student Session*, pages 201 - 210.

8. Poli, R. (2000). Hyperschema Theory for GP with One-Point Crossover, Building Blocks, and Some New Results in GA Theory. In R. Poli, W. Banzhaf, and *et al.*, editors, *Genetic Programming, Proceedings of EuroGP'2000*, Springer-Verlag

9. Radcliffe, N. (1994). The algebra of genetic algorithms. *Annals of Mathematics and Artificial Intelligence*, 10:339-384. http://users.breathemail.net/njr/papers/amai94.pdf

10. Rothlauf F., Goldberg D., Heinzl A. (2002). Network random keys - a tree representation scheme for genetic and evolutionary algorithms. *Evolutionary Computation*, 10(1): 75 - 97.

11. Rowe, J., Vose, M., and Wright, A. (2002). Group properties of crossover and mutation. *Evolutionary Computation*, 10(2): 151-184.

12. Stephens, C. (2001). Some exact results from a coarse grained formulation of genetic dynamics. *Proceedings of the Genetic and Evolutionary Computation Conference (GECCO)*, pages 631-638, Morgan Kaufmann.

13. Vose, M. (1999). The Simple Genetic Algorithm: Foundations and Theory. *MIT Press*.

14. Vose, M. and Wright, A. (2001). Form invariance and implicit parallelism. *Evolutionary Computation*, 9(3):355-370.

Compact Genetic Codes
as a Search Strategy of Evolutionary Processes

Marc Toussaint

Institute for Adaptive and Neural Computation,
University of Edinburgh, 5 Forrest Hill, Edinburgh EH1 2QL, Scotland, UK
mtoussai@inf.ed.ac.uk

Abstract. The choice of genetic representation crucially determines the capability of evolutionary processes to find complex solutions in which many variables interact. The question is how good genetic representations can be found and how they can be adapted online to account for what can be learned about the structure of the problem from previous samples. We address these questions in a scenario that we term indirect Estimation-of-Distribution: We consider a decorrelated search distribution (mutational variability) on a variable length genotype space. A one-to-one encoding onto the phenotype space then needs to induce an adapted phenotypic variability incorporating the dependencies between phenotypic variables that have been observed successful previously. Formalizing this in the framework of Estimation-of-Distribution Algorithms, an adapted phenotypic variability can be characterized as minimizing the Kullback-Leibler divergence to a population of previously selected individuals (parents). Our core result is a relation between the Kullback-Leibler divergence and the description length of the encoding in the specific scenario, stating that compact codes provide a way to minimize this divergence. A proposed class of Compression Evolutionary Algorithms and preliminary experiments with an L-system compression scheme illustrate the approach. We also discuss the implications for the self-adaptive evolution of genetic representations on the basis of neutrality (σ-evolution) towards compact codes.

1 Introduction

The complexity of a problem largely depends on the interactions between variables of a solution. A stochastic search process like evolution will perform well on a complex problem only when the search distribution is adapted to these interactions, i.e., when the search distribution obeys these dependencies between variables. The ability of natural evolution to find highly complex structured organisms, in which many variables interact in determining the fitness, can only be understood when acknowledging that evolution did not pursue an exhaustive search but "learned" to shape the search distribution towards complex, highly structured organisms (see also [25]).

In evolutionary processes the search distribution is determined by the variational operators (mutation and recombination) and, in the case of indirect en-

A.H. Wright et al. (Eds.): FOGA 2005, LNCS 3469, pp. 75–94, 2005.

codings, by the choice of genetic representation. The role of the genetic representation seems to become particularly important when large and regularly or hierarchically structured solutions need to be found. The eye-less gene [8] is an impressive example of a genetic representation specifically designed to induce modular structured variability in nature.

The general questions are how to find genetic representations that induce the desired dependencies on the phenotype space, and how they can be adapted online to account for what can be learned about the structure of the problem from previous samples. There have been various approaches to characterize what a good representation is, considering, for example, all possible representations, Gray vs. binary codes, redundant representations, and recursive encodings [4, 11, 12, 18, 27]. Theoretical approaches concerning the adaptation of the representation based on the specific current population include, for example, Estimation-Of-Distribution Algorithms [17], Walsh analysis [10], and Maximal Entropy principles [28].

Our approach is to consider a specific scenario that we term indirect Estimation-of-Distribution: We assume that the search distribution (mutational variability) on a variable length genotype space is decorrelated. A one-to-one encoding onto the phenotype space then needs to induce a properly structured phenotypic variability. The idea of this scenario is that the encoding receives all responsibilities to induce the structural properties of the phenotypic search distribution, leaving a simple problem of unstructured (decorrelated) adaptation on the genotype level.

We formalize the scenario within the framework of Estimation-of-Distribution Algorithms, where an adapted phenotypic variability can be characterized as minimizing the Kullback-Leibler divergence to a population of previously selected individuals (parents). Our core result is a relation between the Kullback-Leibler divergence and the description length of the encoding in the specific scenario of indirect Estimation-of-Distribution based on a variable length genotype space. The result states that compact genetic codes provide a way to minimize this divergence – and may thus be seen as transferring similar results on Minimum Description Length in the context of modeling, in particular model selection, (e.g., [5, 24]) to the specific domain of evolutionary search based on adaptive representations.

An intuitive way to grasp these results might be the following: Consider a set of good (selected) individuals in the phenotype space. In this parent population there will generally exist dependencies between phenotypic variables of the individuals, measurable as mutual information between them. This information on the dependencies – that stem from selection and have their origin in the structure of the problem – is what should be extracted and exploited for further search. Assume we can map these individuals on variable size strings such that the average string length is minimized. Before the compression there was mutual information between the (phenotypic) variables. After the compression, there should be no mutual information between the (genetic) symbols that describe the individuals because otherwise the mapping would not be a minimum description length compression. Thus, a compression is one way (among others) to map on genotypic representations in which symbols are decoupled (i.e., to map on a factorial code). A compression can also be considered as an implicit analysis of the dependencies that have been present in the parent population because it is able to dissolve them by introducing new symbols. Eventually, the key idea is that *inverting* the compression is a mechanism to induce exactly these dependencies. In other words, when there is noise (decorrelated mutational variability) on the genetic symbols, this should translate to a phenotypic variability that obeys these dependencies.

The following two sections will introduce the theoretical framework, including the indirect induction of a search distribution and the principle of Estimation-of-Distribution Algorithms. Section 4 derives the main results on the relation between compact codes and Estimation-of-Distribution. Section 5 aims to illustrate the approach by proposing a class of Compression Evolutionary Algorithms and presenting preliminary experiments with a Compression GA based on a simple L-system compression scheme (similar to the ideas of Nevill-Manning and Witten [13]). This leads to the discussion, summarizing the main results, but also considering issues on the choice of compression technique, accumulative versus each-time-step compression, and implications for understanding the fully self-adaptive evolution of genetic representations on the basis of neutrality (σ-evolution) towards compact codes.

2   Indirect Induction of Search Distributions

Let P be the search space. A heuristic search scheme is a process in the space L^P of distributions over the search space in which a search distribution $q \in L^P$ is propagated iteratively. In each iteration, samples from q are drawn, evaluated, and the outcome of evaluation is used to design, according to some heuristic, a new search distribution in the next step.

For instance, in ordinary Evolutionary Algorithms (EAs) the search distribution is given by a finite parent population and recombination and mutation operators. This search distribution is sampled, leading to the finite offspring population, which is then evaluated, leading to the selection probability distribution over these offspring. The heuristic to generate the new search distribution in simple EAs is to sample the selection probability distribution, leading to a new parent population which in turn induces a new search distribution. We do not need to specify these operators here explicitly. We develop the theory on the abstract level of search distributions.

In this paper we are interested in indirect codings of search points. In the heuristic search framework, this means that a distribution $\tilde{q} \in L^G$ over another search space G, the *genotype space*, is maintained. The search distribution q over the actual search space P (*phenotype space*) is then given indirectly via a *coding* $\phi : G \to P$,

$$q(x) = \sum_{g \in [x]_\phi} \tilde{q}(g) \, , \quad \text{where } x \in P, \ g \in G, \ [x]_\phi = \{g \in G \mid \phi(g) = x\} \, .$$

We also use the short notation $q = \tilde{q} \circ \phi^{-1}$ for this projection of $\tilde{q}$ under ϕ. The set $[x]_\phi$ of all genotypes mapping to the same phenotype x is an equivalence class under ϕ, also called *neutral set* of x.

Three additional constraints define the "indirect encoding case" considered in this paper: First, we impose that $\phi : G \to P$ shall be bijective (one-to-one). We denote the space of all bijective codings $G \to P$ by Φ. The discussion of self-adaptation in the case of non-injective codings in section 6 will clarify this constraint.

Second, G is the space of variable length strings over some finite alphabet $\mathcal{A}$,

$$G = \bigcup_{l=1}^{\infty} \mathcal{A}^l .$$

It is technically unclear how to define a marginal over the ith symbol (or mutual information between symbols) directly for a variable length distribution $\tilde{q} \in \mathrm{L}^G$. Thus we will consider the decomposition

$$\tilde{q}(g) = \tilde{q}(g|l)\, \tilde{q}(l) , \quad \text{where } g \in G, \; l = \text{length}(g) \in \mathbb{N} .$$

Here, $\tilde{q}(l)$ is a distribution over the genotype length $l \in \mathbb{N}$, and $\tilde{q}(g|l)$ the conditional distribution over a fixed length alphabet $\mathcal{A}^l$. We use the short notation $\tilde{q}_l \equiv \tilde{q}(\cdot|l)$ for this length-conditioned distribution. The marginal $\tilde{q}_l^i$ over the ith symbol ($i \leq l$) and the mutual information $I(\tilde{q}_l) = \sum_i H(\tilde{q}_l^i) - H(\tilde{q}_l)$ can then be defined as usual[1].

As the third constraint we limit the space of possible search distributions in a certain way: We impose that the length-conditioned distributions $\tilde{q}_l$ on the genotype space have to factorize. We denote the set of feasible genotype distributions by

$$\tilde{Q} \subseteq \{\tilde{q} \in \mathrm{L}^G \mid \forall l : \; I(\tilde{q}_l) = 0\} .$$

Putting both together, we have the space $Q \subseteq \mathrm{L}^P$ of feasible search distribution over P as

$$Q = \{\tilde{q} \circ \phi^{-1} \mid \phi \in \Phi, \; \tilde{q} \in \tilde{Q}\} . \tag{1}$$

In summary, indirect induction of the search distribution means that, in order to design a search distribution $q \in \mathrm{L}^P$ we have to pick a bijective coding $\phi \in \Phi$ and a decorrelated distribution $\tilde{q} \in \tilde{Q}$ on the genotype space. In other terms, every feasible search distributions q correspond to a pair $(\phi, \tilde{q})$.

---

[1] For a distribution q over some product space $\mathcal{A} \times \cdots \times \mathcal{A}$, we generally denote the marginal over the ith variable by q^i. The mutual information $I(q) = \sum_i H(q^i) - H(q)$ measures all dependencies between variables of any order (not only pair-wise dependencies)

3 Estimation-of-Distribution

What is a reasonable heuristic to design a search distribution given the results of evaluation of previous sample points? We will follow here the idea of Estimation-Of-Distribution Algorithms [17] which can be described as follows.

We assume that the outcome of evaluation is given as a distribution p over P, typically the parent population[2]. Given that the space of feasible search distributions is limited to Q, a simple heuristic to chose the new search distribution is to pick the one that is most similar to p. Similarity can be measured by the Kullback-Leibler divergence (KLD) $D(p : q) = \mathrm{E}_p \left\{ \log \frac{p(x)}{q(x)} \right\}$ between two distributions, which also captures the structural similarity between two distributions in the sense of the similarity of the different order dependencies (see [1, 23], please note the relations between the KLD, log-likelihood, free energy, and mean energy[3]). Thus, one heuristic to design the new search distribution q' reads

$$q' = \underset{q \in Q}{\operatorname{argmin}} \, D(p : q) \,. \tag{2}$$

We term this specific kind of an EDA *KL-search*. In general, the crucial parameter of KL-search is the choice of the set Q of feasible search distributions. On the one hand, the choice of Q determines the computational cost of the minimization (2) in every step. On the other hand, it determines the algorithm's capability to exploit the structure observable in p.

Some algorithms of the class of EDAs are exact instantiations of KL-search: MIMIC [6] chooses Q to be the set of Markov chains, PBIL [2] chooses Q as the set of factorized distributions. Other EDAs differ from KL-search in the choice of the similarity measure (they use alternatives to the KLD, e.g., BOA [16] takes a Bayesian Dirichlet Metric). But all of them can distinctly be characterized by their choice of Q, which may also be the set of dependency trees [3], Bayesian networks (BOA, [16]), or Bayesian networks with decision trees at each node (hBOA, [15]).

Despite its conceptual simplicity, the minimization required in each iteration of KL-search can be computationally very expensive, depending on the complexity of the distributions in Q. When only simple distributions, like factorized distributions (PBIL) or Markov chains (MIMIC) are allowed, the minimization

---

[2] Generally, in this formalism p is meant to encode any information that we receive from evaluations. Typically, p is non-vanishing only on a finite set of samples (the offspring population) and the values of p might be (in a normalized way) the fitness values of these offspring. Alternatively, p might represent a resampling of such a "fitness distribution over offspring", which corresponds to the parent population

[3] With the definitions of the entropy $H(p) = -\mathrm{E}_p \left\{ \log p(x) \right\}$ and the log-likelihood $\mathcal{L}(q) = \mathrm{E}_p \left\{ \log q(x) \right\}$ we have $D(p : q) = -H(p) - \mathcal{L}(q)$. One could roughly say, "minimizing the KLD means maximizing the log-likelihood *and* the entropy". Further, when defining an energy functional $E(x) = -\log q(x)$, the mean energy $E = \mathrm{E}_p \left\{ E(x) \right\} = -\mathcal{L}(q)$ is the negative log-likelihood while the free energy $F = E - H(p)$ is the KLD

can be calculated directly. For more complex distributions (like Bayesian networks, BOA), the minimization itself requires an iterative procedure.

Finally note that distributions in Q should typically be constrained to have a minimum entropy. In that way, a repeated cycle of entropy decrease (in the course of evaluation) and entropy increase (when picking a new search distribution) ensures exploration and prevents the algorithms from early convergence.

4   Indirect Estimation-of-Distribution via Compression

KL-search proposes how to pick a new search distribution out of Q incorporating the knowledge on the evaluation of previous samples. In section 2 we specified a specific Q that defines the indirect coding case, where a choice of q means to pick a bijective coding ϕ and a factorized distribution $\tilde{q}$ on G. Putting this together, KL-search amounts to a heuristic to pick a coding ϕ such that knowledge on previous evaluations is incorporated. In this section we will derive results on how this heuristic to pick a coding reads more explicitly. We first discuss the simpler fixed length case before addressing the general one:

Fixed length case. Let us first assume that G contains only strings of fixed length l, $G = \mathcal{A}^l$. Then the marginals $\tilde{q}^i$ are straight-forward to define and $I(\tilde{q}) = \sum_i H(\tilde{q}^i) - H(\tilde{q}) = 0$ constrains $\tilde{q}$ to vanishing dependencies (of arbitrary order) between genes.

Combining (2) with the definition (1) of Q for the case of indirect codings, we find

$$D(p:q) = \sum_x p(x) \ln \frac{p(x)}{\sum_{g' \in [x]_\phi} \tilde{q}(g')} = \sum_g \tilde{p}(g) \ln \frac{\tilde{p}(g)}{\tilde{q}(g)} \ . \tag{3}$$

Here we defined $\tilde{p}$ as the back-projection of p onto the coding space G, $\tilde{p} = p \circ \phi$. The last step uses that ϕ is bijective such that there exists exactly one $g \in [x]_\phi$. Next we use that $\tilde{q}$ has to factorize, $\tilde{q}(g) = \tilde{q}(g^1, g^2, ..., g^l) = \tilde{q}^1(g^1)\, \tilde{q}^2(g^2) \cdots \tilde{q}^l(g^l)$,

$$D(p:q) = \sum_g \tilde{p}(g) \ln \frac{\tilde{p}(g)}{\tilde{p}^1(g^1) \cdots \tilde{p}^l(g^l)} + \sum_g \tilde{p}(g) \ln \frac{\tilde{p}^1(g^1) \cdots \tilde{p}^l(g^l)}{\tilde{q}^1(g^1) \cdots \tilde{q}^l(g^l)}$$
$$= I(\tilde{p}) + D(\tilde{p}^{(1)} : \tilde{q}) \ . \tag{4}$$

Here we defined $\tilde{p}^{(1)}(g) = \tilde{p}^1(g^1)\, \tilde{p}^2(g^2) \cdots \tilde{p}^l(g^l)$ as the "factorized reduction" of $\tilde{p}$ (i.e, the product of its marginals, see also [1, 23] on how to generally define the kth order reduction $\tilde{p}^{(k)}$ of $\tilde{p}$ containing all and only the dependencies of order $\leq k$ within $\tilde{p}$).

This result states that, in order to follow the KL-search scheme in the indirect coding case, one should find a coding ϕ and a search distribution $\tilde{q}$ such that $I(\tilde{p}) + D(\tilde{p}^{(1)} : \tilde{q})$ is minimized.

Here, I would like to distinguish two cases. In the first case we assume that $\tilde{Q}$ comprises *all* factorized distributions without a bound on the entropy. In this case, no matter which ϕ is chosen, one can always minimize $D\big(\tilde{p}^{(1)} : \tilde{q}\big)$ down to zero by picking $\tilde{q} = \tilde{p}^{(1)}$. Since $I(\tilde{p})$ is independent of $\tilde{q}$, the minimization (2) can be realized by first optimizing ϕ and then picking $\tilde{q}$:

$$q' = \underset{q \in Q}{\operatorname{argmin}} \, D\big(p : q\big) = (\phi, \tilde{q}) \,, \quad \text{where } \phi = \underset{\phi}{\operatorname{argmin}} \, I(\tilde{p}) \text{ and } \tilde{q} = \tilde{p}^{(1)} \,.$$

We call this procedure (first optimizing ϕ, then picking $\tilde{q}$) the *two step procedure*. Note that $\tilde{p}^{(1)}$ in the last equation depends on the ϕ chosen before.

However, in a realistic algorithm, $\tilde{Q}$ should not comprise all factorized distributions but obey a lower bound on the entropy of these distributions to ensure exploration. (Recall that for $\tilde{q} = \tilde{p}^{(1)}$, and since ϕ is bijective, $H(p) = H(\tilde{p}) = H(\tilde{p}^{(1)}) - I(P_G) = H(\tilde{q}) - I(\tilde{p}) = H(q) - I(\tilde{p})$. Therefore, the entropy of search $H(q)$ would only be greater than $H(p)$ by the amount of $I(\tilde{p})$, which is minimized.) Hence, in the second case, when $\tilde{Q}$ is only a subset of all factorized distributions, $D\big(\tilde{p}^{(1)} : \tilde{q}\big)$ can generally not be minimized to zero and the minimization of (2) can *not exactly* be decomposed in the two steps of first minimizing $I(\tilde{p})$ w.r.t. ϕ, and then, for a fixed ϕ, minimizing $D\big(\tilde{p}^{(1)} : \tilde{q}\big)$ w.r.t. $\tilde{q}$. The exact minimization of (2) remains a coupled problem of finding a pair $(\phi, \tilde{q})$.

For completeness, let us estimate a bound on the "error" made when still adopting the two step procedure of minimization. Let $(\phi^*, \tilde{q}^*)$ be a coding and genotype distribution that indeed minimize (2), and let $(\phi', \tilde{q}')$ be the result of the two step procedure, i.e., ϕ' minimizes $I(\tilde{p}')$ and $\tilde{q}'$ minimizes $D\big(\tilde{p}'^{(1)} : \tilde{q}'\big)$ for the given coding ϕ'. Here, $\tilde{p}' = p \circ \phi'$ and $\tilde{p}^* = p \circ \phi^*$. A rough bound for the error made can be estimated as follows, to be explained in detail below,

$$
\begin{aligned}
D\big(p : q'\big) - D\big(p : q^*\big) &= D\big(\tilde{p}'^{(1)} : \tilde{q}'\big) - D\big(\tilde{p}^{*(1)} : \tilde{q}^*\big) + I(\tilde{p}') - I(\tilde{p}^*) \\
&\leq D\big(\tilde{p}'^{(1)} : \tilde{q}'\big) - D\big(\tilde{p}^{*(1)} : \tilde{q}^*\big) \;\leq\; D\big(\tilde{p}'^{(1)} : \tilde{q}'\big) \\
&= \sum_{i=1}^{l} D\big(\tilde{p}'^{i} : \tilde{q}'^{i}\big) \\
&\leq -l \log \tilde{q}'^{i}(a^i) \;=\; -l \log(1 - \alpha) \;\leq\; l\,\hbar \,.
\end{aligned}
$$

The first inequality stems from the fact that ϕ' minimizes $I(\tilde{p}')$ and thus $I(\tilde{p}') \leq I(\tilde{p}^*)$. Since both, $\tilde{p}'^{(1)}$ and $\tilde{q}'$, are factorized distributions, their Kullback-Leibler divergence decomposes into a sum. For each marginal, when there is a lower bound $\hbar$ on the entropy of $\tilde{q}'^{i}$, the divergence $D\big(\tilde{p}'^{i} : \tilde{q}'^{i}\big)$ is particularly large when $\tilde{p}'^{i}$ has very low entropy. In the worst case, $\tilde{p}'^{i}$ has zero entropy, i.e., is non-zero only for a single symbol $a^i \in \mathcal{A}$. In that case $D\big(\tilde{p}'^{i} : \tilde{q}'^{i}\big) = -\log \tilde{q}'^{i}(a^i)$. In order to minimize $D\big(\tilde{p}'^{i} : \tilde{q}'^{i}\big)$, $\tilde{q}'^{i}$ is chosen to have the form of the typical symbol mutation distribution with mutation rate α, where $\tilde{q}'^{i}(a) = 1 - \alpha$ for $a = a^i$ and $\tilde{q}'^{i}(a) = \frac{\alpha}{|\mathcal{A}|-1}$ for $a \neq a^i$. Then, $H(\tilde{q}^i) = -(1-\alpha)\log(1-\alpha) - \alpha \log \frac{\alpha}{|\mathcal{A}|-1}$. Given the lower bound $\hbar$ on the entropy, the minimal mutation rate α can be chosen to ensure $H(\tilde{q}^i) = \hbar$ and $D\big(\tilde{p}^i : \tilde{q}^i\big) = -\log(1-\alpha) \leq \hbar$. Thus, in the worst case, the "error" made when using the two step procedure instead of the exact minimization of (2) is smaller than $l\,\hbar$.

Variable length case. Let us repeat the above derivations in the general case when $G = \bigcup_{l=1}^{\infty} \mathcal{A}^l$ comprises strings of any length over the alphabet $\mathcal{A}$. The constraint of vanishing mutual information in $\tilde{q}$ now refers to the length-conditioned distributions $\tilde{q}_l$, i.e., we impose $I(\tilde{q}_l) = 0$ while $\tilde{q}(l)$ is unconstrained. (Recall $\tilde{q}(g) = \tilde{q}_l(g)\,\tilde{q}(l)$ where $l = \text{length}(g)$.) Equation (3) now leads to

$$D(p\!:\!q) = \sum_l \sum_g \tilde{p}_l(g)\tilde{p}(l) \ln \frac{\tilde{p}_l(g)\tilde{p}(l)}{\tilde{q}_l(g)\tilde{q}(l)}$$

$$= \sum_l \tilde{p}(l) \sum_g \tilde{p}_l(g) \ln \frac{\tilde{p}_l(g)}{\tilde{q}_l(g)} + \sum_l \tilde{p}(l) \ln \frac{\tilde{p}(l)}{\tilde{q}(l)}$$

$$= \sum_l \tilde{p}(l) \sum_g \tilde{p}_l(g) \ln \frac{\tilde{p}_l(g)}{\tilde{p}_l^{(1)}(g)} + \sum_l \tilde{p}(l) \sum_g \tilde{p}_l(g) \ln \frac{\tilde{p}_l^{(1)}(g)}{\tilde{q}_l(g)} + D\big(\tilde{p}(l) : \tilde{q}(l)\big)$$

$$= \mathrm{E}_l\left\{I(\tilde{p}_l)\right\} + \mathrm{E}_l\left\{D(\tilde{p}_l^{(1)} : \tilde{q}_l)\right\} + D\big(\tilde{p}(l) : \tilde{q}(l)\big) \ , \tag{5}$$

where we introduced $\mathrm{E}_l\left\{\cdot\right\}$ as the expectation over $\tilde{p}(l)$ (which depends on ϕ). The entropy of p can be written as

$$H(p) = -\sum_g p(g) \ln p(g) = -\sum_l \tilde{p}(l) \sum_g \tilde{p}_l(g) \ln[\tilde{p}_l(g)\,\tilde{p}(l)]$$

$$= -\sum_l \tilde{p}(l) \Big[\sum_g \tilde{p}_l(g) \ln \tilde{p}_l(g) + \sum_g \tilde{p}_l(g) \ln \tilde{p}(l)\Big]$$

$$= \sum_l \tilde{p}(l) H(\tilde{p}_l(g)) + H(\tilde{p}(l))$$

$$= \sum_l \tilde{p}(l) \Big[\sum_{i=1}^{l} H(\tilde{p}_l^i) - I(\tilde{p}_l)\Big] + H(\tilde{p}(l))$$

$$= \mathrm{E}_l\left\{\sum_{i=1}^{l} H(\tilde{p}_l^i)\right\} - \mathrm{E}_l\left\{I(\tilde{p}_l)\right\} + H(\tilde{p}(l)) \ . \tag{6}$$

Adding equations (5) and (6) we find

Lemma 1. *In the indirect encoding case, for any $p \in L^P$, any bijective encoding $\phi: G \to P$, and any factorized genotype distribution $\tilde{q} \in \tilde{Q}$, we have*

$$D(p\!:\!q) = \mathrm{E}_l\left\{I(\tilde{p}_l)\right\} + \mathrm{E}_l\left\{D(\tilde{p}_l^{(1)} : \tilde{q}_l)\right\} + D\big(\tilde{p}(l) : \tilde{q}(l)\big) \tag{7}$$

$$= \mathrm{E}_l\left\{\sum_{i=1}^{l} H(\tilde{p}_l^i)\right\} + \mathrm{E}_l\left\{D(\tilde{p}_l^{(1)} : \tilde{q}_l)\right\} + D\big(\tilde{p}(l) : \tilde{q}(l)\big) + H(\tilde{p}(l)) - H(p). \tag{8}$$

The second RHS term in equation (8) is a comparison of only the marginals of $\tilde{p}_l$ and $\tilde{q}_l$, the third term is a comparison of the length distributions, the fourth term is the entropy of the genotype length, which depends on ϕ, and the last term depends only on p, not on ϕ or $\tilde{q}$. The first term in equation (8) is of particular interest here. The following bounds show how it relates to the description length. Note that $\log |\mathcal{A}|$ is the maximal entropy of a marginal:

$$\mathrm{E}_l \left\{ \sum_{i=1}^{l} H(\tilde{p}_l^i) \right\} \leq \log |\mathcal{A}| \, \mathrm{E}_l \left\{ \sum_{i=1}^{l} 1 \right\} = \log |\mathcal{A}| \, \mathrm{E}_l \left\{ l \right\} = L_p \log |\mathcal{A}| \, , \qquad (9)$$

where we introduce $L_p = \mathrm{E}_l \{l\}$ as the expected description length of samples of p in the encoding ϕ. On the other hand, from (6), we get

$$\mathrm{E}_l \left\{ \sum_{i=1}^{l} H(\tilde{p}_l^i) \right\} \geq H(p) - H(\tilde{p}(l)) \, . \qquad (10)$$

Note that for an optimally compact coding $L_p \log |\mathcal{A}| = H(p) - H(\tilde{p}(l))$ [4], and both bounds are exact.

We collect these findings in

Lemma 2. *In the indirect encoding case, the expected description length L_p of samples from p (parents) gives an upper bound on $D(p : q)$,*

$$0 \leq D(p : q) - \mathrm{E}_l \left\{ D(\tilde{p}_l^{(1)} : \tilde{q}_l) \right\} - D(\tilde{p}(l) : \tilde{q}(l)) \leq L_p \log |\mathcal{A}| - H(p) + H(\tilde{p}(l)).$$

For an optimally compact encoding ϕ, these bounds are exact, i.e., we have

$$D(p : q) = \mathrm{E}_l \left\{ D(\tilde{p}_l^{(1)} : \tilde{q}_l) \right\} + D(\tilde{p}(l) : \tilde{q}(l)) \, ,$$

and $D(p : q)$ can easily be minimized by adapting the genotype marginals $\tilde{q}_l^i$ and the length distribution $\tilde{q}(l)$ (which can be set equal to $\tilde{p}(l)$).

Let us briefly summarize and discuss these results by emphasizing certain aspects:

1. Compression is doing "more than we need": Reconsider the exact expressions (7) and (8) of lemma 1. We related the term $\mathrm{E}_l \{\sum_i H(\tilde{p}_l^i)\}$ to the description L_p via the bound (9) and showed that this bound is exact for an optimal compression. One should note though that compression is not the only way to minimize the term $\mathrm{E}_l \{\sum_i H(\tilde{p}_l^i)\}$; it can equally be minimized by reducing the marginal entropies $H(\tilde{p}_l^i)$, which means not to exploit the expressional power of the alphabet. This can better be understood going back to expression (7) involving the mutual information: Ultimately, what matters is to reduce $\mathrm{E}_l \{I(\tilde{p}_l)\}$, i.e. finding a factorial code, which can also be done perfectly with very low marginal

---

[4] Here is a slight difference to the usually considered case of channel capacity: In our case, the length of a genome itself can carry information (even for $\mathcal{A} = \{0\}$) of the amount $H(\tilde{p}(l))$ such that the symbols only need to encode $H(p) - H(\tilde{p}(l))$ entropy. In the channel capacity case, where a continuous stream of symbols is transmitted, the exact bound is $L_p \log |\mathcal{A}| = H(p)$, as for the Shannon-Fano code

entropies, not exploiting the alphabet. By relating it to the description length we showed that a compression is reducing $E_l \{I(\tilde{p}_l)\}$ while *additionally* trying to exploit the alphabet optimally. Thus, compression is doing "more than we need" from the strict point of view of minimizing $D(p:q)$. Clearly, this also means that an optimally compact coding is not the only solution to minimize $D(p:q)$ via indirect induction – but it is one.

2. If there are no further constraints on $\tilde{q}$ (e.g., no entropy bound) then $\tilde{q}_l$ and $\tilde{q}(l)$ can always be set equal to $\tilde{p}_l^{(1)}$ and $\tilde{p}(l)$, thus perfectly minimizing $E_l \left\{ D(\tilde{p}_l^{(1)} : \tilde{q}_l) \right\} + D(\tilde{p}(l) : \tilde{q}(l))$. In this case, we have

$$D(p:q) = E_l \{I(\tilde{p}_l)\} = E_l \left\{ \sum_{i=1}^{l} H(\tilde{p}_l^i) \right\} + H(\tilde{p}(l)) - H(p) \ ,$$

and the problem reduces to finding an encoding that extinguishes the mutual information $E_l \{I(\tilde{p}_l)\}$ or, as discussed, an optimal compression.

3. The two step procedure: If $\tilde{Q}$ is additionally constrained by a bound on the entropy, minimizing $D(p:q)$ remains a coupled problem of reducing $E_l \{I(\tilde{p}_l)\}$ by a proper choice of ϕ and reducing $E_l \left\{ D(\tilde{p}_l^{(1)} : \tilde{q}_l) \right\} + D(\tilde{p}(l) : \tilde{q}(l))$ by a proper choice of $\tilde{q}$, which though depends on ϕ. We discussed this coupled problem already in the fixed length case. The two step procedure of first finding a compact coding ϕ of p and then adapting the marginals of $\tilde{q}$ to those of $p \circ \phi$ is an approximate method for this minimization. In the worst case one may miss a reduction of $D(p:q)$ by an amount $\leq L_p \hbar$, when $\hbar$ is the lower bound on the entropy in each marginal $\tilde{q}_l^i$. Note that a compact coding minimizes this worst case error.

5   Compression EAs

This section aims to provide a perspective on how the ideas on compact codes motivate the design of new Evolutionary Algorithms. We will propose a general scheme to design such algorithms but can present preliminary results only for a special case that is more in the line of conventional GAs rather than EDAs and similar to algorithms proposed earlier. The results are promising. Most importantly though, the experiments exhibit what are crucial aspects to be discussed when the aims are practical implementations of the principle of compact codes.

A straight-forward way to design a new Evolutionary Algorithm, exploiting the idea of compact codes, is to combine any compression technique with any standard EA. Such a *Compression EA* reads

1. Initialize a finite population $p = \{p_1, .., p_\mu\}$.
2. Use a compression technique to find an encoding ϕ that (approximately) minimizes $L_p = \sum_{i=1}^{\mu} \text{length}(\phi^{-1}(p_i))$.

3. Apply standard operators (e.g., mixing or EDA-operators) to generate l offspring genotypes from the μ compressed parent genotypes.

4. Map the l offspring genotypes from G back to P using ϕ.

5. Apply evaluation and selection on these offspring to generate the new parent population p' and repeat from step **2**.

The operators in step **3** may be any standard operators used in Evolutionary Algorithms. They have to be memory-less though, since the encoding will change in each iteration step and thus integrating knowledge from previous time steps becomes futile (cf. the discussion of cumulative compression as an alternative in section 6).

For instance, a most direct implementation of KL-search calculates the length and marginal distributions before resampling them (very similar to the PBIL algorithm, but accounting for variable size strings). This leads to the following *Compression EDA*:

3.a Calculate the length distributions $\tilde{p}(l)$ and the symbol marginals $\tilde{p}^i_l$ for each l from the compressed population $\tilde{p}$.

3.b Set $\tilde{q}(l) = \tilde{p}(l)$ and add entropy by mixing the marginals with the uniform symbol distribution $\mathcal{U}$, $\tilde{q}^i_l = (1 - \alpha)\,\tilde{p}^i_l + \alpha\,\mathcal{U}$.

3.c Take l samples from the distribution $\tilde{q}$ by recursively (i) picking an l from $\tilde{q}(l)$, (ii) $\forall^l_{i=1}$ pick a symbol g^i from $\tilde{q}^i_l$, and (iii) store $g = (g^1, .., g^l)$ as a new offspring genotype.

We must leave it to future work to present results on this Compression EDA. In the following we will investigate a concrete algorithm that is based on conventional mutation operators rather than such an EDA and similar to algorithms proposed earlier [7, 20]. The compression technique is based on L-systems, very simple, but computationally expensive. It is inspired by the work of Nevill-Manning and Witten [13]. Below we will briefly report on some experiences when using Lempel-Ziv compression (i.e., the algorithm of GZIP) instead.

5.1   A L-System Compression GA

The compression. Step **2** of a Compression EA requires to find an encoding ϕ that minimizes the description length of individuals averaged over the current parent population. A simple technique of compression is to recursively analyze the parents for frequent pairs of neighboring symbols and replace such pairs by new symbols (cf. [13]). We realize this scheme with an L-system:

A L-system is a sequence $\Pi = \langle \pi_1, .., \pi_k \rangle$ of k productions π_i. Given the alphabet $\mathcal{A}$, each production $\pi = \langle l : r_1, .., r_m \rangle$ consists of a LHS symbol $l \in \mathcal{A}$ and a sequence $\langle r_1, .., r_m \rangle$ of RHS symbols. The L-system Π defines a mapping ϕ from one sequence s to another by applying all productions π_i, in the given order, on s. Applying a production $\langle l : r_1, .., r_m \rangle$ on s means to replace every l that occurs in s by the sequence $\langle r_1, .., r_m \rangle$. For instance, if the L-system is $\Pi = \langle e : cd, f : dc, c : ab, d : ba \rangle$, then a sequence $\langle ef \rangle$ is mapped to $\phi(\langle ef \rangle) = \langle abbabaab \rangle$. The mapping ϕ can easily be inverted by applying all productions, in reverse order, inversely on a sequence. Inverse application of a production $\langle l:r_1,..,r_m\rangle^{-1}$ on s means to replace every subsequence $\langle r_1,..,r_m\rangle$ that occurs in s by l.

Let $\mathcal{A}$ be the non-negative integer numbers, $\mathcal{A} = \mathbb{N}_0$. Starting with a population $p = \{s_1,..,s_\mu\}$ of $\{0,1\}$-sequences, there is a straight-forward way to construct an L-system that compresses the population by recursively extracting and encapsulating pairs of symbols that occur frequently in the population. This scheme reads

2.a Initialize the L-system as $\Pi = \langle\,\rangle$ and the *new*-symbol as $c = 2$.

2.b Calculate the frequency of every symbol pair that occurs in the population.

2.c For every pair $\langle r_1, r_2\rangle$ of symbols that occurs more often than once, create a new production $\langle c:r_1, r_2\rangle$, append it to the *beginning* of Π, and increment the *new*-symbol $c \leftarrow c + 1$. This is to be done in order, beginning with the pair of highest frequencies, and random order between pairs of same frequency. If there is no such pairs, exit the recursion.

2.d Recode the population $p \leftarrow \{\phi^{-1}(p_1),..,\phi^{-1}(p_\mu)\}$ (effectively, only the new productions will result in replacements).

2.f Repeat from step **2.b**.

The entropy. Following the general scheme of a Compression EA, we next need to specify a mixing operator in step **3**. This operator is supposed to induce the necessary entropy in $\tilde{q}$. We discussed earlier that this entropy should exceed the entropy $H(p)$ of the parent population to ensure further exploration, and we handled this issue theoretically by putting a lower bound on the entropy of feasible search distributions in the minimization of the Kullback-Leibler divergence.

However, first experiments showed quickly that there is an interesting problem when using the above compression in a straight-forward manner. Actually, the compression can be considered as too good and eventually violating the constraint of the lower bound on entropy: In the case of a *finite population*, the compression scheme often leads to a *full compression*, where in the end every individual is represented by a different *single* symbol. The compressed population thus is only a set of μ different single symbols, and the description length was minimized down to 1. The mutual information indeed vanishes for the full compression. However, on this representation it is impossible to induce more entropy than the parent population had (which is, if they are disjoint, $\log\mu$). The parent population already has maximal entropy under all distributions over only one symbol (since $\tilde{q}(l) = \tilde{p}(l)$ is fixed) of the alphabet $\{1,..,\mu\}$. Thus, the full compression violates the constraint of the lower bound on entropy of $\tilde{q}$ and is thus infeasible.

A solution to this problem with the above compression scheme would be to stop compression at some level, maybe at the price that the mutual information is not fully extinguished, but allowing for the addition of entropy. We follow this approach by choosing the level of compression stochastically or, equivalently, to add entropy on different levels of compression.

Concretely, the algorithm stochastically decides whether to apply one-point mutations on each level of compression, i.e., in each recursion of the compression scheme. We insert the following step in the compression recursion:

2.e For each individual, decide with a probability α whether to apply a one-point mutation. A one-point mutation randomly picks a location and, with equal probabilities, deletes it, replaces it with a new symbol in $\{0, .., c-1\}$, or inserts such a new symbol.

This competes the description of the algorithm which omits an additional step **3**.

The hierarchical XOR problem. The fitness function we consider is the Hierarchical XOR (HXOR) function [7, 26]. For a string $s \in \{0,1\}^n$ we first define a boolean function $h(s) \in \{0,1\}$, determining whether s is "valid" or not: Let $n = \text{length}(s)$ be the string length, $l \in \mathbb{N}_0$ such that $n/2 \le 2^l < n$ ($l = \lfloor \log_2(n-1) \rfloor$), and $L = s_{1:2^l}$ and $R = s_{2^l+1:n}$ the left and right parts of the string when cut at location 2^l. Then, for $n \ge 2$, we define

$$h(s) = 1 \iff \left[n = 2^{l+1} \wedge h(L) = 1 \wedge h(R) = 1 \wedge L = \bar{R} \right] ,$$

and $h(s) = 1$ if $n = 1$. Here $\bar{R}$ is the bit-wise negation of the right part. The last condition means that the bit-wise XOR between left and right part must be true for each bit. For instance, the strings for which $h(s) = 1$ are, up to length 16, $\langle 0 \rangle$, $\langle 01 \rangle$, $\langle 0110 \rangle$, $\langle 0110\,1001 \rangle$, $\langle 0110\,1001\,1001\,0110 \rangle$, and their bit-wise negations. Based on h, we define a fitness function $H(s) \in \mathbb{N}$, for $n \ge 2$,

$$H(s) = H(L) + H(R) + \begin{cases} n & \text{if } h(s) = 1 \\ 0 & \text{else} \end{cases} ,$$

and $H(s) = 1$ if $n = 1$. To normalize and put a limit on the string length, we define the lth HXOR function $H_l(s) \in [0,1]$: If s is longer than 2^l, let $s' = s_{1:2^l}$ and otherwise $s' = s$. Then,

$$H_l(s) = \frac{1}{2^l(l+1)} H(s') .$$

The normalization is given by the highest possible value $2^l(1+l)$ of $H(s)$ for a length 2^l string. There exist two global optima of H_l, namely the two "valid" strings of length l for which $h(s) = 1$.

Parameters. For the experiments, we use simple (μ,l)-selection with population sizes $\mu = 30$ and $l = 100$ and one elitist. The mutation probability (as indicated in step **2.e** of the algorithm) is $\alpha = 0.1$. The population was initialized with all $\langle 01 \rangle$ individuals.

Results. We tested the algorithm on the HXOR problem. Figure 1A displays the fitness trajectories for 20 runs on the 1024-bit HXOR problem. All runs consistently found the optimal 1024-bit string. It took on average 519 generations to reach this global optimum (with standard deviation ± 168 generations).

Figure 1B displays the average first hitting generation and standard deviation for different length HXOR problems, up to a problem size of 8192 bits. The

Table 1. A L-system found to compress a population of identical solutions to the 1024-bit HXOR. The full L-system is composed of 27 productions and reads ⟨C:10, D:1C, E:0D, F:CD, G:0F, H:EG, I:CH, J:CG, K:EI, L:EJ, M:LI, N:JK, O:NM, P:JM, Q:KO, R:PO, S:KR, T:QS, U:QP, V:UT, W:SV, X:US, Y:PT, Z:PW, a:XY, b:VZ, c:ba⟩. In the table, we expanded the RHS of these productions and, for brevity, only displayed the first 15 productions

C:10
D:110
E:0110
F:1 0110
G:01 0110
H:01 1001 0110
I:1001 1001 0110
J:1001 0110

K:0110 1001 1001 0110
L:0110 1001 0110
M:0110 1001 0110 1001 1001 0110
N:1001 0110 0110 1001 1001 0110
O:1001 0110 0110 1001 1001 0110 0110 1001 0110 1001 1001 0110
P:1001 0110 0110 1001 0110 1001 1001 0110
etc...

algorithm found the optimum in all runs for all problem sizes and the variance of the first hitting generation is relatively small for a stochastic search scheme. The diagram also shows that the number of generations needed to find an optimum seems to grow linearly with the problem size.

To give an impression on the codings developed by the L-system compression scheme, table 1 displays an L-system found to compress a population of identical solutions to the 1024-bit HXOR problem. As discussed above, this compression implements a *full compression* such that eventually every individual is represented by a single symbol c. We find that some productions represent the typical modules of HXOR solutions (like E:0110 or J:1001 0110) while others represent parts of these modules. Recall that the order in which productions are added to the L-system is stochastic (cf. step **2.c**). Consequently, the coding is not strictly hierarchical as a human might have designed it (or DEVREP, see below) and the L-system comprises more pair-productions than minimally necessary to encode the sequence of length 1024 (the minimum would be 19 productions).

Generally, the performance of the algorithm – in terms of the generations needed – is of the same order as the DEVREP algorithm presented by de Jong [7], which is the only previous algorithm we are aware of capable of solving large HXOR problems. ([7] reported only on results for the 64-bit and 1024-bit HXOR problem, where about $2.3 \cdot 10^7$ bit evaluations were needed in the single 1024-bit run presented.) The DEVREP algorithm is tailored to hierarchical problems, where different hierarchy levels are explicitly distinguished and mutational variations allowed only within a specific hierarchy level. It should be clear that plain variable length GAs perform extremely poorly on the HXOR problem because of its complex deceptiveness when search is performed on an atomic representation (see [7] for experiments).

Notes on GZIP. The compression scheme used is computationally very expensive. We argue below that a proper solution might be a cumulative approach to find compressions which are not recalculated in each generation. Alternatively, one might want to use efficient standard compression techniques, for example the

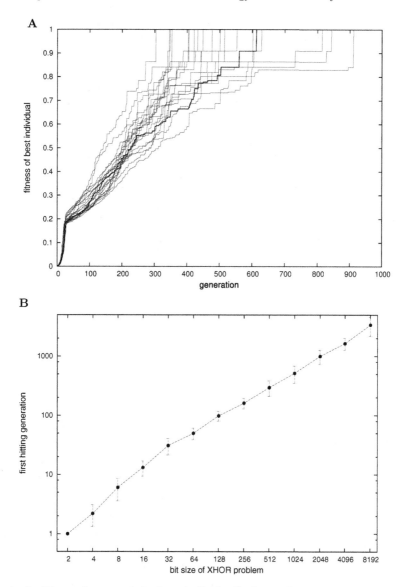

Fig. 1. A. The trajectory of the best individual's fitness for 20 runs on the 1024-bit HXOR problem. One of the runs is drawn bold. **B.** First hitting generations, averaged over 20 runs for each of the different HXOR problem sizes

Lempel-Ziv compression as used by GZIP. We performed some experiments also with this compression algorithm.

However, LZ-compression has some instructive practical drawbacks. Let us first consider to compress a single sequence, mutate it, and decompress it. The symbol mutations of the compressed string have to obey some constraints such

Table 2. The first line is the original string, composed of 15×abcd. The lines below show 20 variations, generated by compressing the string with LZ-compression, applying a single one-point mutation, and decompressing again. Dots indicate that the symbol has not varied compared to the original string

```
abcdabcdabcdabcdabcdabcdabcdabcdabcdabcdabcdabcdabcdabcdabcd
.a...a...a...a...a...a...a...a...a...a...a...a...a...a...a..
......................................d.ddabcdabcd
..................................................abdabcdabcd
........ddabcdabcdabcddbcdd.dddbcddbcddbcddbcddbcddbcdabcdabcd
...................cabcdabcdac.caccdaccdaccdabcdabcd
......aabadabcdabaabcdab.dab.dabcdabcdab.aba..a...a..a.
........bcdabcdabcdbbcdbcdbcd.bcd.bcd.bcd.bcd.bcdabcdabcd
d...d...d...d...d...d...d...d...d...d...d...d...d...d...d...
......ab..a.......ab......a...a...........ab..ab..a...a...a.
.........................................................
.......................................cdacdaccdabcdabcdabcdabcd
......bc..b.......bc......b...b............bc..bc..b...b...b.
....................cc.bcd.bcd.bcdabcdabcdabcdabcd
...........abcddaacdaabcd..........abcddaabcd..a...a...a..aacdaacd
...a...a...a...a...a...a...a...a...a...a...a...a
....dcd.cddbcddbcddbcddcdd..d...d...d...d...d...d...d...
.................................................bcdabcdab
..................dabcabcdabcdadabca....adcdadcdadcdabcdabcd
.........................................................
................bcd....................bcd.bcd.b...b...b.
```

that the compressed sequence remains a valid string that can be decompressed by the LZ algorithm (the possible symbol range at each location is different). Table 2 displays the result of 20 different one-point mutations on the compressed representation. Here, a second issue about LZ-compression becomes apparent. The mutations at the beginning of the string are less likely to be severe or modular than at the end of the string. The reason is that LZ-compression is a 1-pass scheme which builds up codes while parsing the string. In the beginning, no codes have been build up yet and the string remains largely uncompressed. This a priori asymmetry is undesirable in the context of evolutionary exploration. A possible trick is to concatenate the same string several times, compress the concatenation, mutate only the part which represents the last copy of the string, and decompress.

When implementing an algorithm, we exploited this observation. To compress and mutate a single individual, we concatenated all individuals of the population and additionally appended the individual of interest into a long sequence and compressed it. In this manner, LZ-compression first uses the whole population sequence to build up codes that are then used to compress the last individual. Then we mutated the compressed representation of the last individual and decompressed it. A computationally very expensive scheme, again.

Experiments with the HXOR problem showed a very high variance in performance between runs. For the 128-bit HXOR problem it frequently occurred that runs never find the optimum while others find it within a few hundred generations. Generally, the experiments with LZ-compression gave some important insight in the relevance of how the codes are build up. They do however not support the practical use of conventional LZ-compression for evolutionary algorithms.

6 Discussion

Summary. The basis on which we developed the theoretical analysis was the indirect Estimation-of-Distribution scenario, where a distribution over P is estimated via a distribution over a variable length genotype space G and a bijective encoding $\phi : G \to P$. The genotype distribution is restricted to factorize. Thus the problem of estimating a distribution is split into "decoding" the structural aspects of the distribution (via compression) and then estimating the remaining structure-less factorized distribution. The main result are Lemmas 1 and 2 relating the KLD subject to minimization to the description length of the encoding.

The class of Compression EAs proposed in the last section are straightforward combinations of compression techniques with standard EA operators. An algorithm based on a simple type of L-system compression performs reliably well on the hierarchically deceptive HXOR function, which we tested up to a problem size of 8192 bits. The number of generations needed to find an optimum seems to grow linearly with the problem size.

However, the main aim of giving this explicit example for a Compression EA was to introduce to the following discussion of two aspects that seem to be important (i) when thinking about future, computationally efficient Compression EAs and (ii) when considering the implications of the presented theory for the understanding of natural evolution and the self-adaptive evolution of genetic representations.

Compression techniques. Generally, to compress data one needs to analyze dependencies in the data and introduce new symbols for dependent features. Standard string compression algorithms, such as LZ-compression as well as the L-system compression we investigated, basically search for contiguous patterns in the original or partly compressed string. This was successful for the HXOR problem since the dependencies can (on various hierarchy levels) be described in terms of contiguous patterns. For many other hard optimization problems (e.g., MAXSAT) the dependencies will typically be between arbitrary variables and hardly detectable for standard string compression techniques. The question arises what kind of compression techniques are suitable for a specific class of problems. In its generality, we must leave this unanswered. Theoretically, it is straight-forward to design a specific factorial code for any distribution which is explicitly given, e.g., by a graphical model. How to construct such codes from data is yet an open issue.

Cumulative compression. The L-system compression we used has properties that proved beneficial in the experiments: unlike LZ-compression it is unbiased w.r.t. which modules are encapsulated and it can develop arbitrary hierarchies. To re-compute the compression from scratch at every generation also has advantages: the encoding is always adapted to the current population, i.e. the currently available information about the problem, and the stochasticity of the compression scheme leads to more diversity in exploration at different generations.

Clearly though, recomputing the compression at every generation is computationally very expensive. The general scheme of a Compression EA should thus be modified to develop the compact encoding in a cumulative manner rather than recomputing it at every generation. In the case of L-system compression, cumulative could mean that the L-system is persistent over the generations and at each generation only a single new production is added or an old, unused production deleted. Assuming that the inherent structure of found solutions will not evolve too fast, the cumulative scheme would still allow to provide a good compression of the current population.

Besides being computationally much more effective, the cumulative approach has another interesting perspective. Since the encoding is incrementally build up during evolution, the encoding becomes a variable that integrates information about the successful solutions over more than one generation. This is comparable to strategy parameters in Evolution Strategies which, for instance, integrate the average movement of the population over the recent history, assuming that this search direction will also be profitable in the future [9]. Such schemes have hardly been transferred to ordinary GAs because the notion of "proceeding in the same direction" makes no sense on the hypercube. The notion does make sense though on the more abstract level of "proceeding by incorporating the same structural dependencies that have previously been successful" – where the notion of "search direction" is replaced by the notion of "structural properties of search". The encoding becomes the variable which is capable to integrate such information. The objective of compactness indicates how an adequate permanent adaptation can be achieved.

Self-adaptive σ-evolution of compact representations. In this work we addressed a *bijective* encoding ϕ that is adapted explicitly (externally) at each generation. This approach is complementary to a formalism we proposed earlier to describe the self-adaptive evolution of genetic representations [21, 22]. The proper way to formalize the self-adaptive case is to consider a *fixed* but *non-injective* genotype-phenotype mapping. In that case there exists a variety (neutral set) of different genotypes that map to the same phenotype. The "choice of genetic representation" here means which genotype from the neutral set is chosen to encode the phenotype. The specific example investigated in [20] clarifies the relation between the two complementary frameworks: A genotype in the self-adaptive scenario may be the *tuple* (g_0, Π) of a (compact) string g_0 (termed axiom, or egg cell) and an L-system Π (termed genome). The global genotype-phenotype mapping from the space of such tuples to phenotype is clearly fixed. But different genotypes, involving different L-systems Π, may map to the same phenotype and thus induce structurally completely different phenotypic variabilities. In this scenario, the evolution of genetic representations can be understood as moves in the neutral set, e.g., neutral reorganizations of the L-system Π.

In [22] it was show that the self-adaptive evolution of genetic representations is driven by an implicit selection which discriminates phenotypically equivalent genotypes by the quality of the phenotypic variability that they induce (which is related to the effective fitness [14, 19]). More precisely, the selectional advantage of different genetic representations of the same phenotype is proportional to the negative *Kullback-Leibler divergence* between the phenotypic variability and the Boltzmann fitness distribution (assuming a lower bound on phenotypic entropy, see [22] for details).

The objective of minimizing the Kullback-Leibler divergence is thus the key to transfer the results we derived here to the self-adaptive scenario, now stating that there is an effective selection pressure on the description length of a genetic representation. Indeed, a tendency towards compact representations was observed in experiments on σ-evolution [20]. It was previously argued that its origin is the advantage of mutational robustness when every genetic symbol underlies a constant mutation rate. We can now add another origin, namely that compact representations are *structurally* more favorable – meaning that the phenotypic variability they induce allows us to reduce the Kullback-Leibler divergence and follow the Estimation-of-Distribution principle. It is yet open to which degree both effects contributed to the evolution of compact codes.

Acknowledgment

I would like to thank the German Research Foundation (DFG) for their funding of the Emmy Noether fellowship TO 409/1-1, allowing me to pursue this research.

References

1. S. Amari. Information geometry on hierarchy of probability distributions. *IEEE Transactions on Information Theory*, 47(5):1701–1711, 2001.
2. S. Baluja. Population-based incremental learning: A method for integrating genetic search based function optimization and competitive learning. Technical Report CMU-CS-94-163, Comp. Sci. Dep., Carnegie Mellon U., 1994.
3. S. Baluja and S. Davies. Using optimal dependency-trees for combinatorial optimization: Learning the structure of the search space. In *Proc. of Fourteenth Int. Conf. on Machine Learning (ICML 1997)*, pages 30–38, 1997.
4. L. Barbulescu, J.-P. Watson, and D. Whitley. Dynamic representations and escaping local optima: Improving genetic algorithms and local search. In *Seventeenth National Conference on Artificial Intelligence (AAAI)*, pages 879–884, 2000.
5. A. Barron, J. Rissanen, and B. Yu. The minimum description length principle in coding and modeling. *IEEE Transactions on Information Theory*, 44:2743–2760, 1998.
6. J. S. de Bonet, C. L. Isbell, Jr., and P. Viola. MIMIC: Finding optima by estimating probability densities. In M. C. Mozer, M. I. Jordan, and T. Petsche, editors, *Advances in Neural Information Processing Systems*, volume 9, page 424. The MIT Press, 1997.
7. E. D. de Jong. Representation development from Pareto-Coevolution. In *2003 Genetic and Evolutionary Computation Conference (GECCO 2003)*, 2003.
8. G. Halder, P. Callaerts, and W. Gehring. Induction of ectopic eyes by targeted expression of the eyeless gene in Drosophila. *Science*, 267:1788–1792, 1995.
9. N. Hansen and A. Ostermeier. Completely derandomized self-adaption in evolutionary strategies. *Evolutionary Computation*, 9:159–195, 2001.
10. R. B. Heckendorn and A. H. Wright. Efficient linkage discovery by limited probing. *Evolutionary Computation*, 2004. Accepted for publication.

11. G. S. Hornby and J. B. Pollack. The advantages of generative grammatical encodings for physical design. In *Proc. of 2001 Congress on Evolutionary Computation (CEC 2001)*, pages 600–607. IEEE Press, 2001.

12. G. E. Liepins and M. D. Vose. Representation issues in Genetic Algorithms. *Journal of Experimental and Theoretical Artificial Intelligence*, 2, 1990.

13. C. G. Nevill-Manning and I. H. Witten. Identifying hierarchical structure in sequences: A linear-time algorithm. *Journal of Artificial Intelligence Research*, 7:67–82, 1997.

14. P. Nordin and W. Banzhaf. Complexity compression and evolution. In L. Eshelman, editor, *Genetic Algorithms: Proc. of Sixth International Conf. (ICGA 1995)*, pages 310–317. Morgan Kaufmann, Pittsburgh, 15-19 1995.

15. M. Pelikan and D. E. Goldberg. Hierarchical BOA solves Ising spin glasses and MAXSAT. In *Genetic and Evolutionary Computation Conference 2003 (GECCO-2003)*, pages pp. 1271–1282. Springer-Verlag, 2003.

16. M. Pelikan, D. E. Goldberg, and E. Cantú-Paz. Linkage problem, distribution estimation, and Bayesian networks. *Evolutionary Computation*, 9:311–340, 2000.

17. M. Pelikan, D. E. Goldberg, and F. Lobo. A survey of optimization by building and using probabilistic models. Technical Report IlliGAL-99018, Illinois Genetic Algorithms Laboratory, 1999.

18. F. Rothlauf and D. E. Goldberg. Redundant representations in Evolutionary Computation. *Evolutionary Computation*, 11:381–415, 2003.

19. C. R. Stephens and J. M. Vargas. Effective fitness as an alternative paradigm for evolutionary computation I: General formalism. *Genetic Programming and Evolvable Machines*, 1:363–378, 2000.

20. M. Toussaint. Demonstrating the evolution of complex genetic representations: An evolution of artificial plants. In *2003 Genetic and Evolutionary Computation Conference (GECCO 2003)*, pages 86–97, 2003.

21. M. Toussaint. The evolution of genetic representations and modular neural adaptation, April 2003. PhD thesis, Institut für Neuroinformatik, Ruhr-Universiät-Bochum, Germany. Published with the Logos Verlag Berlin (2004), ISBN 3-8325-0579-2, 173 pages.

22. M. Toussaint. On the evolution of phenotypic exploration distributions. In C. Cotta, K. De Jong, R. Poli, and J. Rowe, editors, *Foundations of Genetic Algorithms 7 (FOGA VII)*, pages 169–182. Morgan Kaufmann, 2003.

23. M. Toussaint. Notes on information geometry and evolutionary processes, 2004. Los Alamos pre-print nlin.AO/0408040.

24. P. M. B. Vitányi and M. Li. Minimum Description Length induction, Bayesianism, and Kolmogorov complexity. *IEEE Trans. Inform. Theory*, IT-46:446–464, 2000.

25. G. P. Wagner and L. Altenberg. Complex adaptations and the evolution of evolvability. *Evolution*, 50:967–976, 1996.

26. R. A. Watson and J. B. Pollack. Hierarchically consistent test problems for genetic algorithms: Summary and additional results. In *Late breaking papers at the Genetic and Evolutionary Computation Conference*, pages 292–297, 1999.

27. D. Whitley, S. Rana, and R. Heckendorn. Representation issues in neighborhood search and evolutionary algorithms. In *Genetic Algorithms and Evolution Strategy in Engineering and Computer Science*, pages 39–58. John Wiley & Sons Ltd., 1997.

28. A. H. Wright, R. Poli, C. R. Stephens, W. B. Langdon, and S. Pulavarty. An Estimation of Distribution Algorithm based on maximum entropy. In *2004 Genetic and Evolutionary Computation Conference (GECCO 2004)*, pages 343–354. Springer, Berlin, 2004.

Asymptotic Convergence of Some Metaheuristics Used for Multiobjetive Optimization

Mario Villalobos-Arias[1,*], Carlos A. Coello Coello[2], and Onésimo Hernández-Lerma[1]

[1] CINVESTAV-IPN
Department of Mathematics
A. Postal 14-740
México, D.F. 07000, Mexico
{mava,ohernand}@math.cinvestav.mx
[2] CINVESTAV-IPN
Evolutionary Computation Group
Depto. de Ingeniería Eléctrica
Sección de Computación
Av. Instituto Politécnico Nacional No. 2508
Col. San Pedro Zacatenco
México, D. F. 07300, Mexico
ccoello@cs.cinvestav.mx

Abstract. This paper presents the asymptotic convergence analysis of Simulated Annealing, an Artificial Immune System and a General Evolutionary Algorithm for multiobjective optimization problems. In the case of a General Evolutionary Algorithm, we refer to any algorithm in which the transition probabilities use a uniform mutation rule. We prove that these algorithms converge if elitism is used.

1 Introduction

In nature, most problems have several objectives which we aim to optimize simultaneously. Such problems are called "multiobjective", and their solution requires a suitable definition of optimality (usually called "Pareto optimality"). Such problems normally have not one, but an infinite set of possible solutions, which represent possible trade-offs among the objectives (such solutions constitute the so-called "Pareto optimal set").

Diverse metaheuristics have been adopted to solve multiobjective optimization problems (MOP) [2]. In this paper, we study three of them: simulated annealing (SA) [10, 15], artificial immune systems (AIS) [14] and evolutionary algorithms (EA) [9, 6]. For these metaheuristics that use a uniform mutation rule (see end of Section 3.1) we show that the associated Markov chain converges geometrically to its stationary distribution, but not necessarily to the optimal solution set of the multiobjective optimization problem. Convergence to the optimal solution set is ensured if elitism (whose definition is provided in this paper) is used.

---

* Permanent Address: Escuela de Matemática, Universidad de Costa Rica, San José, Costa Rica
mvillalo@cariari.ucr.ac.cr

A.H. Wright et al. (Eds.): FOGA 2005, LNCS 3469, pp. 95–111, 2005.

Metaheuristics such as those indicated in this paper, have become a standard tool to solve both single-objective and multiobjective optimization problems. In the single-objective case, the convergence of a metaheuristic is reasonably well-understood [17]. However, when dealing with multiobjective optimization problems, there is not much work available in the literature, except for extremely particular cases (see for example [16]).

The remainder of this paper is organized as follows. Section 2 introduces the problem of our interest. The three specific algorithms studied in this paper are introduced in Section 3. In Section 4 we present some basic definitions related to Markov chain theory. Our main results (i.e., the corresponding proofs) are presented in Section 5. Section 6 provides our conclusions and some possible paths of future research.

2   The Multiobjective Optimization Problem

Let X be a set and $F : X \longrightarrow \mathbb{R}^d$ a given vector function with components $f_i : X \longrightarrow \mathbb{R}$ for each $i \in \{1, \ldots, d\}$. The multiobjective optimization problem (MOP) we are concerned with is to find $x^* \in X$ such that

$$F(x^*) = \min_{x \in X} F(x) = \min_{x \in X} [f_1(x), \ldots, f_d(x)], \tag{1}$$

where the minimum is understood in the sense of the standard Pareto order in which two vectors in $\mathbb{R}^d$ are compared as follows.

If $\boldsymbol{u} = (u_1, \ldots, u_d)$ and $\boldsymbol{v} = (v_1, \ldots, v_d)$ are vectors in $\mathbb{R}^d$, then

$$\boldsymbol{u} \preceq \boldsymbol{v} \iff u_i \leq v_i \ \forall \, i \in \{1, \ldots, d\}.$$

This relation is a partial order. We also write $\boldsymbol{u} \prec \boldsymbol{v} \iff \boldsymbol{u} \preceq \boldsymbol{v}$ and $\boldsymbol{u} \neq \boldsymbol{v}$.

Definition 1: A point $x^* \in X$ is called a *Pareto optimal solution* for the MOP (1) if there is no $x \in X$ such that $F(x) \prec F(x^*)$. The set

$$\mathcal{P}^* = \{x \in X \ : \ x \text{ is a Pareto optimal solution}\}$$

is called the *Pareto optimal set*, and its image under F, i.e.

$$F(\mathcal{P}^*) := \{F(x) \ : \ x \in \mathcal{P}^*\},$$

is called *Pareto front*.

In the remainder of the paper we will use the following well–known "scalarization" result.

Proposition 1: If $\boldsymbol{x}^* \in X$ is a solution of the weighted problem:

$$\min_{\boldsymbol{x} \in X} \sum_{s=1}^{d} w_s f_s(\boldsymbol{x}), \text{ where } w_s \geq 0 \ \forall s \in \{1, \ldots, n\} \text{ and } \sum_{s=1}^{d} w_s = 1,$$

then $\boldsymbol{x}^* \in \mathcal{P}^*$.

Proof. See, for instance, [13, p.78].

Now we introduce some notation that will be used later on. Let

$$\Sigma_{opt} := \{x \in X : \textstyle\sum_{s=1}^{d} f_s(x) = \Sigma_m\},$$

where

$$\Sigma_m := \min_{x \in X} \sum_{s=1}^{d} f_s(x). \tag{2}$$

Then, by Proposition 1, the Pareto optimal set $\mathcal{P}^*$ contains Σ_{opt}, i.e.

$$\Sigma_{opt} \subset \mathcal{P}^*. \tag{3}$$

As we are concerned with computational aspects, in the remainder of the paper we will assume that the set X in (1) is *finite*. For an EA and the AIS, in which the elements are represented by strings of length l with 0 or 1 at each entry, we take $X = \mathbb{B}^l$, with $\mathbb{B} = \{0, 1\}$. For SA we only assume that X is finite.

3   Algorithms

3.1   Evolutionary Algorithms

Evolutionary algorithms are techniques that use a population which evolves over time (i.e., generations) applying some operations to the current population to obtain the next one. Some of these operations are

- mutation
- selection
- crossover
- reordering

The type of EAs we are interested in are modeled as Markov chains with transition probabilities that use uniform mutation and possibly other operations. This mutation is applied with a certain parameter or probability p_m, which is positive and less than $1/2$, i.e.

$$p_m \in (0, 1/2). \tag{4}$$

Some examples of this type of EAs are the following:

- genetic algorithms (see [9]),
- evolution strategies (see [18]),
- evolutionary programming (see [8, 7]).

These types of algorithms can be modeled as a Markov chain $\{X_k : k \geq 0\}$ whose state space S is the set of all possible populations of n individuals, each one represented by a bit string of length l. Hence $S = (\mathbb{B}^l)^n = \mathbb{B}^{nl}$, where $\mathbb{B} = \{0, 1\}$ and so S is the set of all possible vectors of n entries, each of which is a string of length l with 0 or 1 at each entry.

Let $i \in S$ be a state, so that i can be represented as

$$i = (i_1, i_2, \ldots, i_n),$$

where each i_s is a string of length l of 0's and 1's.

The chain's transition probability is given by

$$P_{ij} = \mathbb{P}(X_{k+1} = j \mid X_k = i).$$

Thus the transition matrix is of the form

$$P = (P_{ij}) = LM, \tag{5}$$

where M is the transition matrix corresponding to the mutation operation and L represents the other operations.

Note that these matrices are stochastic, i.e. $L_{ij} \geq 0$, $M_{ij} \geq 0$ for all i, j, and for each $i \in S$

$$\sum_{j \in S} L_{ij} = 1 \text{ and } \sum_{j \in S} M_{ij} = 1. \tag{6}$$

The Mutation Probability

The mutation probability is very important in the convergence analysis of the EA. To calculate it from state i to state j we use that the individual i_s is transformed into the individual j_s applying *uniform mutation* (i.e. a flip mutation, with probability p_m, is applied to each entry of i_s) then each entry of i_s is transformed into the corresponding one of j_s with probability $1 - p_m$ or p_m depending on if the corresponding entries are equal or different, as in the following scheme.

$$
\begin{array}{cc}
 & 1 \ \ 2 \ \cdots \ n \\
i & \boxed{i_1}\boxed{i_2}\boxed{\cdots}\boxed{i_n}
\end{array}
$$

$$\text{mutation} \downarrow \ \downarrow \ \cdots \ \downarrow$$

$$
\begin{array}{cc}
j & \boxed{j_1}\boxed{j_2}\boxed{\cdots}\boxed{j_n}
\end{array}
$$

Thus, for each individual in the population the mutation probability can be calculated as

$$p_m^{H(i_s,j_s)} (1 - p_m)^{l - H(i_s,j_s)} \ \forall s \in \{1, \ldots, n\},$$

where $H(i_s, j_s)$ is the Hamming distance between i_s and j_s. It follows that the mutation probability from i to j is:

$$M_{ij} = \prod_{s=1}^{n} p_m^{H(i_s,j_s)} (1 - p_m)^{l - H(i_s,j_s)} \tag{7}$$

3.2   The Simulated Annealing Algorithm

Kirkpatrick et al. [10] and Černy [15] proposed an optimization algorithm based on some analogies with an annealing process in which a crystal is produced. This led to the development of an algorithm called "Simulated Annealing" which is a heuristic search technique that has been quite successful in combinatorial optimization problems (see [1] and [11] for details).

The simulated annealing algorithm generates a succession of possible solutions of the optimization problem. These possible solutions are the states of a Markov chain and the "energy" of a state is the evaluation of the possible solution that it represents.

The temperature is simulated with a sequence of positive control parameters c_k. A transition of the Markov chain occurs in two steps, given the value c_k of the control parameter. First, if the current state is i, a new state j is generated with a certain probability $G_{ij}(c_k)$, defined below. Then an "acceptance rule" $A_{ij}(c_k)$ is applied to j. Our main result hinges on a suitable selection of the acceptance rule, which we now discuss.

The generation probability. For each state i, let S_i be a subset of $S \setminus \{i\}$ called the neighborhood of i. We shall assume that the number of elements in S_i is the same, say Θ, for all $i \in S$, and also that the neighbor relation is symmetric, that is, $j \in S_i$ if and only if $i \in S_j$. Then, denoting by χ_{S_i} the indicator function of S_i (i.e. $\chi_{S_i}(j) := 1$ if $j \in S_i$ and 0 otherwise), we define the generation probability

$$G_{ij}(c_k) := \frac{\chi_{S_i}(j)}{\Theta} \quad \text{for all } i, j \in S. \tag{8}$$

The acceptance probability. This probability value is crucial for the behavior of the simulated annealing algorithm.

The idea of this acceptance rule is that any new state that improves the actual state will be accepted with probability 1 and the others are accepted with certain probability that tends to zero as time goes to infinity.

When dealing with MOPs there are different options to define the acceptance rule. For instance, Serafini [20] proposes to use the L_∞–Tchebycheff norm given by

$$A'_{ij}(c) = \min \left\{ 1, \exp \left(\max_{s \in \{1,\dots,d\}} \frac{\lambda_s(f_s(i) - f_s(j))}{c} \right) \right\},$$

where the λ_s are given positive parameters.

On the other hand, Ulungu and coworkers [21, 22, 24, 23] use

$$A''_{ij}(c) := \min \left\{ 1, \exp \left(\sum_{s=1}^{d} \frac{\lambda_s(f_s(i) - f_s(j))}{c} \right) \right\}$$

$$= \exp \left\{ - \left(\sum_{s=1}^{d} \frac{\lambda_s(f_s(j) - f_s(i))}{c} \right)^{+} \right\}. \tag{9}$$

where as usual, a^+ denotes the positive part of a number $a \in \mathbb{R}$, namely

$$a^+ := \begin{cases} a & \text{if } a > 0, \\ 0 & \text{otherwise.} \end{cases}$$

Here, we will use the acceptance probability presented in [20]:

$$A_{ij}(c) := \prod_{s=1}^{d} \min \left\{ 1, \exp \left(\frac{f_s(i) - f_s(j)}{c} \right) \right\},$$

which can be expressed in the simpler form

$$A_{ij}(c) = \exp\left(-\frac{\sum_{s=1}^{d}(f_s(j) - f_s(i))^+}{c}\right). \tag{10}$$

For the last two acceptance rules, we have shown somewhere else that the SA for MOP converges (see [25]).

The transition probability. Having the generation and the acceptance probabilities, we can now define the *transition probability* from i to j as

$$P_{ij}(c_k) := \begin{cases} G_{ij}(c_k)A_{ij}(c_k) & \text{if } i \neq j, \\ 1 - \sum_{l \in S, l \neq i} P_{il}(c_k) & \text{if } i = j, \end{cases} \tag{11}$$

where A_{ij} is as in (10) (or as in (9)).

3.3 Artificial Immune System

The Artificial Immune System (AIS) algorithm is a technique that, as its name indicates, simulates in a computer certain aspects of an immune system. When an antigen enters our immune system, it is immediately detected and generates a response from the immune system. As a consequence, antibodies are generated by the immune system. Antibodies are molecules that play the main role in the immune response. They are capable of adhering to the antigens in order to neutralize and mark them for elimination by other cells of the immune system. Successful antibodies are cloned and hypermutated. This is called the *clonal selection principle* and has been the basis for developing the algorithm on which we base the work reported in this paper [4].

For our mathematical model, we will consider the AIS (based on clonal selection theory [4]) for multiobjective optimization proposed in [3]. From here on, we will refer to this approach using the same name adopted by the authors of this algorithm: "Multi-objective Immune System Algorithm" (MISA for short). Next, we will focus our discussion only on the aspects that are most relevant for its mathematical modelling. For a detailed discussion on this algorithm, readers should refer to [3].

In MISA the antigens are simulated with a population of strings of 0's and 1's. The population is divided in two parts, a primary set and a secondary set; the primary set contains the "best" individuals (or elements) of the population. The transition of one population to another is made by means of two mutation rules and a reordering operation. First, the elements of the primary set are copied several times, then in each of these copies a fixed number of bits are mutated, at random. Regarding the secondary set, a uniform mutation with parameter p_m is applied. This parameter is positive and less than $1/2$, i.e. $p_m \in (0, 1/2)$.

After that, the elements are reordered, moving the "best" individuals to the primary set. MISA can be modeled with a Markov chain $\{X_k : k \geq 0\}$, with state space $S = \mathbb{B}^{nl}$, where $\mathbb{B} = \{0,1\}$. In this case a individual can be represented as:

$$i = (i^1, i^2) = (i_1, i_2, \dots, i_{n_1}; i_{n_1+1}, \dots, i_n),$$

i^1 represents the primary set and i^2 the secondary.

3.4   Using Elitism

In our case, when dealing with MOPs, we say that we are using *elitism* in an algorithm if we use an extra set, called the *elite* set, in which we put the "best" elements
(nondominated elements of the state in our case) found. This elite set usually does not
participate in the evolution (although, there are multi-objective evolutionary algorithms
that use the elite set in the selection process, such as the Strength Pareto Evolutionary
Algorithm [27]), since it is used only to store the nondominated elements.

After each transition we apply an *elitism operation* that accepts a new state if there
is an element in the population that improves some element in the elite set (i.e., if there
is an element in the population that dominates, in the Pareto sense, some element in the
elite set).

If we are using elitism, the representation of the states changes to the following
form:

$$\hat{i} = (i^e; i) = (i_1^e, \cdots, i_r^e; i_1, \cdots, i_n),$$

where $i_1^e, \cdots, i_r^e$ are the members of the elite set of the state, r is the number of elements
in the elite set and we assume that the cardinality of $\mathcal{P}^*$ is greater than or equal to r. In
addition we assume that $r \leq n$.

Note that in general $i_1^e, \cdots, i_r^e$ are not necessarily the "best" elements of the state $\hat{i}$,
but after applying the elitism operation in i^e they are the "best" elements of the state.

Let $\hat{P}$ be the transition matrix associated with the new states. If all the elements in
the elite set of a state are Pareto optimal, then any state that contains an element in the
elite set that is not a Pareto optimal will not be accepted, i.e.

$$\text{if } \{i_1^e, \cdots, i_r^e\} \subset \mathcal{P}^* \text{ and } \{j_1^e, \cdots, j_r^e\} \not\subset \mathcal{P}^* \text{ then } \hat{P}_{ij} = 0. \tag{12}$$

4   Markov Chain Theory

We provide here some standard definitions and results.

We first introduce the definition of convergence of an algorithm, which uses the
following notation: if $V = (v_1, v_2, \ldots, v_n)$ is a vector, then $\{V\}$ denotes the set of
entries of V, i.e.

$$\{V\} = \{v_1, v_2, \ldots, v_n\}.$$

Definition 2: Let $\{X_k : k \geq 0\}$ be the Markov chain associated to an algorithm. We
say that the algorithm converges with probability 1 if

$$\mathbb{P}(\{X_k\} \subset \mathcal{P}^*) \to 1 \text{ as } k \to \infty.$$

In the case in which we are using elitism we replace X_k by X_k^e, the elite set of the
state (i.e. if $X_k = i$ then $X_k^e = i^e$).

The next result gives an upper bound on the rate of convergence of P^k as $k \to \infty$.
We will use it to show the existence of the stationary distribution in Theorem 2.

Lemma 1: Let N be the cardinality of S, and let P_{ij}^k be the entry ij of P^k. Suppose that there exists an integer $\nu > 0$ and a set J of $N_1 \geq 1$ values of j such that

$$\min_{\substack{1 \leq i \leq N \\ j \in J}} P_{ij}^\nu = \delta > 0.$$

Then there are numbers $\pi_1, \pi_2, \ldots, \pi_{N_1}$ such that

$$\lim_{k \to \infty} P_{ij}^k = \pi_j \ \forall i = 1, \ldots, N, \ \forall j \in J, \text{ with } \pi_j \geq \delta > 0,$$

and $\pi_1, \pi_2, \ldots, \pi_{N_1}$ form a set of stationary probabilities. Moreover

$$|P_{ij}^k - \pi_j| \leq (1 - N_1 \delta)^{\frac{k}{\nu} - 1} \ \forall \, i = 1, \ldots, N, \ \forall \, j \in J, \ \forall \, k = 1, 2, \ldots.$$

Proof. See, for example, [5, p. 173].

We will need some properties of the limiting distribution, which we present next. Recall that a probability distribution q is called the *limiting distribution* of a Markov chain with transition probability P if

$$q_i = \lim_{k \to \infty} \mathbb{P}(X_k = i | X_0 = j) \text{ for all } i, j \in S.$$

If such a limiting distribution q exists and $a_i(k) = \mathbb{P}(X_k = i)$, for $i \in S$, denotes the distribution of X_k, then

$$\lim_{k \to \infty} a_i(k) = q_i \text{ for all } i \in S.$$

Moreover, q is an *invariant* (or *stationary*) distribution of the Markov chain, which means that

$$q = q \, P; \tag{13}$$

that is, q is a left eigenvector of P with eigenvalue 1. A converse to this result (which is true for *finite* Markov chains) is given in Lemma 2 below.

Observe that (13) trivially holds if q is a probability distribution satisfying

$$q_i P_{ij} = q_j P_{ji} \ \forall i, j \in S. \tag{14}$$

This equation is called the *detailed balance equation*, and (13) is called the *global balance equation*.

Lemma 2:[12, p.19] Let P be the transition matrix of a finite, irreducible and aperiodic Markov chain. Then the chain has a unique stationary distribution q (that is q is the unique distribution that satisfies (13)) and, in addition, q is the chain's limiting distribution.

Definition 3: Let X be as in problem (1). We say that X is *complete* if for each $x \in X \setminus \mathcal{P}^*$ there exists $x^* \in \mathcal{P}^*$ such that $F(x^*) \preceq F(x)$.

For instance, if X is finite then X is complete.

Let $i, j \in S$ be two arbitrary states, we say that i *leads* to j, and write $i \to j$, if there exists an integer $k \geq 1$ such that $P_{ij}^k > 0$. If i does not lead to j, then we write $i \not\to j$.

We call a state i *inessential* if there exists a state j such that $i \to j$ but $j \not\to i$. Otherwise the state i is called *essential*.

We denote the set of essential states by E and the set of inessential states by I. Clearly,

$$S = E \cup I.$$

We say that P is in *canonical form* if it can be written as

$$P = \begin{pmatrix} P_1 & 0 \\ R & Q \end{pmatrix}.$$

Observe that P can be put in this form by reordering the states, that is, the essential states at the beginning and the inessential states at the end. In this case, P_1 is the matrix associated with the transitions between essential states, R with transitions from inessential to essential states, and Q with transitions between inessential states.

Note that P^k has a Q^k in the position of Q in P, i.e.

$$P^k = \begin{pmatrix} P_1^k & 0 \\ R_k & Q^k \end{pmatrix},$$

where R_k is a matrix that depends of P_1, Q and R.

Now we present some results that will be essential in the proof of Theorem 3.

Lemma 3: Let P be a stochastic matrix, and let Q be the submatrix of P associated with transitions between inessential states. Then, as $k \to \infty$,

$$Q^k \to 0 \text{ elementwise geometrically fast.}$$

Proof. See, for instance, [19, p.120]. ∎

As a consequence of Lemma 3 we have the following.

Corollary 1: For any initial distribution,

$$\mathbb{P}(X_k \in I) \to 0 \text{ as } k \to \infty.$$

Proof. For any initial distribution vector p_0, let $p_0(I)$ be the subvector that corresponds to the inessential states. Then, by Lemma 3,

$$\mathbb{P}(X_k \in I) = p_0(I)'Q^k \mathbf{1} \to 0 \text{ as } k \to \infty. \quad\blacksquare$$

5   Main Results

In this section we present some recent results on the convergence of the algorithms introduced in Section 3, for multiobjective optimization problems (MOPs).

5.1   Convergence of Simulated Annealing

Following the ideas of Laarhoven, Aarts and Korst in [1, 11] we developed a convergence proof of SA for MOPs, which is presented in the following Theorem.

Theorem 1: Let $P(c)$ be the transition matrix associated with the SA algorithm defined by (8), (10), (11) and, moreover, suppose that $G(c)$ is irreducible. Then:

(a) The Markov chain has a stationary distribution $q(c)$ whose components are given by

$$q_i(c) = \frac{1}{N_0(c)} \exp \left(-\frac{\sum_{s=1}^{d} f_s(i)}{c} \right), \tag{15}$$

where

$$N_0(c) = \sum_{j \in S} \exp \left(-\frac{\sum_{s=1}^{d} f_s(j)}{c} \right) \tag{16}$$

(b) For each $i \in S$

$$q_i^* := \lim_{c \searrow 0} q_i(c) = \frac{1}{|\Sigma_{opt}|} \chi_{\Sigma_{opt}}(i),$$

where $|\Sigma_{opt}|$ denotes the number of elements in Σ_{opt}.

(c) The SA algorithm converges with probability 1.

These results remain valid if (10) is replaced with (9).

Proof of Theorem 1.

(a) Since G is irreducible, using Lemma 2 it can be seen that the Markov chain is irreducible and aperiodic (see [1, p.39]). Hence, by Lemma 2 there exists a unique stationary distribution. We now use (8) and (11) to see that (14) holds for all $i \neq j$. First note that

$$q_i(c)P_{ij}(c) = q_i(c)G_{ij}(c)A_{ij}(c)$$
$$= \begin{cases} \frac{1}{\Theta} q_i(c)A_{ij}(c) & \text{if } j \in S_i \\ 0 & \text{if } j \notin S_i. \end{cases}$$

Similarly,

$$q_j(c)P_{ji}(c) = q_j(c)G_{ji}(c)A_{ji}(c)$$
$$= \begin{cases} \frac{1}{\Theta} q_j(c)A_{ji}(c) & \text{if } i \in S_j \\ 0 & \text{if } i \notin S_j. \end{cases}$$

Thus, since $i \in S_j$ if and only if $j \in S_i$, to obtain (14) we only have to prove that

$$q_i(c)A_{ij}(c) = q_j A_{ji}(c).$$

But this follows from (10), (15) and using that for any real numbers $a_1, a_2, \ldots, a_n$, $b_1, b_2, \ldots, b_n$, we have

$$\sum_{k=1}^{n}(a_k - b_k) + \left(\sum_{k=1}^{n}(b_k - a_k)\right)^+ = \left(\sum_{k=1}^{n}(a_k - b_k)\right)^+ ,$$

$$\sum_{k=1}^{n}(a_k - b_k) + \sum_{k=1}^{n}(b_k - a_k)^+ = \sum_{k=1}^{n}(a_k - b_k)^+ .$$

because

$$q_i(c)A_{ij}(c) =$$

$$= \frac{1}{N_0(c)} \exp\left(-\frac{\sum_{s=1}^{n} f_s(i)}{c}\right) \exp\left(-\frac{\sum_{s=1}^{n}(f_s(j) - f_s(i))^+}{c}\right)$$

$$= \frac{1}{N_0(c)} \exp\left(-\frac{\sum_{s=1}^{n} f_s(j)}{c}\right)$$

$$\exp\left(-\frac{\sum_{s=1}^{n}(f_s(i) - f_s(j)) + \sum_{s=1}^{n}(f_s(j) - f_s(i))^+}{c}\right)$$

$$= \frac{1}{N_0(c)} \exp\left(-\frac{\sum_{s=1}^{n} f_s(j)}{c}\right) \exp\left(-\frac{\sum_{s=1}^{n}(f_s(i) - f_s(j))^+}{c}\right)$$

$$= q_j(c)A_{ji}(c).$$

This shows that (14) holds, which in turn yields part (a) in Theorem 1. (Note that this proof, with obvious changes, remains valid if the acceptance probability is given by (9) rather than (10)).
(b) Note that for each $a \le 0$

$$\lim_{x \searrow 0} e^{\frac{a}{x}} = \begin{cases} 1 & \text{if } a = 0, \\ 0 & \text{otherwise.} \end{cases} \tag{17}$$

Now, by (2), (15) and (16)

$$q_i(c) = \frac{\exp\left(-\frac{\sum_{s=1}^{n} f_s(i)}{c}\right)}{\sum_{j \in S} \exp\left(-\frac{\sum_{s=1}^{n} f_s(j)}{c}\right)}$$

$$= \frac{\exp\left(\frac{\Sigma_m - \sum_{s=1}^{n} f_s(i)}{c}\right)}{\sum_{j \in S} \exp\left(\frac{\Sigma_m - \sum_{s=1}^{n} f_s(j)}{c}\right)}$$

$$= \frac{\exp\left(\frac{\Sigma_m - \sum_{s=1}^{n} f_s(i)}{c}\right)}{\sum_{j \in S} \exp\left(\frac{\Sigma_m - \sum_{s=1}^{n} f_s(j)}{c}\right)} \left(\chi_{\Sigma_{opt}}(i) + \chi_{S - \Sigma_{opt}}(i)\right)$$

$$= \frac{1}{\sum_{j \in S} \exp\left(\frac{\Sigma_m - \sum_{s=1}^{n} f_s(j)}{c}\right)} \chi_{\Sigma_{opt}}(i)$$

$$+ \frac{\exp\left(\frac{\Sigma_m - \sum_{s=1}^{n} f_s(i)}{c}\right)}{\sum_{j \in S} \exp\left(\frac{\Sigma_m - \sum_{s=1}^{n} f_s(j)}{c}\right)} \chi_{S - \Sigma_{opt}}(i).$$

Now let $c \searrow 0$. Then, by (17), the second term of the latter sum goes to 0, whereas the denominator of the first term goes to $|\Sigma_{opt}|$. Hence

$$\lim_{c \searrow 0} q_i(c) = \frac{1}{|\Sigma_{opt}|} \chi_{\Sigma_{opt}}(i) + 0 = q_i^*,$$

which completes the proof of part (b).

(c) By (b) and Lemma 2

$$\lim_{c \searrow 0} \lim_{k \to \infty} \mathbb{P}\{X_k = i\} = \lim_{c \searrow 0} q_i(c) = q_i^*,$$

and so by (3)

$$\lim_{c \searrow 0} \lim_{k \to \infty} \mathbb{P}\{X_k \in \mathcal{P}^*\} \geq \lim_{c \searrow 0} \lim_{k \to \infty} \mathbb{P}\{X_k \in \Sigma_{opt}\} = 1. \tag{18}$$

Thus

$$\lim_{c \searrow 0} \lim_{k \to \infty} \mathbb{P}\{X_k \in \mathcal{P}^*\} = 1,$$

and (c) follows.   ∎

5.2 Convergence of Evolutionary Algorithms

In this subsection we present convergence results for the EA for solving MOPs, in which we show that the use of elitism is necessary to guarantee the convergence of this kind of algorithms.

The first result is related to the existence of a stationary distribution for the Markov chain of the EA.

Theorem 2: Let P be the transition matrix of an EA. Then P has a stationary distribution π such that

$$|P_{ij}^k - \pi_j| \leq \left(1 - 2^{nl} p_m^{nl}\right)^{k-1} \quad \forall i, j \in S \ \forall k = 1, 2, \dots. \tag{19}$$

Moreover, π has all entries positive.

Theorem 2 states that P^k converges geometrically to π. Nevertheless, in spite of this result, the convergence of the EA to the Pareto optimal set cannot be guaranteed. In fact, from Theorem 2 and using the fact that π has all entries positive, we immediately deduce the following.

Corollary 2: The EA does not converge.

To ensure convergence of the EA we need to use elitism.

Theorem 3: The EA using elitism converges.

The next lemma will be used in the proof of Theorem 2.

Lemma 4: Let P be the transition matrix of the EA. Then

$$\min_{i,j \in S} P_{ij} = p_m^{nl} > 0 \ \forall i, j \in S, \tag{20}$$

and therefore P is primitive.

Proof. By (4) we have

$$p_m < \frac{1}{2} < 1 - p_m.$$

Thus, from (7),

$$M_{ij} = \prod_{s=1}^{n} p_m^{H(i_s,j_s)}(1 - p_m)^{l-H(i_s,j_s)}$$

$$> \prod_{s=1}^{n} p_m^{H(i_s,j_s)} p_m^{l-H(i_s,j_s)} \quad = \quad \prod_{s=1}^{n} p_m^{l}$$

$$= p_m^{nl}$$

On the other hand, by (5) and (6)

$$P_{ij} = \sum_{s \in S} R_{is} M_{sj}$$

$$\geq p_m^{nl} \sum_{s \in S} R_{is}$$

$$= p_m^{nl} > 0,$$

To verify (20), observe that P_{ij} attains the minimum in (20) if i has 0 in all entries and j has 1 in all entries. Thus the desired conclusion follows. ∎

Proofs

Proof of Theorem 2. Because (20) holds for all $j \in S$ we have that $J = S$, $N_1 = N = 2^{nl}$ and $\nu = 1$. Thus, by Lemma 1, P has a stationary distribution π with all entries positive and we get (19). ∎

Despite the fact that Theorem 3 is an extension of a result originally presented by Rudolph [17], our proof is more general. Additionally, we do not have to make any assumptions regarding the existence of a single optimal point (i.e., our proof is simpler), due to the use of essential and inessential states.

Proof of Theorem 3. By Corollary 1, it suffices to show that the states that contain elements in the elite set that are not Pareto optimal are inessential states. To this end, first note that $X = \mathbb{B}^l$ is complete, because it is finite.

Now suppose that there is a state $\hat{i} = (i^e; i)$ in which the elite set contains elements $i_{s_1}^e, \ldots, i_{s_k}^e$ that are not Pareto optimal. Then, as X is complete, there are elements, say $j_{s_1}^e, \ldots, j_{s_k}^e \in \mathcal{P}^*$, that dominate $i_{s_1}^e, \ldots, i_{s_k}^e$, respectively.

Take $\hat{j} = (j^e; j)$ such that all Pareto optimal points of i^e are in j^e and replace the other elements of i^e with the corresponding $j^e_{s_1}, \ldots, j^e_{s_k}$. Thus, all the elements in j^e are Pareto optimal.

Now let

$$j = (j^e_1, \ldots, j^e_r, \underbrace{i^e_{s_1}, \ldots, i^e_{s_1}}_{n-r \text{ copies}}).$$

By Lemma 4 we have $i \to j$. Hence, with positive probability we can pass from (i^e, i) to (i^e, j), and then we apply the elitism operation to pass from (i^e, j) to (j^e, j). This implies that $\hat{i} \to \hat{j}$. On the other hand, using (12), $\hat{j} \nrightarrow \hat{i}$ and therefore $\hat{i}$ is an inessential state.

Finally, from Corollary 1 we have

$$\mathbb{P}(\{X^e_k\} \subset \mathcal{P}^*) = \mathbb{P}(X_k \in E) = 1 - \mathbb{P}(X_k \in I) \to 1 - 0 = 1$$

as $k \to \infty$.

This completes the proof of Theorem 3.   ∎

5.3   Convergence of an Artificial Immune System Algorithm

A previous proof for a version of MISA was presented in [26], in which some constraints were imposed on the way in which one could go from one state to another. Here, we present a proof of a more general version of MISA. The idea is the same for the EA, and is presented in the next lemma.

Lemma 5: If any state in MISA has in its elite set an element that is not a Pareto optimal, this state is an inessential state.

Proof. Note that $X = \mathbb{B}^l$ is complete, because it is finite.

Let $\hat{i} = (i^e; i^1, i^2)$ be a state in which the elite set contains elements that are not Pareto optimal.

1. From i^1, a set of clones is generated. Next, a fixed number of (randomly chosen) string positions of these clones are mutated. Then we change the initial positions in all the strings of the clones (there exists a positive probability of doing this). The set obtained from this previous process is called $ClonesM(i^1)$.
2. Since a uniform mutation is applied to i^2, we change whatever is necessary in all the elements within this set, so that we can obtain the worst element of $ClonesM(i^1)$. As before, there exists a positive probability of doing this, so that none of these elements enters the primary set.
3. Then, all the elements are rearranged and we select the nondominated elements and they are placed in j^1. Now, let j^2 contain a number of individuals of the remainder of the elements available, until completing N (N is the population size).
4. When we apply elitism we obtain the set j^e.
5. To the clones of j^1, we mutate the same initial string positions. Then $ClonesM(j^1) \subseteq ClonesM(i^1)$. Therefore, the best elements of $ClonesM(j^1)$ will be in j^1 again. When we apply elitism to the elements of j^1, we do not modify the set j^e.

6. Let $j_{s_1}^e, \ldots, j_{s_k}^e$ be the elements of j^e that are not Pareto optimal. As X is complete, there exist elements $i_{s_1}^\star, \ldots, j_{s_k}^\star \in \mathcal{P}^*$ that dominate $j_{s_1}^e, \ldots, j_{s_k}^e$, respectively.

7. Now, since we apply uniform mutation to j^2, we can obtain from $j_1^2, \ldots, j_k^2$ to $i_{s_1}^\star, \ldots, j_{s_k}^\star$ respectively, and the other elements of j^2 are left as they were before.

8. Like $ClonesM(j^1)$ and $\{j_{k+1}, \ldots, j_{n_2}\}$ had already been modified j^e when applying elitism, we will not modify again j^e. Thus, the only part of j^e that is modified will be $i_{s_1}^\star, \ldots, j_{s_k}^\star$ and they will replace the nondominated elements of j^e.

9. Finally, let $i^\dagger$ be the resulting state of this process. Using the previous process, we can go from $\hat{i}$ to $i^\dagger$ ($\hat{i} \to i^\dagger$), but as in $i^{\dagger e}$ there are only Pareto optimal solutions, from (12) $P_{i^\dagger \hat{i}} = 0$ ($i.e$ $i^\dagger \not\to \hat{i}$). This proves that $\hat{i}$ is an inessential state.   ∎

From Lemma 5, the convergence of MISA is easily obtained as follows.

Theorem 4: The MISA algorithm using elitism converges.

Proof. From Lemma 5 and Corollary 1 we have

$$\mathbb{P}(\{X_k^e\} \subset \mathcal{P}^*) = \mathbb{P}(X_k \in E) = 1 - \mathbb{P}(X_k \in I) \to 1 - 0 = 1$$

as $k \to \infty$. This completes the proof.   ∎

6   Conclusions and Future Work

We have presented the convergence proofs of three meta-heuristics that have been used for solving MOPs: simulated annealing, an artificial immune system (based on clonal selection theory), and a general evolutionary algorithm.

It is worth noting that in the case of the general EA, our convergence proof extends previous proofs of convergence presented for genetic algorithms used for single-objective optimization (e.g., [17]). Actually, our proof is valid for a more general class of evolutionary algorithms that use uniform mutation.

Regarding the artificial immune system, the proof included here, together with some of our previous work [26], constitute the only attempts currently known to prove convergence of such metaheuristic.

Finally, regarding simulated annealing, our proof relies on previous work by Laarhoven, Aarts and Korst [1, 11], but it constitutes (to the best of our knowledge), the first proof of convergence of simulated annealing in multiobjective optimization problems.

As part of our future work, we intend to extend these results to a more general case in which not even uniform mutation is required. We also plan to analyze other types of heuristics used for multiobjective optimization, and to try to determine bounds of convergence for such algorithms.

Acknowledgments

The first author acknowledges support from the Universidad de Costa Rica through a scholarship to pursue graduate studies at the Department of Mathematics of

CINVESTAV-IPN. The second author acknowledges support from NSF-CONACyT project No. 42435-Y. The last author acknowledges partial support from CONACyT grant 37355-E.

References

1. E.H. Aarts and J.H. Korst. *Simulated Annealing and Boltzmann Machines: A Stochastic Approach to Combinatorial Optimization and Neural Computing*. Wiley, Chichester, UK, 1989.
2. Carlos A. Coello Coello, David A. Van Veldhuizen, and Gary B. Lamont. *Evolutionary Algorithms for Solving Multi-Objective Problems*. Kluwer Academic Publishers, New York, 2002.
3. Nareli Cruz Cortés and Carlos A. Coello Coello. Multiobjective Optimization Using Ideas from the Clonal Selection Principle. In Erick Cantú-Paz et al., editor, *Genetic and Evolutionary Computation—GECCO 2003. Proceedings, Part I*, pages 158–170. Springer. Lecture Notes in Computer Science Vol. 2723, July 2003.
4. Leandro Nunes de Castro and F. J. Von Zuben. Learning and Optimization Using the Clonal Selection Principle. *IEEE Transactions on Evolutionary Computation*, 6(3):239–251, 2002.
5. J. L. Doob. *Stochastic Processes*. Wiley, New York, 1953.
6. David B. Fogel. *Evolutionary Computation. Toward a New Philosophy of Machine Intelligence*. The Institute of Electrical and Electronic Engineers, New York, 1995.
7. Lawrence J. Fogel. *Artificial Intelligence through Simulated Evolution*. Wiley, New York, 1966.
8. Lawrence J. Fogel. *Artificial Intelligence through Simulated Evolution. Forty Years of Evolutionary Programming*. Wiley, New York, 199.
9. David E. Goldberg. *Genetic Algorithms in Search, Optimization, and Machine Learning*. Addison-Wesley, Reading, MA, 1989.
10. S. Kirkpatrick, C.D. Gellatt, and M.P. Vecchi. Optimization by Simulated Annealing. *Science*, 220:671–680, 1983.
11. P. Laarhoven and E.H. Aarts. *Simulated Annealing: Theory and Applications*. D. Reidel, Boston, MA, 1987.
12. G.F. Lawler. *Introduction to Stochastic Processes*. Chapman & Hall/CRC, Boca Raton, FLA, 1995.
13. Kaisa M. Miettinen. *Nonlinear Multiobjective Optimization*. Kluwer, Boston, MA, 1998.
14. Leandro Nunes de Castro and Jonathan Timmis. *An Introduction to Artificial Immune Systems: A New Computational Intelligence Paradigm*. Springer-Verlag, 2002.
15. V. Černy. A Thermodynamical Approach to the Traveling Salesman Problem: An Efficient Simulation Algorithm. *Journal of Optimization Theory and Applications*, 45(1):41–51, 1985.
16. G. Rudolph and A. Agapie. Convergence Properties of Some Multi-objective Evolutionary Algorithms. In *Proceedings of the 2000 Conference on Evolutionary Computation*, volume 2, pages 1010–1016, Piscataway, NJ, July 2000. IEEE Press.
17. Günther Rudolph. Convergence Analysis of Canonical Genetic Algorithms. *IEEE Transactions on Neural Networks*, 5:96–101, 1994.
18. Hans-Paul Schwefel. *Evolution and Optimum Seeking*. Wiley, New York, 1995.
19. E. Seneta. *Non-Negative Matrices and Markov Chains*. Springer-Verlag, New York, second edition, 1981.
20. Paolo Serafini. Simulated Annealing for Multiple Objective Optimization Problems. In G.H. Tzeng, H.F. Wang, U.P. Wen, and P.L. Yu, editors, *Proceedings of the Tenth International Conference on Multiple Criteria Decision Making: Expand and Enrich the Domains of Thinking and Application*, volume 1, pages 283–292, Berlin, 1994. Springer-Verlag.

21. E.L. Ulungu. *Optimisation combinatoire multicritere: Determination de l'ensemble des solutions efficaces et methodes interactives.* PhD thesis, Faculté des Sciences, Université de Mons-Hainaut, Mons, Belgium, 1993.
22. E.L. Ulungu, J. Teghem, and Ph. Fortemps. Heuristics for Multi-Objective Combinatorial Optimization by Simulated Annealing. In J. Gu, G. Chen, Q. Wei, and S. Wang, editors, *Multiple Criteria Decision Making: Theory and Applications. Proceedings of the 6th National Conference on Multiple Criteria Decision Making*, pages 228–238, Windsor, UK, 1995. Sci-Tech.
23. E.L. Ulungu, J. Teghem, Ph. Fortemps, and D. Tuyttens. MOSA Method: A Tool for Solving Multiobjective Combinatorial Optimization Problems. *Journal of Multi-Criteria Decision Analysis*, 8(4):221–236, 1999.
24. E.L. Ulungu, J. Teghem, and Ch. Ost. Efficiency of Interactive Multi-Objective Simulated Annealing Through a Case Study. *Journal of the Operational Research Society*, 49:1044–1050, 1998.
25. M. Villalobos-Arias, C. A. Coello Coello, and O. Hernández-Lerma. Asymptotic Convergence of a Simulated Annealing Algorithm for Multiobjective Optimization Problems. Technical Report EVOCINV-02-2004, Evolutionary Computation Group at CINVESTAV, Sección de Computación, Departamento de Ingeniería Eléctrica, CINVESTAV-IPN, México, D.F., March 2004. available at:
   http://delta.cs.cinvestav.mx/~ccoello/2004.html.
26. M. Villalobos-Arias, C. A. Coello Coello, and O. Hernández-Lerma. Convergence Analysis of a Multiobjective Artificial Immune System Algorithm. In *Artificial Immune Systems: Third International Conference (ICARIS 2004). Proceedings*, pages 226–235, Catania, Sicily, Italy, September 2004. Springer. Lecture Notes in Computer Science Vol. 3239.
27. Eckart Zitzler and Lothar Thiele. Multiobjective Evolutionary Algorithms: A Comparative Case Study and the Strength Pareto Approach. *IEEE Transactions on Evolutionary Computation*, 3(4):257–271, November 1999.

Running Time Analysis of a Multiobjective Evolutionary Algorithm on Simple and Hard Problems

Rajeev Kumar and Nilanjan Banerjee

Department of Computer Science and Engineering
Indian Institute of Technology Kharagpur
Kharagpur, WB 721 302, India
rkumar@cse.iitkgp.ernet.in, nilanb@cs.umass.edu

Abstract. In this paper, we suggest a multiobjective evolutionary algorithm based on a restricted mating pool (REMO) with a separate archive for storing the remaining population. Such archive based algorithms have been used for solving real-world applications, however, no theoretical results are available. In this paper, we present a rigorous running time complexity analysis for the algorithm on two simple discrete pseudo boolean functions and on the multiobjective knapsack problem which is known to be NP-complete. We use two well known simple functions LOTZ (Leading Zeros: Trailing Ones) and a quadratic function. For the knapsack problem we formalize a $(1 + \epsilon)$-approximation set under a constraint on the weights of the items. We then generalize the idea by eliminating the constraints based on a principle of partitioning the items into blocks and analyze REMO on it. We use a simple strategy based on partitioning of the decision space into fitness layers for the analysis.

1 Introduction

Evolutionary algorithms are emerging as a powerful tool to solve NP-hard combinatorial optimization problems. EAs use a randomized search technique with a *population* of individuals. The genetic operators used by EAs do not apply, in general, any problem-specific knowledge, however, special genetic operators may be designed by incorporating domain knowledge to expedite the search for some applications. In the multiobjective scenario, EAs often find effectively a set of diverse and mutually competitive solutions. Some results for solving computationally hard problems [1, 2] using multiobjective EAs are available in the literature – e.g., m-dimensional knapsack [3], minimum spanning tree [4, 5], partitioning of high-dimensional patterns spaces [6], code-book design [7], communication network topology design [8], and network design [9].

The EA operators like mutation and crossover imitate the process of natural evolution. The underlying principles of the operators are simple, but nevertheless, EAs exhibit complex behavior which is difficult to analyze theoretically. Hence, though there are numerous empirical reports on the application of EAs, work on their theoretical analysis is rare. However, besides empirical findings, theoretical analysis is essential to understand the performance and behavior of such heuristics. Some work, in this direction, has recently started, e.g., [10–12].

In case of single objective optimization, many results have been obtained on the analysis of evolutionary algorithms. Results on the time bounds of algorithms in the

A.H. Wright et al. (Eds.): FOGA 2005, LNCS 3469, pp. 112–131, 2005.

discrete search space [13] as well as continuous search space [14] are available. Some analysis on special functions using $(1 + 1)$ EA has been done - linear functions [15], ONE-MAX function [16], unimodal function [13, 17], and pseudo-boolean quadratic functions [18]. Most of the work above analyzed evolutionary algorithms with mutation as the only genetic operator. However, a proof that crossover is essential is presented in [19].

The analysis of the multiobjective case, however, is more difficult than its single objective counterpart since it involves issues like the size of Pareto-set, diversity of the obtained solutions and convergence to the Pareto-front [20, 21]. Consequently, results on theoretical analysis of multiobjective evolutionary algorithms are few. Rudolph [22, 23] and Rudolph and Agapie [24] have studied multiobjective optimizers with respect to their limit behavior. Laumanns et al. pioneered in deriving sharp asymptotic bounds for two-objective toy functions [10, 11]. Recently, Giel [25] and Thierens [26] derived bounds for another simple function.

Most of the work done earlier deals with analysis of simple problems. However, analysis of a multiobjective evolutionary algorithm on a simple variant of the 0-1 knapsack problem was started by Laumanns et al. [12]. They analyzed the expected running time of two multiobjective evolutionary algorithms 'Simple Evolutionary Multiobjective Optimizer (SEMO)' and 'Fair Evolutionary Multiobjective Optimizer (FEMO)' for a *simple* instance of the multiobjective 0-1 knapsack problem. The considered problem instance has two profit values per item and cannot be solved by one-bit mutations. In the analysis, the authors make use of two general upper bound techniques, the decision space partition method and the graph search method. The paper demonstrates how these methods, which have previously only been applied to algorithms with one-bit mutations, are equally applicable for mutation operators where each bit is flipped independently with a certain probability. However, the work takes care of only a very special instance of the knapsack and cannot be extended to any knapsack problem.

None of the work involves analysis of real-world combinatorial optimization problems using multiobjective EAs. In this work, we continue such an analysis for the well-known bi-objective 0–1 knapsack problem [27–30]. In the most general case, the Pareto-optimal set for the knapsack can be exponentially large in the input size. Therefore, we first, formulate a $(1 + \epsilon)$-approximate set for the 0 - 1 knapsack, followed by a rigorous analysis of the expected time to find the solution-set. We also carry out the expected running time analysis on two simple pseudo-boolean functions, namely the Leading Zeros: Trailing Ones (LOTZ) [10] and Quadratic Function (QF) [25].

We suggest a simple multiobjective optimizer based on an archiving strategy which is well adapted to work efficiently for problems where the Pareto-optimal points are Hamming neighbors (i.e., having a Hamming distance of 1). We call our algorithm Restricted Evolutionary Multiobjective Optimizer (REMO). The algorithm uses a special archive of two individuals which are selected based on a special fitness function which selects two individuals with the largest Hamming distance. Such a mechanism assures that the individuals selected for mutation are more likely to produce new individuals. However, this assumption holds for functions where the optimal set consists of individuals which are Hamming neighbors of each other and the distribution of the optimal points is fairly uniform.

The rest of the paper is organized as follows. Section 2 describes the related work in the field of theoretical analysis of evolutionary algorithms. Section 3 includes a few definitions pertaining to multiobjective optimization. Section 4 describes our algorithm REMO. Section 5 presents the analysis of the REMO on the LOTZ and the Quadratic function. Section 6 formulates the linear functions, the knapsack problem and its $(1+\epsilon)$-approximate set. The analysis of the algorithm on the knapsack problem is given in section 7. Finally, conclusions are drawn in section 8.

2   Related Work

2.1   Problems Analyzed

The work of Beyer et al. [31] revolves around how long a particular algorithm takes to find the optimal solutions for a given class of functions. The motivation to start such an analysis was to improve the knowledge of the randomized search heuristics on a given class of functions. Rudolph [22, 23] and Rudolph and Agapie [24] studied multiobjective optimizers with respect to their limit behavior. They aimed to find whether a particular algorithm converges if the number of iterations goes to infinity.

In the single objective case, the working of EAs have been analyzed for many analytical functions - linear functions [15]; unimodal functions [13]; and quadratic functions [18]. Some recent work has been done on sorting and shortest-path problems by recasting them as combinatorial problems [32]. A study is done to evaluate the blackbox complexity of problems too [33].

All the above work used a base-line simple $(1+1)$ EA. The analysis of $(1+1)$ EA was done using the method of partitioning of the decision space in accordance with the complexity of the problem into fitness layers as one of the methods [15]. For all such work, the only genetic operator used was mutation. However, Jansen and Wegener [19] analyzed the effectiveness of crossover operator, and showed that crossover does help for a class of problems.

For multiobjective optimization, the analysis of the asymptotic expected optimization time has been started by Laumanns et al. [10]. They presented the analysis of population-based EAs (SEMO and FEMO) on a bi-objective problem (LOTZ) with conflicting objectives. They extended this work by introducing another pseudo-boolean problem (Count Ones Count Zeros: COCZ), another algorithm Greedy Evolutionary Multiobjective Optimizer (GEMO), and scaling the LOTZ and COCZ problems to larger number of decision variables and of objectives [11]. A similar analysis was performed by Giel [25] and Thierens [26] on another bi-objective problem (Multiobjective Count Ones: MOCO) and the quadratic function that we use for our algorithm. These authors designed simple toy functions to understand the behavior of simple EAs for multiobjective problems. In [12] the authors solve a special instance of the multiobjective knapsack problem.

2.2   Algorithms Analyzed

The single objective optimizer basically yields a single optimal solution so $(1+1)$ EAs have been successfully used and analyzed for different functions. However, in the multiobjective case, an optimizer should return a set of *incomparable or equivalent* solutions. Hence a population-based EA is preferred. For this purpose, Laumanns et al. proposed a base-line population based EA called Simple Evolutionary Multiobjective Optimizer (SEMO). Another strategy used is a multi-start variant of $(1 + 1)$ EA [11].

These algorithms have an unbounded population. Individuals are added or removed from the population based on some selection criterion. Laumanns et al. introduced two other variants of SEMO called FEMO (Fair EMO) and GEMO (Greedy EMO) which differ in their selection schemes. The algorithms do not have any defined stopping criterion and are run till the desired representative approximate set of the Pareto-front is in the population [11].

There is another group of algorithms which use an explicit or implicit archiving strategy to store the best individuals obtained so far. This approach has proven to be very effective in finding the optimal Pareto-front at much reduced computational cost, e.g., NSGA-II [34], PAES [35], PCGA [36] and SPEA2 [37]. Also, many real-world problems have effectively been solved using such a strategy. However, there exists no analysis of such algorithms. In this work, we propose and use an archive-based EA.

Another issue in archive-based EAs is the size of the archive and the mating pool. If we restrict the number of individuals used for mating in the population to a constant, the expected waiting time till the desired individual is selected for mutation, is considerably reduced. But, for such an algorithm an efficient selection strategy to choose the proper individuals from the archive to the population needs to be devised. This is further discussed while formulating and analyzing the REMO algorithm

3   Basic Definitions

In the multiobjective optimization scenario there are m incommensurable and often conflicting objectives that need to be optimized simultaneously. We formally define some terms below which are important from the perspective of MOEAs. We follow [3, 23–25, 36, 38] for some of the definitions.

Definition 1. Multiobjective Optimization Problem (MOP): *A general Multiobjective Optimization problem includes a set of n decision variables $x = (x_1, x_2, ..., x_n)$, a set of m objective functions $F = \{f_1, f_2, ..., f_m\}$ and a set of k constraints $C = \{c_1, c_2, ..., c_k\}$. The objectives and the constraints are functions of the decision variables.* The goal is to:

Maximize/Minimize: $F(x) = \{f_1(x), f_2(x), ..., f_m(x)\}$
subject to satisfaction of the constraints:
$C(x) = \{c_1(x), c_2(x), ..., c_k(x)\} \leq 0$ for
$X = (x_1, x_2, ..., x_n) \in Y$
$F = (f_1, f_2, ..., f_m) \in G$

where X is the decision vector, F is the objective vector, Y denotes the decision space and G is a function space.

The m objectives may be mutually conflicting in nature. In some formalizations the k constraints defined above are treated as objective functions, thus, making the problem constraint-free, and vice-versa the objectives may be treated as constraints to reduce the dimensionality of the objective-space.

Definition 2. Quasi Order, Partial Order: *A binary relation $\preceq$ on a set* **F** *is called a quasi order if it is both reflexive and transitive. The pair $(F, \preceq)$ is called a partially ordered set if it is an antisymmetric quasi order.*

The Pareto dominance relationship in multiobjective evolutionary algorithms are partial orders (posets). The reason for the Pareto dominance relation to be a partial order is that there might be a number of individuals in the population which are mutually *incomparable or equivalent* to each other. An ordering is not defined for them.

Definition 3. $\preceq_q$: *Let Y be the decision space and let $(G, \preceq)$ be a poset of objective values. Let $f : Y \to G$ be a mapping. Then f induces a quasi-order $\preceq_q$ on Y (a set of binary vectors) by the following definitions for a minimization problem:*

$$x \prec_q y \text{ iff } f(x) \prec f(y)$$
$$x =_q y \text{ iff } f(x) = f(y)$$
$$x \preceq_q y \text{ iff } x \prec_q y \lor x =_q y .$$

The above definition introduces the concept of Pareto-dominance and Pareto optimality. Pareto-dominance can be defined as follows:

In a maximization problem of m objectives $o_1, o_2, ..., o_m$, an individual objective vector F_i is partially less than another individual objective vector F_j (symbolically represented by $F_i \prec F_j$) iff $(\forall_{o_i})(f_i^{o_i} \leq f_j^{o_i}) \land (\exists_{o_j})(f_i^{o_j} < f_j^{o_j})$, where $f_i^{o_i}$ and $f_i^{o_j}$ are components of F_i and F_j respectively.

Then F_j is said to dominate F_i. If an individual is not dominated by any other individual, it is said to be non-dominated. We use the notion of Pareto-optimality if $F = (f_1, ..., f_m)$ is a vector-valued objective function. Pareto dominance is formally defined in the next definition.

Definition 4. Pareto Dominance: *A vector $f_m = \{f_1^m, ..., f_k^m\}$ is said to dominate a vector $f_n = \{f_1^n, ..., f_j^n\}$ (denoted by $f_m \prec f_n$) iff f_n is partially less than f_m, i.e., $\forall i \in \{1, ..., k\}, f_m^i \leq f_n^i \land \exists i \in \{1, ..., k\} : f_i^m < f_i^n$.*

Definition 5. Pareto Optimal Set: *A set $A \subseteq Y$ (where Y denotes the entire decision space) is called a Pareto optimal set iff*

$$\forall a \in A: \text{ there does not exist } x \in Y: a \prec x .$$

In most practical cases it is not possible to generate the entire Pareto-optimal set. This might be the case when the size of the set is exponentially large in the input size. Thus, we confine our goal to attain an approximate set. This approximate set is usually polynomial in size. Since in most cases the objective functions are not bijective, there are a number of individuals in the decision space which are mapped to the same objective function. Hence, one might define an approximate set by selecting only one individual corresponding to an objective function. This is usually done in the case of single objective optimization problem.

Definition 6. Approximate Set: *A set $A_p \subseteq A$ (Pareto-optimal Set) is called an approximate set if there is no individual in A_p which is weakly dominated by any other member of A_p.*

Another strategy that might be used to attain an approximate set is to try and obtain an inferior Pareto front. Such a front may be inferior with respect to the distance from the actual front in the decision space or the objective space. If the front differs from the actual optimal front by a distance of ϵ in the objective space, then, the dominance relation is called a $(1 + \epsilon)$-dominance.

Definition 7. $(1 + \epsilon)$-Domination: *For decision vectors $X_1, X_2 \in X$ where $X_1 = (x_{11}, x_{12}, ..., x_{1i})$ and $X_2 = (x_{21}, x_{22}, ..., x_{2i})$, we say that X_2 $(1 + \epsilon)$-dominates X_1, denoted by $X_1 \preceq^{1+\epsilon} X_2$, if $f(x_{1i}) \leq (1+\epsilon) \cdot f(x_{2i})$ for all objectives to be maximized, and $f(x_{2i}) \leq (1 + \epsilon) \cdot f(x_{1i})$ for all objectives to be minimized.*

The $(1 + \epsilon)$-dominance is transitive. The optimal set created by applying the above dominance relation is called a $(1 + \epsilon)$-approximate set.

Definition 8. $(1+\epsilon)$-Approximate Set: *A set $A_{1+\epsilon}$ is called a $(1+\epsilon)$-approximate set of the Pareto set if for all elements a_p in the Pareto-set there exists an element $a' \in A_{1+\epsilon}$ such that $a_p \preceq^{1+\epsilon} a'$.*

Definition 9. δ-Sampling of the Pareto-Set: *If P denotes the Pareto-optimal set, then a δ-sampling of P is a set $P' \subset P$, such that no two individuals in P' is within a distance of δ units in the objective space (assuming some metric in the objective space).*

One might also attain an approximate set by taking a proper subset of the Pareto-optimal set. A strategy used to get such a subset is called sampling. A δ-sampling is a special form of sampling in which no two individuals are within a distance of δ in the objective space. The reason is to reduce the output length of the algorithm and reduce the problem to P-space.

Definition 10. Running Time of an EA: *The running time of an EA searching for an approximate set is defined as the number of iterations of the EA loop until the population is an approximate set for the considered problem.*

4   Algorithm

The algorithm suggested in this paper uses a restricted mating pool or population P of only two individuals and a separate archive A for storing all other points that are likely to be produced during a run of the algorithm. The two individuals to be chosen are selected based on a special function called **Handler**. With a probability of $\frac{1}{2}$ a function $Fit(x, P \cup A) = H(x)$ is evaluated where $H(x)$ is the number of hamming neighbors of x in $P \cup A$. The two individuals with the smallest Fit-values are selected into the population P and the rest are transferred to the archive A. Such a selection strategy assures that we select an individual for mating that has a higher probability of producing a new offspring in the next run of the algorithm. Such a selection strategy has been found to improve the expected running time for simple functions like LOTZ. The intuition behind such a selection scheme is that for simple functions whose optimal set is uniformly distributed over the front and whose individuals are Hamming neighbors of each other. However, for the other half of the cases the handler function selects the two individuals at random. This is similar to the selection mechanism in SEMO [10]. This is done because in certain functions it is probable that if the two individuals selected are always those with the largest hamming distance, the algorithm might not be able to explore all the bit vectors in the optimal set. This happens for harder problems like the 0 - 1 knapsack. The algorithm takes ϵ ($\epsilon \geq 0$) as an input parameter and produces as its output a $(1 + \epsilon)$-approximate set of the Pareto-optimal set. The algorithms in its main loop creates an individual uniformly at random and adds the individual into the population if it is not weakly dominated by any other individual in the population and it is not dominated by any other individual. All those individuals which are dominated by the new individual and if the new individual does not dominate the individual in the P and A are removed from P and A. Such a strategy is adopted so that if any individual from the $(1 + \epsilon)$-set is created it is never removed from the population. Note that if $\epsilon = 0$ and we aim to find the entire Pareto set then we only need to check whether there is some individual which is dominated by the newly created individual and remove it from P and A.

Restricted Evolutionary Multiobjective Optimizer (REMO)

---

1. Input Parameter: ϵ, $\epsilon \geq 0$, if we desire a$(1 + \epsilon)$-approximate set of the Pareto-optimal set.
2. Initialize two sets $P = \phi$ and $A = \phi$, where P is the mating pool and A is an archive.
3. Choose an individual x uniformly at random from $Y = \{0, 1\}^n$.
4. $P = \{x\}$.
5. **loop**
6. Select an individual y from P at random.
7. Apply mutation operator on y by flipping a single randomly chosen bit and create y'.
8. $P = P \setminus \{l \in P \mid l \prec y' \wedge l$ does not $(1 + \epsilon)$-dominate $y'\}$.
9. $A = A \setminus \{l \in A \mid l \prec y' \wedge l$ does not $(1 + \epsilon)$-dominate $y'\}$.
10. **if** there does not exist $z \in P \cup A$ such that $y' \prec z$ or $f(z) = f(y')$ **then** $P = P \cup \{y'\}$.
11. **end if.**
12. **if** cardinality of P is greater than 2 **then**
13. **Handler Function**
14. **end if.**
15. **end loop.**

Handler Function

---

1. Generate a random number R in the interval $(0, 1)$.
2. **if** $R > \frac{1}{2}$ **then** Step 3 **else** Step 5.
3. For all the members of $P \cup A$ calculate a fitness function $Fit(x, P \cup A) = H(x)$ where $H(x)$ denotes the number of hamming neighbors of x in $P \cup A$.
4. Select two individuals with the minimum $Fit(x, P \cup A)$ values into P and put the rest of the individuals in the archive A. In case of equal $Fit(x, P \cup A)$ values the selection is made at random.
5. Select two individuals at random from $P \cup A$.

---

5 Analysis of REMO on Simple Functions

5.1 Leading Ones Trailing Zeros

The Leading Ones (LO), Trailing Zeros (TZ) and the LOTZ problems can be defined as follows where the aim is to maximize both the objectives:

$$LO : \{0,1\}^n \rightarrow N \qquad\qquad LO(x) = \sum_{i=1}^{n} \prod_{j=1}^{i} x_j$$

$$TZ : \{0,1\}^n \rightarrow N \qquad\qquad TZ(x) = \sum_{i=1}^{n} \prod_{j=i}^{n} (1 - x_j)$$

$$LOTZ : \{0,1\}^n \rightarrow N^2 \qquad\quad LOTZ(x) = (LO(x), TZ(x))$$

Proposition 1. The Pareto front (optimal set of points in the objective space) for LOTZ can be represented as a set $S = \{(i, n - i) \mid 0 \le i \le n\}$ and the Pareto set consists of all bit vectors belonging to the set $P = \{ 1^i 0^{n-i} \mid 0 \le i \le n \}$[10].

Proof. The proof is the same as given in [10].

Analysis: The analysis of the function above is divided into two distinct phases. Phase 1 ends with the first Pareto-optimal point in the population P, and Phase 2 ends with the entire Pareto-optimal set in $P \cup A$. We assume $\epsilon = 0$, thus, we aim to find the entire Pareto-optimal set.

Theorem 2. The expected running time of REMO on LOTZ is $O(n^2)$ with a probability of $1 - e^{-\Omega(n)}$.

Proof. We partition the decision space into fitness layers defined as (i, j), $(0 \le i, j \le n)$ where i refers to the number of Leading-ones and j is the number of Trailing-zeros in a chromosome. The individuals in one particular fitness layer are incomparable to each other. A parent is considered to climb up a fitness layer if it produces a child which dominates it. In Phase 1 a mutation event is considered a success if we climb up a fitness layer. If the probability of a success S, denoted by $P(S) \ge p_i$ then the expected waiting time for a success $E(S) \le \frac{1}{p_i}$.

For LOTZ, in Phase 1 the population cannot contain more than one individual for REMO because a single bit flip will create a child that is either dominating or is dominated by its parent and the algorithm does not accept weakly dominated individuals. The decision space is partitioned into fitness layers as defined above. Phase 1 begins with a initial random bit vector in P. Let us assume that after T iterations in Phase 1 the individual $A(i, j)$ is in the population P. The individual can climb up a fitness layer (i, j) by a single bit mutation if it produces the child $(i + 1, j)$ or $(i, j + 1)$. The probability of flipping any particular bit in the parent is $\frac{1}{n}$, thus the probability associated with such a transition is $\frac{2}{n}$. The factor of 2 is multiplied because we could either flip the leftmost 0 or the rightmost 1 for a success. Therefore, the expected waiting time for such a successful bit flip is at most $\frac{n}{2}$. If we assume that Phase 1 begins with the worst individual $(0, 0)$ in the population then algorithm would require at most n successful mutation steps till the first Pareto-optimal point is found. Thus, it takes $\sum_{i=1}^{i=n} \frac{n}{2} = \frac{n^2}{2}$ expected number of steps for the completion of Phase 1.

To prove that the above bound holds with an overwhelming probability let us consider that the algorithm is run for n^2 steps. The expected number of successes for these n^2 steps is at least $2n$. If S denotes the number of successes, then by Chernoff's bounds:

$$P[S \leq (1 - \tfrac{1}{2}) \cdot 2n] = P[S \leq n] \leq e^{-\frac{n}{4}} = e^{-\Omega(n)}$$

Hence, the above bound for Phase 1 holds with a probability of $1 - e^{-\Omega(n)}$ which is exponentially close to 1.

Phase 2 begins with an individual of the form $I = (i, n-i)$ in P. A success in Phase 2 is defined as the production of another Pareto-optimal individual. The first successful mutation in Phase 2 leads to production of the individual $I_{+1} = (i + 1, n - i - 1)$ or $I_{-1} = (i - 1, n - i + 1))$ in the population P. The probability of such a step is given by $\frac{2}{n}$. Thus, the waiting time till the first success occurs is given by $\frac{n}{2}$. If we assume that after the first success I and I_{-1} are in P (without loss of generality), then the Pareto-optimal front can be described as two paths from $1^{i-1}0^{n-i+1}$ to 0^n and 1^i0^{n-i} to 1^n. At any instance of time T, let the individuals in P be represented by $L = (l, n - l)$ and $K = (k, n - k)$ where $0 \leq k < l \leq n$. As the algorithm would have followed the path from $(i - 1, n - i + 1)$ to $(k, n - k)$ and $(i, n - i)$ to $(l, n - l)$ to reach to the points L and K, it is clear that at time T all the individuals of the form $S = (j, n - j)$ with $l < j < k$ have already been found and form a part of the archive A. Moreover, the handler function, assures that L and K are farthest apart as far as Hamming distance is concerned. At time T the probability of choosing any one individual for mutation is $\frac{1}{2}$. Let us assume, without loss of generality, that the individual selected is $(k, n - k)$. The flipping of the left most 0 produces the individual $K_{+1} = (k + 1, n - k - 1)$ and the flipping of the rightmost 1 produces the individual $K_{-1} = (k - 1, n - k + 1)$. Since, the algorithm does not accept weakly dominated individuals and K_{+1} is already in A, the production of K_{-1} can only be considered as a success. Thus the probability of producing another Pareto-optimal individual at time T is $\frac{1}{2n}$. Thus, the expected waiting time of producing another Pareto-optimal individual is at most $2n$. Since, no solutions on the Pareto-optimal front is revisited in Phase 2, it takes a maximum of $n + 1$ steps for its completion. The special case hold when the individual 0^n or 1^n appears in the population. Such individuals represent the end points of the Pareto-front and their mutation do not produce any individual that can be accepted. Moreover, such individuals always form a part of the population P since they have to be a part of the pair of individuals which have the maximum Hamming distance. However, if such an individual is a part of P the probability bound still holds as the probability of choosing the right individual (the individual which is not 0^n or 1^n) for mutation is still $\frac{1}{2}$ and the probability of successful mutation is $\frac{1}{2n}$ and expected running time bound holds. Therefore, REMO takes $O(n^2)$ for Phase 2.

Now, we consider Phase 2 with $4n^2$ steps. By arguments similar to Phase 1, it can be shown by Chernoff's bounds that the probability of the number of successes in Phase 2 being less than n, is $e^{-\Omega(n)}$.

Altogether considering both the Phases, REMO takes n^2 steps to find the entire Pareto-optimal set for LOTZ.

For the bound on the expected time we have not assumed anything about the initial population. Thus, we notice that the above bound on the probability holds for the next n^2 steps. Since the lower bound on the probability that the algorithm will find the entire

Pareto set is more than $\frac{1}{2}$ (in fact exponentially close to 1) the expected number of times the algorithm has to run is bounded by 2.

Combining the results of both the phases 1 and 2 yields the bounds in the theorem.

□

Comments: The above bound only considers the number of iterations that the algorithm needs to find the Pareto set. However, if all the evaluations of the loop is considered, the handler function can take a time $O(n^2)$ to find the pair of individuals with the least Fit value and hence, the entire algorithm may take time of $O(n^4)$.

5.2 Quadratic Function

We use a continuous quadratic function which has been widely used in *empirical* analysis of multiobjective evolutionary optimizers and adapt it to the discrete boolean decision space exactly as done in [25]. The function in the continuous decision space is $((x - a)^2, (x - b)^2)$ where the aim is the minimization of both the objectives. The function in the discrete boolean decision space is described in the following manner exactly as in [25]:

$$QF : \{0, 1\}^n \to N^2.$$
$$\textbf{if } \|x\| = \sum_{i=1}^{n} x_i$$
$$QF : (\|x\| - a)^2, (\|x\| - b)^2)$$

The idea is to just test the efficiency of REMO in solving problems like those of QF and in fact the analysis follows a line similar to that of [25].

Proposition 2. *Without loss of generality we assume that $a > b$. The Pareto-optimal front of QF can be represented by the set $F = \{(i^2, (i - (b - a))^2) \mid a \leq i \leq b\}$. The Pareto-optimal points of QF are those bit vectors where $a \leq \|x\| \leq b$.*

Proof. The proof is the same as given in [25].

Theorem 2. *The running time of REMO on QF is $O(n \log n)$ for any value of a and b.* (We assume $\epsilon = 0$, thus we aim to find the entire Pareto-optimal set.)

Proof. We partition the analysis into two phases. The analysis turns out to be very similar to that done in [25] but it is much simpler due to the local mutation operator used in REMO. Phase 1 ends with the first Pareto-optimal point in P and the second phase continues till all the Pareto-optimal bit vectors are in $P \cup A$.

It can be proven that in Phase 1 there can be a maximum of 1 (similar to [25] individuals in $P \cup A$. Thus, the archive A is empty. This is because a single bit mutation of a parent with $\|x\| < a$ or $\|x\| > b$ will produce an individual which is dominated by or dominates its parent. We partition the decision space into sets with individuals having the same number of ones. Let us consider a bit vector represented as I_d where d represents the number of ones in the individual. A single bit mutation of I_d is considered to be a success if the number of ones increases (decreases) when $d < a(d > b)$.

Therefore, a success S requires the flipping of any one of the d 1-bits ($n - d$ 0-bits) when $d < a$ ($d > b$). The probability of a successful mutation $P(S) = \frac{d}{n} (or \frac{n-d}{n})$. The expected waiting time of S, given by $E(S) \le \frac{n}{d} (or \frac{n}{n-d})$. Hence, the total expected time till the first Pareto optimal point arrives in the population is at most $\sum_{d=1}^{n} \frac{n}{d} = nH_n = n \log n + \Theta(2n) = O(n \log n)$ (where H_n is the nth Harmonic number) by the linearity of expectations.

Phase 2 starts with the assumption that $b - a > 1$ or else there would be no second phase. The number of individuals in the population is bounded by 2. The selection mechanism ensures that they are the bit vectors that are most capable of producing new individuals. The Pareto-front can be visualized as a path of individuals with number of ones varying from a to b or b to a. Let us represent any individual with $a < \|x\| < b$ as I_k where k represents the number of ones in the bit vector. Such a bit vector can be produced either by an individual with $k + 1$ ones or $k - 1$ ones. The associated probability for such a successful mutation is at least $\frac{k+1}{2n}$ and $\frac{n-k+1}{2n}$ respectively. Hence, the expected waiting time till the I_k^{th} Pareto optimal point is in the population (assuming that its parent is in the population) is $E(I_k) \le \frac{2n}{k+1}$ and $\frac{2n}{n-k+1}$ for the two cases above. Thus, the total expected time till all the Pareto points are in $P \cup A$ is at most $\sum_{k=a}^{b} E(I_k) \le \sum_{k=a}^{b} \frac{2n}{k+1} \le \sum_{k=0}^{b-a} \frac{2n}{k+1} = 2nH_{b-a}$ where H_n stands for the nth harmonic number.

Therefore, the expected time for Phase 2 is at most $2ne \log(b - a) + \theta(2ne) = O(n \log(b - a))$.

Altogether both Phases take a total of $O(n \log n + n \log(b - a))$ time. Since a and b can have a maximum value of n the running time for REMO on QF is $O(n \log n)$ which is the same as proven in [25]. $\square$

6   Linear Functions and Knapsack Problem

6.1   Linear Functions

A bi-objective linear function is:

$$F(x) = (f_1(x) = \sum_{i=1}^{n} w_i x_i, f_2(x) = \sum_{i=1}^{n} w_i' x_i)$$
where $w_i > 0, w_i' > 0$

The aim of a bi-objective problem may be to maximize or minimize both the objectives; in such a case the problem reduces to a single objective one. In our study, therefore, we take the case of simultaneously maximizing f_1 and minimizing f_2. Thus, the problem is formulated with two mutually conflicting objectives.

In this section, we show that for the function $F(x)$ the number of Pareto-optimal points can range from $n+1$ to 2^n. Thus, for any arbitrary values of the weights w and w' the Pareto-optimal set can be exponential in n. Throughout this section we investigate the case where the bits of the individuals are arranged in their decreasing value of $\frac{w}{w'}$. Thus, $\frac{w_1}{w_1'} \ge \frac{w_2}{w_2'} \ge ... \ge \frac{w_n}{w_n'}$.

Lemma 1. For any bi-objective linear function $F(x) = (f_1, f_2)$ the set $A_1 = \{1^i 0^{n-i}\}$ where $0 \le i \le n$ represents a set of Pareto-optimal points.

Proof. Let us consider an individual K^* in the decision space (which does not belong to A_1) and an individual $K \in A_1$ which is of the form $(1^k 0^{n-k})$ for $0 \le k \le n$. If the set of 1-bits of K^* is a subset of the set of 1-bits in K it is clear that K and K^* are incomparable.

However, if the set of 1-bits of K^* is not a subset of K, let S denote the set of common bit positions that are set to one in both K^* and K with $x_1 = \sum_{i \in S} w_i$. Let S_1 denote the set of bit positions that are set to one in K but not in K^* and $y_1 = \sum_{i \in S_1} w_i'$. If the individual K^* has to dominate K, in the best case $f_2(K^*)$ is at most $f_2(K)$. Since all the bits are arranged in the decreasing order of $\frac{w_i}{w_i'}$, $f_1(K^*)$ is at most $x_1 + \frac{w_{k+1}}{w_{k+1}'} y_1 \le x_1 + \frac{w_k}{w_k'} y_1$. Now, $f_1(K)$ is at least $x_1 + \frac{w_k}{w_k'} y_1$. Hence, $f_1(K) > f_1(K^*)$. Therefore, K^* cannot dominate K.

Now we need to prove that two individuals in A_1 are mutually incomparable to each other. Let us consider another individual $I = 1^i 0^{n-i}$ in A_1 where $0 \le i \le n$. If $i < k$, $f_1(I) < f_1(K)$ and $f_2(I) < f_2(K)$, implying that I and K are incomparable (by definition 2). A similar argument holds for $i > k$, hence proving the lemma. $\square$

Example 1. Let us consider a linear function with three weights. The w of the components are $W = \{20, 15, 19\}$ and the weights W' are $W' = \{8, 7, 12\}$. Clearly $\frac{W_1}{W_1'} > \frac{W_2}{W_2'} > \frac{W_3}{W_3'}$. Let us consider the bit vector 110. This individual has the first and second weights set to 1. Individuals whose bits set to 1 are a subset of the above individuals, for example 100, will have both w and w' less than 110, and hence is incomparable to it. Individuals which are not a subset of 110, like for example 101 will be dominated by 110 or is incomparable to it. As an example, $w(110) = 35$, $w'(110) = 15$ and $w(101) = 39$, $w'(101) = 20$, these two individuals are incomparable. This hold for any individual which is not a subset of 110.

Proposition 3. The size of the Pareto-optimal set for the most general case of a linear function $F(x)$ lies between $n + 1$ and 2^n.

Proof. It is clear from lemma 1 that the lower bound on the number of Pareto-optimal individuals for $F(x)$ is $n + 1$. Moreover, the upper bound holds for cases where all the bit vectors are Pareto-optimal. We next show that there are examples which fit into the above bounds.

Case 1: Let us consider a linear function such that $w_1 > w_2 > w_3 > ... > w_n$ and $w_1' < w_2' < w_3' < ... < w_n'$. Each Pareto-optimal point is of the form $X = 1^i 0^{n-i}$ where $0 \le i \le n$. It is clear that individuals of the form X represent a Pareto-optimal solution because it contains the i largest weights of f_1 and the i smallest weights of f_2. Flipping the left-most 0-bit of X to 1 or the right-most 1-bit to 0 creates an individual which is incomparable to X. Moreover, any individual with a 0 followed by a 1 cannot be Pareto-optimal as it can be improved in both objectives by simply swapping the bits. The Pareto-optimal set of such a function thus contains $n + 1$ individuals.

Case 2: For the other extreme case, let us consider a linear function for which $\frac{w_1}{w_1'} = \frac{w_2}{w_2'} = \frac{w_3}{w_3'} = ... = \frac{w_n}{w_n'}$ and $w_1 > w_2 > w_3 > ... > w_n$. It is clear that for such a function all the points in the decision space $\{0,1\}^n$ are Pareto-optimal. Thus, the total number of Pareto points for this case is 2^n.   □

Consequently, for any randomized algorithm the expected runtime to find the entire Pareto-optimal set for the above case of bi-objective linear functions is $\Omega(2^n)$.

6.2 Knapsack Problem

Next, we show that the above problem of the conflicting objectives for a linear function can be interpreted as the 0 - 1 Knapsack problem.

Definition 11. 0–1 Knapsack Problem: The *knapsack problem* with n items is described by the knapsack of size b and three sets of variables related to the items: decision variables $x_1, x_2, ..., x_n$; positive weights $W_1, W_2, ..., W_n$; and profits $P_1, P_2, ..., P_n$; where, for each $1 \leq i \leq n$, x_i is either 0 or 1. The W_i and P_i represent the weight and profit, as integers, of the ith item respectively.

The single-objective knapsack problem can be formally stated as:

Maximize $P = \sum_{i=1}^{n} P_i x_i$
subject to $\sum_{i=1}^{n} W_i x_i \leq b$,
where $x_i = 0$ or 1

In order to recast the above single-objective problem along with one constraint on weights of the items into a bi-objective problem, we use the formulation similar to the linear function described above. Thus, a bi-objective knapsack problem of n items with two conflicting objectives (maximizing profits and minimizing weights) is formulated as:

Maximize $P = \sum_{j=1}^{n} P_j x_j$ and   Minimize $W = \sum_{j=1}^{n} W_j x_j$

Therefore, if the set of items is denoted by I, the aim is to find all sets $I_i \subseteq I$, such that for each I_j there is no other set which has a larger profit than profit (I_j) for the given weight bound $W(I_j)$. Thus, it is equivalent to finding the collections of items with the maximum profit in a knapsack with capacity $W(I_j)$.

The $0 - 1$ knapsack problem, in the optimization form above is NP-complete.

In the following section we formalize a $(1 + \epsilon)$- approximate set for the above knapsack problem. We work under the assumption that the items are arranged in a strictly decreasing value of $\frac{P}{W}$.

Lemma 2. Let us define a set X_j^i as the set of i most efficient items (efficiency defined by $\frac{P}{W}$) among the j smallest weight items if the sum of the weights of the j items is less than the $(j + 1)^{st}$ item. Let A_1 be the set of all such X_j^i and the following constraint be imposed on weights of the items: $\forall X_j^i$, if $\{I_1^j, ..., I_i^j\}$ represents the set of items in X_j^i, then $W_{I_{i+1}} < \epsilon \cdot \sum_{k=1}^{k=i} W_{I_k}$ for $i < j$. Now, if A_2 represents a singleton set, $\{0^i 10^{n-(i+1)}\}$ where the 1 is set at the position of the lightest item. Then, $A_1 \cup A_2$ represents a $(1 + \epsilon)$-approximation of the Pareto-optimal set for the knapsack problem.

Example 2. Let us consider a knapsack with three items. The profits of the items is given by $P = \{40, 20, 10\}$ and the weights by $W = \{20, 11, 6\}$. The knapsack satisfies the constraint on the weights (given in Lemma 2) for $\epsilon = 0.6$. X_1 is a trivial case as it contains the lightest item. For X_2^i, $1 \leq i \leq 2$, we consider the second and the third items. Clearly $W_3 < \epsilon \cdot W_2$ and hence it satisfies the constraint. For X_3^i, $1 \leq i \leq 3$ we have all the items. Since, $W_2 < \epsilon \cdot W_1$ and $W_3 < \epsilon \cdot (W_1 + W_2)$, the knapsack weights satisfy the given constraint.

Proof. Let (P^o, W^o) represent any arbitrary Pareto-optimal solution for the knapsack problem. We need to prove, that corresponding to such a solution we can always find a solution (P', W') in $A_1 \cup A_2$ such that $(P^o, W^o) \preceq^{1+\epsilon} (P', W')$. If W^o is the item with the smallest weight then the element in set A_2 weakly dominates (P^o, W^o) and hence $(1 + \epsilon)$-dominates (P^o, W^o). Now, let us consider a more general case. Let π define the permutation of the items which reorders them according to their weights. Corresponding to any W^o, we aim to find an item $I_{\pi(j)}$ such that $W_{\pi(j)} \leq (1+\epsilon) \cdot W^o < W_{\pi(j+1)}$. If such an item cannot be found then $j = n + 1$. It is clear that for this $j + 1$, an $X_{j+1}^i \in A_1$, a set of items $I_1^{j+1}, I_2^{j+1}, ..., I_i^{j+1}$ (i can be equal to j), can always be found such that $\sum_{k=1}^{k=i} W_{I_k} \leq (1 + \epsilon) \cdot W^o < \sum_{k=1}^{k=i+1} W_{I_k}$. We claim that $W^o \leq \sum_{k=1}^{k=i} W_{I_k} \leq (1 + \epsilon) \cdot W^o$. This holds because of the constraint on the weights imposed in the lemma. It is clear from the constraint that if $W_{I_{i+1}} \leq \epsilon \cdot \sum_{k=1}^{i} W_{I_k}$ and adding of the $(i + 1)^{st}$ item into the knapsack increases the weight of the knapsack above $(1+\epsilon) \cdot W^o$, the sum of the weights of the items in X_{j+1}^i is at least W^o. Since, the sum of the weights in the set X_{j+1}^i obeys the weight bound $(1 + \epsilon) \cdot W^o$, with respect to the weights, $(P^o, W^o) \preceq^{1+\epsilon} (P', W')$. Now, we know that $W' \geq W^o$. Let S_c denote the items common in both the solution (P', W') and (P^o, W^o) and S_{un} denote the set of items in the solution (P', W') and not in (P^o, W^o). Let $A = \sum_{I_m \in S_c} P_{I_m}$ and $B = \sum_{I_m \in S_{uc}} W_{I_m}$. Now, $W^o < W_{\pi(j+1)}$, therefore (P^o, W^o) contains items in the set X_{j+1} (which is the set of the first $j + 1$ lightest items). Since X_{j+1}^i is the set of the i most efficient items in X_{j+1} (efficiency defined as $\frac{P}{W}$), $P^o \leq A + B \cdot \frac{P_{i+1}}{W_{i+1}} \leq A + B \cdot \frac{P_i}{W_i} \leq P'$, thus proving that $(P^o, W^o) \preceq^{1+\epsilon} (P', W')$, hence the lemma.   □

In the next lemma, we extend the results obtained in lemma 2 to formalize a $(1+\epsilon)$-approximate set for any knapsack.

Lemma 3. Let $I = \{I_1, I_2, ..., I_n\}$ represent the set of items of the knapsack. We partition the set I into m blocks ($1 \leq m \leq n$), $B_1, B_2, ..., B_m$ which satisfy the following conditions:

1. Each block consists of a set of items and all pairs of blocks are mutually disjoint. However, $B_1 \cup B_2... \cup B_m = I$.
2. The items in each block satisfy the weight constraint described in lemma 3 and are arranged in their strictly decreasing $\frac{P_i}{W_i}$ ratio in the block.

Sets A_1 and A_2, similar to those defined in lemma 2 are defined for every block B_m. Let $A^m = A_1^m \cup A_2^m$ for the mth block. If S denotes the set of items formed by taking one set from every A^m, then, the collection of all such S sets, represented by S_{comp}, denotes a $(1 + \epsilon)$-approximation of the Pareto-optimal set for any knapsack problem. If $m = n$ the set reduces to a power set of the item-set.

Example 3. Let us consider a knapsack of four items, $\{I_1, I_2, I_3, I_4\}$. The profits of the items are given by $P = \{40, 20, 10, 15\}$ and the weights are given by $W = \{20, 11, 6, 40\}$. The items of the above knapsack can be divided into blocks $B_1 = \{I_1, I_2, I_3\}$ and $B_2 = \{I_4\}$. The block B_1 satisfies the constraint as it is the same set of items described in example 2 and block B_2 satisfies the constraint trivially as it has only one item. Hence, the blocks describe a valid partitioning.

Proof. Let us consider any Pareto-optimal collection of items U with objective values (P^o, W^o). We aim to prove that corresponding to every U we can find a solution V of the form S (defined above) which $(1 + \epsilon)$-dominates U.

We partition both solutions U and V into blocks of items as defined above. Let us consider a particular block of items B_i, and denote the set of items in U and V in the block as B_i^U and B_i^V respectively. Since, the block consists of items which satisfy all the conditions of lemma 3, the solution represented by B_i^U $(1 + \epsilon)$-dominates B_i^V (irrespective of whether B_i^U represents a Pareto-optimal solution for the items in B_i). It is clear that the above argument holds for every block. Now, weight(U) = $\sum_{i=1}^m W(B_i)$ and profit(U) = $\sum_{i=1}^m P(B_i)$. Since, for every block B_i, weight($B_i(V)$) $< (1 + \epsilon) \cdot$ weight($B_i(U)$) and $(1 + \epsilon) \cdot$ profit($B_i(V)$) > profit($B_i(U)$), weight(V) < $(1 + \epsilon) \cdot$ weight(U) and $(1 + \epsilon) \cdot$ profit(V) > profit(U) by a simple summation of the profits and weights of items in every block. Therefore, $U \preceq^{1+\epsilon} V$, proving the lemma.
□

Lemma 4. The total number of individuals in the $(1 + \epsilon)$-approximate set (defined in Lemma 3) is upper bounded by $(\frac{n}{m})^{2m}$, where m is the number of blocks (m defined in lemma 3) and n is the number of items in the knapsack, if $n \neq m$.

Proof. Let the number of individuals in the kth block be n_k. The number of sets of the form X_j^i in the kth block is of the order $O(n_k^2)$. Thus, the total number of sets possible for all the blocks is upper bounded by $(n_1 \cdot n_2 \cdot \ldots \cdot n_m)^2$ if m is not a constant. Now, $(n_1 \cdot n_2 \cdot \ldots \cdot n_m)^{\frac{1}{m}} \leq \frac{\sum_{i=1}^m n_i}{m}$ (Arithmetic mean $\geq$ Geometric Mean). Thus, $(n_1 \cdot n_2 \cdot \ldots \cdot n_m)^2 \leq (\frac{\sum_{i=1}^m n_i}{m})^{2m} = O((\frac{n}{m})^{2m})$, if $m \neq n$. However, if $n = m$, by the partitioning described in Lemma 3, all the bit vectors represent the $(1 + \epsilon)$-approximate set for the knapsack. Hence, in such a case the total number of individuals in the set is 2^n.
□

7 Analysis of REMO on Knapsack Problem

Lemma 5. If P_{knap} is the sum of all the profits of the items in the knapsack, then the total number of individuals in the population and archive is at most $P_{knap} + 1$.

Proof. It is clear that at any time the population and archive consists of individuals which are incomparable to each other as in the case of SEMO. We aim to prove that any two individuals of the population and archive will have distinct profit values. We try to prove the claim by contradiction. Let us assume that there are two individuals a and b in $P \cup A$ which have the same profit values. As the algorithm does not accept any weakly dominated individuals either weight$(a) >$ weight (b) or weight$(a) <$ weight(b). However, this contradicts our initial assumption that the population consists of individuals which are incomparable to each other. Hence, a and b have distinct profit values. As all the items have integer profits and the total profit can be zero, the total number of individuals are bounded by the sum of the profits $P_{knap} + 1$. $\qquad\Box$

Theorem 1. The expected running time for REMO to find a $(1 + \epsilon)$-approximate set of the Pareto-front (formalization given in lemma 3) for any instance of the knapsack problem with n items is $O(\frac{n^{2m+1} P_{knap}}{m^{2m+1}})$ where m is the number of blocks into which the items can be divided (blocks are defined in lemma 3) and P_{knap} refers to the sum of the profits of the items in the knapsack and $n \neq m$. Moreover, the above bound holds with an overwhelming probability. It is worth noting that the expected waiting time is a polynomial in n if the sum of the *profits* and the *number of partitions* m is a polynomial in n, which in turn depends on the ϵ value. It is difficult to derive any general relationship between ϵ and m, since it depends on the distribution of the weights. However, if the ϵ is large then the number of items in a particular block is likely to be large and hence the number of blocks will be small and the running time can be polynomial in n.

Proof. We divide the analysis into two phases. The first phase continues till 0^n is in the population or the archive. Phase 2 continues till all the vectors in set S is in $P \cup A$. (We take ϵ as an input.) In the first phase, the aim is to have a all $0s$ string in the population. We partition the decision space into fitness layers. A fitness layer i is defined as a solution which has the i smallest weight items in the knapsack. At any point of time in phase 1, there is a maximum of one solution Z which has the i smallest weights in the knapsack. Removal of any one of these items from Z reduces the weight of the knapsack and hence produces a solution which is accepted. With a probability of $\frac{1}{2}$ (as in the handler function), an individual is selected from the archive at random. Therefore, the probability of selection of Z for mutation is $\frac{1}{4(P_{knap}+1)}$ ($P_{knap} + 1$ is the bound on the population size, by lemma 5). If we flip the 1-bit corresponding to the heaviest item in Z (which occurs with a probability of $\frac{1}{n}$), the mutated individual is accepted. Thus, the probability of producing an accepted individual is $\frac{1}{4n(P_{knap}+1)}$. Therefore, the expected waiting time till a successful mutation occurs is at most $4n(P_{knap} + 1)$. Since, a maximum of n successes in Phase 1 assures that 0^n is in the population, the expected time for completion of Phase 1 is $4n^2(P_{knap} + 1)$. Therefore, Phase 1 ends in $O(n^2 P_{knap})$ time. If we consider $8n^2(P_{knap} + 1)$ steps the expected number of successes is at least $2n$. By Chernoff's bound the probability of number of successes being less than n is at most $e^{-\Omega(n)}$.

The second phase starts with the individual 0^n in the population. An individual that is required to be produced can be described as a collection of items $I_{coll} = C^1 \cup C^2 \cup$

$\dots \cup C^m$, where, C^k is either X_j^i or one item (of the smallest weight) in the kth block. If $X_j^{i^k}$ refers to the set X_j^i in the kth block, it is clear to see that $H(0^n, X_j^{1^k}) = 1$, where H refers to the Hamming distance. Thus, by a single bit flip of 0^n we can produce a new desired point. Since, the algorithm accepts bit vectors which are $(1+\epsilon)$-approximations of the Pareto-optimal points, the point generated will be taken into the population and will never be removed. It is also clear that $H(X_j^{i-1^k}, X_j^{i^k}) = 1$. Hence, corresponding to every bit vector R which belongs to the $(1+\epsilon)$-approximate set of the Pareto front there is a bit vector in the population or the archive which by a single bit flip can produce R. Thus, from individuals like I_{coll}, we can produce another desired individual (which belongs to the $(1+\epsilon)$-approximate set) by a single bit flip. With a probability of $\frac{1}{2}$ (Handler function), two individuals in the population are chosen at random. Therefore, the probability of the individual like I_{coll} being chosen into the population is $\frac{1}{2(P_{knap}+1)}$ ($P_{knap} + 1$ is the bound on the population size, proven in lemma 5). The probability of I_{coll} being chosen for mutation is thus $\frac{1}{4(P_{knap}+1)}$. Flipping a desired 1-bit or 0-bit in any of the m blocks of I_{coll} will produce a desired individual. Thus, the probability that I_{coll} will produce another desired individual is $\frac{m}{4n(P_{knap}+1)}$. The expected number of waiting steps for a successful mutation is thus, at most $\frac{4n(P_{knap}+1)}{m}$. If R' is the new individual produced, since, no items have equal $\frac{P_i}{W_i}$, there cannot be any individual in the population or the archive which weakly dominates R'. Hence, R' will always be accepted. As every individual in the $(1+\epsilon)$-approximate set can be produced from some individual in the population or the archive by a single bit mutation, the total time taken to produce the entire $(1+\epsilon)$-approximate set is upper bounded by $\frac{4n^{2m+1}(P_{knap}+1)}{m^{2m+1}}$ (total size of the $(1+\epsilon)$-approximate set is bounded by $O((\frac{n}{m})^{2m})$ by lemma 5).

If we consider a phase of $\frac{8(P_{knap}+1)n^{2m+1}}{m^{2m+1}}$ steps, then by Chernoff's bounds, the probability of there being less than $\frac{n^{2m}}{m^{2m}}$ successes is bounded by $e^{-\Omega(n)}$. Altogether both the phases take a total of $O(\frac{P_{knap}n^{2m+1}}{m^{2m+1}})$ for finding the entire $(1+\epsilon)$- approximate set.

For the bound on the expected time we have not assumed anything about the initial population. Thus, we notice that the above bound on the probability holds for the next $\frac{P_{knap}n^{2m+1}}{m^{2m+1}}$ steps. Since the lower bound on the probability that the algorithm will find the entire $(1+\epsilon)$-approximate set is more than $\frac{1}{2}$ (in fact exponentially close to 1) the expected number of runs is bounded by 2.

Combining the results of both the phases yields the bound in the theorem. It is worth noting that for cases when the value of the number of partitions into which the set of items have to be partitioned is a *constant* the expected running time is a polynomial in n.   □

Comments: An important point to note in the analysis of the algorithm is that for the knapsack problem the selection of the individual in the Handler function is done at random. However, this occurs with a probability of 0.5. It is likely that the specialized Fit function is able to find individual which is a subset of the approximate set in time which is faster than the random selection. Note that $1^i 0^{n-i}$ is a subset of the approximate solution which can be efficiently found by the algorithm as in the case of LOTZ. Thus, though we find a worst case upper bound by considering a total random selection, the actual running time bound may be much better with the Fit selection scheme. Moreover, it is also worth noting that the algorithm is adapted to get a $(1 + \epsilon)$ as well as an optimal set for with $\epsilon = 0$.

8   Conclusions

In this paper, an archive based multiobjective evolutionary optimizer (REMO) is presented and a rigorous runtime complexity analysis is carried out of the algorithm on simple discrete boolean functions (the LOTZ function [11] and quadratic function [25]) and a NP-Complete problem (0 - 1 knapsack problem). The key feature of REMO is its special restricted population for mating and a separate archive. Such algorithms have been widely used in solving real world applications. The idea is to restrict the mating pool to a constant c. The value of 2 for c is sufficient for most simple functions. In case of certain functions a single individual population with a similar selection scheme as REMO may suffice. However, two bit vectors may be required for functions where the Pareto front can be reached via two paths as is the case of the quadratic function. However, for more complicated functions like the knapsack it is better to use a more random selection mechanism since the specialized selection may lead to the algorithm getting trapped in a local optima.

Acknowledgements

The authors would like to thank Ingo Wegener for valuable discussions during the course of this work. The authors thank Oliver Giel for his invaluable help in proving some lemmas in the knapsack problem. The authors would also like to thank Lothar Schmitt, Kenneth Jong, Alden Wright and anonymous reviewers for suggestions and corrections.

The part of the work was done while Nilanjan Banerjee was at University of Dortmund during summer 2003; his visit was financed by Professor Ingo Wegener's chair and the German Academic Exchange Service (DAAD). Rajeev Kumar acknowledges support from the Ministry of Human Resource Development, Government of India, project during the period of this work.

References

1. Garey, M.R., Johnson, D.S.: Computers and Interactability: A Guide to the Theory of NP-Completeness. San Francisco, LA: Freeman (1979)
2. Hochbaum, D.: Approximation Algorithms for NP-Hard Problems. Boston, MA: PWS (1997)
3. Zitzler, E., Thiele, L.: Multiobjective evolutionary algorithms: a comparative case study and the strength pareto approach. IEEE Trans. Evolutionary Computation **3** (1999) 257–271
4. Knowles, J.D., Corne, D.W.: A Comparison of Encodings and Algorithms for Multiobjective Minimum Spanning Tree Problems. In: Proc. Congress on Evolutionary Computation (CEC-01). Volume 1. (2001) 544–551

5. Kumar, R., Singh, P.K., Chakrabarti, P.P.: Improved quality of solutions for multiobjective spanning tree problem using evolutionary algorithm. In: Proc. Int. Conf. High Performance Computing (HiPC-04), LNCS 3296. (2004) 494–503

6. Kumar, R., Rockett, P.I.: Multiobjective genetic algorithm partitioning for hierarchical learning of high-dimensional pattern spaces: A learning-follows-decomposition strategy. IEEE Trans. Neural Networks **9** (1998) 822–830

7. Kumar, R.: Codebook design for vector quantization using multiobjective genetic algorithms. In: Proc. PPSN/SAB Workshop on Multiobjective Problem Solving from Nature. (2000)

8. Kumar, R., Parida, P.P., Gupta, M.: Topological design of communication networks using multiobjective genetic optimization. In: Proc. Congress Evolutionary Computation (CEC-2002). (2002) 425–430

9. Kumar, R., Banerjee, N.: Multicriteria network design using evolutionary algorithm. In: Proc. Genetic and Evolutionary Computations Conference (GECCO-03), LNCS 2023. (2003) 2179–2190

10. Laumanns, M., Thiele, L., Zitzler, E., Welzl, E., Deb, K.: Running time analysis of multiobjective evolutionary algorithms on a discrete optimization problem. In: Proc. Parallel Problem Solving from Nature (PPSN VII), LNCS 2439. (2002) 44–53

11. Laumanns, M., Thiele, L., Zitzler, E.: Running time analysis of evolutionary algorithms on pseudo-boolean functions. IEEE Trans. Evolutionary Computation **8** (2004) 170–182

12. Laumanns, M., Thiele, L., Zitzler, E.: Running time analysis of evolutionary algorithms on a simplified multiobjective knapsack problem. Natural Computing **3** (2004) 37–51

13. Droste, S., Jansen, T., Wegener, I.: On the Optimization of Unimodal Functions with the (1+1) Evolutionary Algorithm. In: Proc. Parallel Problem Solving from Nature (PPSN-V), LNCS 1498. (1998) 13–22

14. Jagersküpper, J.: Analysis of simple evolutionary algorithm for minimization in euclidean spaces. In: Proc. 30th Int. Colloquium Automata, Languages and Programming (ICALP 2003), LNCS 2719. (2003) 1068–1079

15. Droste, S., Jansen, T., Wegener, I.: On the Analysis of the (1+1) Evolutionary Algorithm. Theoretical Computer Science **276** (2002) 51–81

16. Garnier, J., Kallel, L., Schoenauer, M.: Rigorous hitting times for binary mutations. Evolutionary Computation **7** (1999) 167–203

17. Rudolph, G.: How mutation and selection solve long path problems in polynomial expected time. Evolutionary Computation **4** (1996) 207–211

18. Wegener, I., Witt, C.: On the analysis of a simple evolutionary algorithm on quadratic pseudo-boolean functions. J. Discrete Algorithms (2002)

19. Jansen, T., Wegener, I.: The analysis of evolutionary algorithms: a proof that crossover really can help. Algorithmica **34** (2002) 47–66

20. Coello, C.A.C., Veldhuizen, D.A.V., Lamont, G.B.: Evolutionary Algorithms for Solving Multiojective Problems. Boston, MA: Kluwer (2002)

21. Deb, K.: Multiobjective Optimization Using Evolutionary Algorithms. Chichester, UK: Wiley (2001)

22. Rudolph, G.: Convergence Properties of Evolutionary Algorithms. Hamburg, Germany: Verlag Dr. Kovač (1997)

23. Rudolph, G.: Evolutionary search for minimal elements in partially ordered finite sets. In: Proc. Annual Conference on Evolutionary Programming. (1998) 345–353

24. Rudolph, G., Agapie, A.: Convergence properties of some multiobjective evolutionary algorithms. In: Proc. Congress on Evolutionary Computation. (2000) 1010–1016

25. Giel, O.: Runtime analysis for a simple multiobjective evolutionary algorithm. Tech-Report, Dept. Computer Science, Univ. Dortmund, Germany (2003)

26. Thierens, D.: Convergence time analysis for the multi-objective counting ones problem. In: Proc. Conf. Evolutionary Multiobjective Optimization (EMO-03), LNCS 2632. (2003) 355–364

27. Asho, I.: Interactive Knapsacks: Theory and Applications. Ph.D. Thesis, Tech Report No.: A-2002-13, Department of Computer and Information Sciences, University of Tampere (2002)

28. Frieze, A., Clarke, M.: Approximation algorithms for m-dimensional 0-1 knapsack problem: Worst case and probabilistic analysis. European J. Operations Research **15** (1984) 100–109

29. Erlebach, T., Kellerer, H., Pferschy, U.: Approximating Multiobjective Knapsack Problems. Management Science **48** (2002) 1603–1612

30. Ibarra, O.H., Kim, C.E.: Fast approximation algorithms for the knapsack and sum of subset problem. J. ACM **22** (1984) 463–468

31. Beyer, H.G., Schwefel, H.P., Wegener, I.: How to Analyse Evolutionary Algorithms? Theoretical Computer Science **287** (2002) 101–130

32. Scharnow, J., Tinnefeld, K., Wegener, I.: Fitness landscapes based on sorting and shortest path problems. In: Proc. Parallel Problem Solving From Nature (PPSN VII), LNCS 2439. (2002) 54–63

33. Droste, S., Jansen, T., Tinnefeld, K., Wegener, I.: A new framework for the valuation of algorithms for black-box optimization. In: Proc. Foundations of Genetic Algorithms Workshop (FOGA VII). (2002) 197–214

34. Deb, K., Others: A Fast Non-Dominated Sorting Genetic Algorithm for Multiobjective Optimization: NSGA-II. In: Proc. Parallel Problem Solving from Nature (PPSN-VI), LNCS. (2000) 849–858

35. Knowles, J.D., Corne, D.W.: Approximating the Non-Dominated Front Using the Pareto Achieved Evolution Strategy. Evolutionary Computation **8** (2000) 149–172

36. Kumar, R., Rockett, P.I.: Improved Sampling of the Pareto-front in Multiobjective Genetic Optimization by Steady-State Evolution: A Pareto Converging Genetic Algorithm. Evolutionary Computation **10** (2002) 283–314

37. Zitzler, E., Laumanns, M., , Thiele, L.: SPEA2: Improving the Strength Pareto Evolutionary Algorithm. In: Proc. Evolutionary Methods for Design, Optimization and Control with Applications to Industrial Problems (EUROGEN). (2001)

38. Zitzler, E., Thiele, L., Laumanns, M., Fonseca, C.M., da Fonseca, V.G.: Performance assessment of multiobjective optimizers: An analysis and review. IEEE Trans. Evolutionary Computation **7** (2003) 117–132

Tournament Selection, Iterated Coupon-Collection Problem, and Backward-Chaining Evolutionary Algorithms

Riccardo Poli

Department of Computer Science, University of Essex, UK
rpoli@essex.ac.uk

Abstract. Tournament selection performs tournaments by first sampling individuals uniformly at random from the population and then selecting the best of the sample for some genetic operation. This sampling process needs to be repeated many times when creating a new generation. However, even upon iteration, it may happen not to sample some of the individuals in the population. These individuals can therefore play no role in future generations. Under conditions of low selection pressure, the fraction of individuals not involved in any way in the selection process may be substantial. In this paper we investigate how we can model this process and we explore the possibility, methods and consequences of not generating and evaluating those individuals with the aim of increasing the efficiency of evolutionary algorithms based on tournament selection. In some conditions, considerable savings in terms of fitness evaluations are easily achievable, without altering in any way the expected behaviour of such algorithms.

1 Introduction

Tournament selection is one of the most popular forms of selection in evolutionary algorithms (EAs). In its simplest form, a group of n individuals is chosen randomly uniformly from the current population, and the one with the best fitness is selected (e.g., see (Bäck et al., 2000)). The parameter n is called the *tournament size* and can be used to vary the selection pressure exerted by this method (the higher n the higher the pressure to select above average quality individuals).

Different selection methods, including tournament selection, have been analysed mathematically in depth in (Blickle and Thiele, 1995, 1997; Motoki, 2002). The main emphasis of previous research has been the evaluation of the changes produced by selection on the fitness distribution of the population. The proportion of individuals of a population that is not selected during the selection phase is one of the quantities that have been used to characterise the behaviour of selection algorithms. This quantity is called the *loss of (fitness) diversity*. Under the implicit assumption that the population is wholly diverse (each individual has

A.H. Wright et al. (Eds.): FOGA 2005, LNCS 3469, pp. 132–155, 2005.

a unique rank), the loss of diversity p_d for tournament selection was estimated in (Blickle and Thiele, 1995, 1997) as

$$p_d = n^{-\frac{1}{n-1}} - n^{-\frac{n}{n-1}},$$

and later calculated exactly in (Motoki, 2002) as

$$p_d = \frac{1}{M} \sum_{k=1}^{M} \left(1 - \frac{k^n - (k-1)^n}{M^n}\right)^M,$$

where M is the population size.

If one assumed that selection only is used or that we use selection to form a mating pool[1], the creation of a new generation would require exactly M selection steps. These are exactly the conditions assumed in the work mentioned above. However, in this paper we do not make this assumption. Instead, we consider the case where each genetic operator directly invokes the selection procedure to provide a sufficient number of parents for its application (e.g., twice in case of crossover). So, there are situations where more than M selection steps are required to form a new generation. More precisely, in a generational selecto-recombinative algorithm, where crossover is performed with probability p_c and reproduction is performed with probability $1 - p_c$, the number of selection steps required to form a new generation is a stochastic variable with mean

$$M(1 - p_c) + \rho M p_c = M[1 + (\rho - 1)p_c],$$

where $\rho = 1$ for a crossover operator which returns two offspring after each application, and $\rho = 2$ if only one offspring is returned[2]. The two-offspring version of crossover is more efficient in terms of tournaments required, and also, since $\rho = 1$, the number of selection steps required to form a new generation is not stochastic and is simply M. For brevity in the following we will use the definition $\alpha = [1 + (\rho - 1)p_c]$.

So, because in each tournament we need n individuals, tournament selection requires drawing $n\alpha M$ individuals uniformly at random (with resampling) from the current population. An interesting side effect of this process is that not all individuals in a particular generation are necessarily sampled within the $n\alpha M$ draws, and this is particularly true for small values of the tournament size n.

---

[1] The mating pool is an intermediate population which gets created by using selection only and from which other operations, such as reproduction and crossing over, draw individuals uniformly at random

[2] In the following we will ignore the (potential) stochasticity of the number of selection steps required to create a new generation. This is justifiable for various reasons: a) it simplifies the analysis (but without significant loss in terms of accuracy of the results obtained, as empirically verified), b) when $\rho = 1$ (two-offspring crossover or mutation only algorithm) there is no stochasticity (and so the analysis is exact), c) even with $\rho = 2$ (one-offspring crossover) it is possible to slightly modify the evolutionary algorithm in such a way that there is no stochasticity

For example, let us imagine to run an evolutionary algorithm starting from a random population containing 4 individuals, which we will denote as 1, 2, 3 and 4. Let us assume that we are creating the next generation using tournament selection with tournament size 2 and mutation only. Then, the creation of the first individual will require randomly picking two individuals from the current population (say individuals 1 and 4) and selecting the best for mutation. We repeat the processes to create the second, third and fourth new individuals. It is not unconceivable that in so doing maybe individual 3 was never involved in any tournament.

It is absolutely crucial, at this stage, to stress the difference between *not sampling* and *not selecting* an individual in a particular generation. The latter refers to an individual which was involved in one or more tournaments, but did not win any, and this is exactly what previous work on tournament selection has concentrated on. The former, instead, refers to an individual which did not participate in any tournament at all, simply because it was not sampled during the creation of the required αM tournament sets. It is individuals such as this that are the focus of this paper. Therefore, the results in this paper are orthogonal to those appeared in the previous work mentioned above and are not limited by uniqueness assumptions.

Continuing with our argument, in general, how many individuals should we expect not to take part in any of αM tournaments? As will be shown in the next section, an answer comes straight from the literature on the coupon collector problem. However, before we explain the connection in more detail, we may want to reflect briefly on why this effect is important.

In general those individuals that do not get sampled by the selection process have no influence whatsoever on future generations. However, these individuals use up resources, e.g., memory, but also, and more importantly, CPU time for their creation and evaluation. For instance, individual 3 in the previous example was randomly generated and had its fitness evaluated in preparation for selection, but neither its fitness nor its genetic make up could have any influence on future generations. So, one might ask, why did we generate such an individual in the first place? And what about generations following the first two? It is entirely possible that an individual in generation two got created and evaluated, but was then neglected by tournament selection, so it had no effect whatsoever on generations 3, 4, etc. Did we really need to generate and evaluate such an individual? If not, what about the parents of such an individual: did we need them? What sort of saving could we obtain by not creating unnecessary individuals in a run?

In this paper we intend to analyse the relationship between tournament selection and the coupon collector's problem, and attempt to answer all of the questions above and more. In particular, we want to rethink the way evolutionary algorithms are run for the purpose of making best use of the available resources *without altering in any way the expected behaviour of such algorithms.* As will become clear in the next sections, in some conditions, saving of 20% fitness evaluations or, in fact, even more are easily achievable.

2   Coupon Collection and Tournament Selection

In the *coupon collector problem*, every time a collector buys a certain product, a coupon is given to him. The coupon is equally likely to be any one of N types. In order to win a prize, the collector must have at least one coupon of each type. The question is: how many products will the collector have to buy on average before he can expect to have a full set of coupons? The answer (Feller, 1971) is obtained by considering that the probability of obtaining a first coupon in one trial is 1 (so the expected waiting time is just 1 trial), the probability of obtaining a second coupon (distinct from the first one) is $\frac{N-1}{N}$ (so the expected waiting time is $\frac{N}{N-1}$), the probability of obtaining a third coupon (distinct from the first two) is $\frac{N-2}{N}$ (so the expected waiting time is $\frac{N}{N-2}$), and so on. So, the expected number of trials to obtain a full set of coupons is

$$E_N = 1 + \frac{N}{N-1} + \frac{N}{N-2} + \cdots N = N \times \left(\frac{1}{N} + \frac{1}{N-1} + \cdots 1 \right) = N \log N + O(N).$$

It is well known that the $N \log N$ limit is sharp. For example, if X is a random variable representing the number of coupons collected, for any constant c

$$\lim_{N \to \infty} \Pr\{X > N \log N + cN\} = 1 - e^{-e^{-c}}.$$

So, for $c \approx 3$ this probability is less than 5%.

How is the process of tournament selection related to the coupon collection problem? We can imagine that the M individuals in the current population are $N = M$ distinct coupons and that tournament selection will draw (with replacement) $n\alpha M$ times from this pool of coupons. Because of the sharpness of the coupon-collector limit mentioned above, if $M \log M + cM < n\alpha M$, i.e., if $n\alpha > \log M + c$ for some suitable positive constant c, then we should expect tournament selection to sample all individuals in the population most of the time. However, for sufficiently small tournament sizes or for sufficiently large populations the probability that there will be individuals not sampled by selection becomes significant.

So, how many different coupons (individuals) should we expect to have sampled at the end of the $n\alpha M$ trials? In the coupon collection problem, the expected number of trials necessary to obtain a set of x distinct coupons is

$$E_x = 1 + \frac{N}{N-1} + \frac{N}{N-2} + \cdots \frac{N}{N-x+1} = N \log \frac{N}{N-x} + O(N).$$

By setting $E_x = n\alpha M$, $N = M$ and ignoring terms of order $O(N)$, from this we obtain an estimate for the number of distinct individuals sampled by selection

$$x \approx M \left(1 - e^{-n\alpha} \right). \tag{1}$$

This indicates that the expected proportion of individuals *not* sampled in the current population varies approximately like a negative exponential of the tournament size.

This approximation is quite accurate. However, we can calculate the expected number of individuals neglected after performing $n\alpha M$ trials directly. We first calculate the probability that one individual is not involved in one trial as $1 - 1/M$. Then the expected number of individuals not involved in any tournaments is simply

$$M(1 - 1/M)^{n\alpha M} = M\left(\frac{M}{M-1}\right)^{-n\alpha M},$$

which also varies like a negative exponential of the tournament size.

As shown in Figure 1 for $\alpha = 1$ (two-offspring crossover or no crossover), typically for $n = 2$ over 13% of the population is neglected, for $n = 3$ this drops to 5%, for $n = 4$ this is 2%, and becomes negligible for bigger values of n.

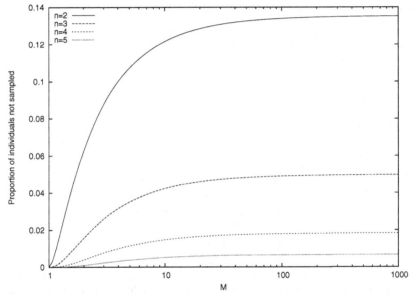

Fig. 1. Proportion of individuals not sampled in one generation by tournament selection for different tournament and population sizes assuming two-offspring crossover or mutation only are used

This simple analysis suggests that saving computational resources by avoiding the creation and evaluation of individuals which will not be sampled by the tournament selection process appears to be possible only for relatively low selection pressures. However, tournament sizes in the range 2–5 are quite common in practice, particularly when attacking hard, multi-modal problems which require extensive exploration of the search space before zooming the search onto any particular region. By rethinking how we perform selection, in this paper we will show how we can achieve substantial computational savings in *any evolutionary algorithm* based on tournament selection on *any problem* where low selection pressure is appropriate *without changing* in any way the course of a run! We will start exploring how we can achieve all this in the next section.

3   Iterated Coupon Collector Problem

Now, let us consider a new game, that we will call the *iterated coupon collection problem*, where the coupon set changes at regular intervals, but the number of coupons available, N, remains constant. Initially the collector is given a (possibly incomplete) set of $m(0)$ old coupons. Each old coupon allows the collector to draw n new coupons. So, he can perform a total of $nm(0)$ trials, which will produce a set of $m(1)$ distinct coupons from the new set. The coupon set now changes, and the player performs $nm(1)$ trials to gather as many as possible new distinct coupons. And so on. Interesting questions here are: what happens to $m(t)$ as t grows? Will it reach a limit? Will it oscillate? In which way will the values of n, $m(0)$ and N influence its behaviour?

Before we answer these questions let us motivate our analysis a bit. How is this new problem related to evolutionary algorithms and tournament selection? The connection is simple (we will assume $\alpha = 1$ for the sake of clarity). Suppose an oracle told us which individuals in a particular generation, G, are not involved in any way in future generations, because selection will not sample them. Then we could concentrate on the other individuals in the population, creating and evaluating only them. Let $m(0)$ be the number of such individuals (these are like the initial set of old coupons given to the player). Clearly, in order to create such individuals, we will need to know who their parent(s) were. This will require running $m(0)$ tournaments to select such parents. In each tournament we randomly pick n individuals from generation $G - 1$ (each distinct individual in that generation is equivalent to a coupon in the new coupon set). After, $nm(0)$ such trials we will be in a position to determine (without the need for an oracle) which individuals in generation $G - 1$ will contribute to future generations, we can count them and denote this number with $m(1)$ [3]. So, again, we can concentrate on these individuals only. They are the equivalent of the new set of coupons the collector has gathered. We can now perform $nm(1)$ trials to determine (again without the need for an oracle) which individuals in generation $G - 2$ (the new coupon set) will contribute to future generations, we can count them and denote this number with $m(2)$ and so on until we reach the initial random generation. There the game stops. So, effectively, *the iterated coupon collector problem is a model for the sampling behaviour of tournament selection over multiple generations in a generational evolutionary algorithm.*

Knowing the sequence $m(t)$ for a particular evolutionary algorithm would tell us how much we could save by not creating and evaluating individuals which will not be sampled by selection. Naturally, we will not have an oracle to help us choose G and to give us $m(0)$. For now, while we concentrate on understanding more about the iterated coupon collector problem, we could think of G as the

---

[3] Because at this stage we are only interested in knowing the number of individuals playing an active role in generation $G - 1$, there is no need to determine the winners of the tournaments: we just need to know who was involved in which tournament. So, we do not even need to evaluate fitness, and, therefore, we do not need to know the genetic makeup of any individual

number of generations we are prepared to run our evolutionary algorithm for, and we might imagine that $m(0) = M$ (the whole population).

In the classical coupon collection problem, the shopper will typically perform as many trials as necessary to gather a full collection of coupons. As we have seen before, however, it is quite easy to estimate how many distinct coupons one should expect at the end of any given *fixed* number of trials. Because the iterated coupon collection game starts with a known number of trials, we can calculate the *expected value* of $m(1)$. However, we cannot directly apply the theory in the previous section to gather information about $m(2)$. This is because $m(1)$ is a stochastic variable, so in order to estimate $m(2)$ we would need to know the probability distribution of $m(1)$ not just its expected value.

Exact probabilistic modelling can be obtained by considering the coupon collection game as a Markov chain (Feller, 1971), where the state of the chain is the number of distinct coupons collected. The transition matrix for the chain can easily be constructed by noticing that the chain can be in state k (i.e., the collector has k distinct coupons) at the next time step only if either it was already in state k and a new coupon has been acquired that is a duplicate (which happens with probability $\frac{k-1}{N}$) or it was in state $k-1$ and the coupon just acquired is a new one (which, of course, happens with probability $\frac{N-k+1}{N}$). So, the number of distinct individuals in the previous generation sampled when randomly picking individuals for tournament selection can be described by the following Markov transition matrix:

$$A = \frac{1}{M} \begin{pmatrix} 0 & 0 & 0 & 0 \cdots 0 \\ M & 1 & 0 & 0 \cdots 0 \\ 0 & M-1 & 2 & 0 \cdots 0 \\ 0 & 0 & M-2 & 3 \cdots 0 \\ \vdots & \vdots & \vdots & \vdots \ddots \vdots \\ 0 & 0 & 0 & 0 \cdots M \end{pmatrix}.$$

The process always starts from state 0, which can be represented using the state probability vector $e_0 = (1\,0\,0\,0\cdots 0)^T$. So, the probability distribution over the states after x time steps (i.e., coupon draws or random samples from the population) is given by $A^x e_0$, which is simply the first column of the matrix A^x.

So, if we are interested only in $m(0)$ individuals in generation G, the probability distribution of $m(1)$ (the number of distinct individuals we need to know from generation $G-1$) is given by $A^{n\alpha m(0)} e_0$ [4]. For example, if the population size is $M = 3$, the tournament size is $n = 2$, we use a two-offspring version of crossover ($\alpha = 1$) and we are interested in $m(0) = 1$ individuals, then the probability distribution of $m(1)$ is represented by the following probability vector

---

[4] This, of course, gives us also the probability distribution over the number of draws, $n\alpha m(1)$, we will need to make from generation $G-2$ in order to fully determine the $m(1)$ individuals we want to know at generation $G-1$

$$\left(\frac{1}{3}\right)^2 \begin{pmatrix} 0\,0\,0\,0 \\ 3\,1\,0\,0 \\ 0\,2\,2\,0 \\ 0\,0\,1\,3 \end{pmatrix}^2 \begin{pmatrix} 1 \\ 0 \\ 0 \\ 0 \end{pmatrix} = \frac{1}{3} \begin{pmatrix} 0 \\ 1 \\ 2 \\ 0 \end{pmatrix}.$$

If we were interested in $m(0) = 2$ individuals at generation G, the probability distribution over the number $m(1)$ of unique individuals sampled would be $A^4 e_0 = (0.0\ 0.0370\ 0.5185\ 0.4444)^T$. Finally, if we were interested in the whole population ($m(1) = M = 3$), the distribution would be $A^6 e_0 = (0\ 0.0041\ 0.2551\ 0.7407)^T$, which reveals that, in these conditions, even when building a whole generation there are still more than 1 in 4 chances of not sampling the whole population at the previous generation. Of course, if $m(0) = 0$, the probability vector for $m(1)$ is e_0, i.e., $m(1) = 0$.

Although this example is trivial, it reveals that for any given $m(0)$ we can compute a distribution over $m(1)$. That is, we can define a new *Markov chain to model the iterated coupon collector problem*. In this chain a state is exactly the same as in the coupon-collector chain (i.e., the number of distinct coupons sampled), *except that now a time step corresponds to a complete set of draws from the new coupon set rather than just the draw of one coupon*. The transition matrix B for this new chain can be obtained very simply: column i of B is $A^{\alpha i} e_0$. That is

$$B = \left(e_0 | A^\alpha e_0 | A^{2\alpha} e_0 | \cdots | A^{M\alpha} e_0 \right).$$

For instance, for the case $M = 3$, $n = 2$ and $\alpha = 1$ considered above

$$B = \begin{pmatrix} 1 & 0 & 0 & 0 \\ 0 & 0.3333 & 0.0370 & 0.0041 \\ 0 & 0.6667 & 0.5185 & 0.2551 \\ 0 & 0 & 0.4444 & 0.7407 \end{pmatrix}.$$

Of course, once the transition matrix is defined, the chain can be iterated to compute the probability distributions of $m(2)$, $m(3)$ and so on, as back as necessary to reach generation 0.

In general B is block diagonal of the form

$$B = \left(\begin{array}{c|c} 1 & \mathbf{0}^T \\ \hline \mathbf{0} & C \end{array} \right),$$

where $\mathbf{0}$ is a column vector containing M zeros and C is a $M \times M$ stochastic matrix. Clearly B is not ergodic (from state 0 we cannot reach any state other than 0 itself), so we cannot expect a unique limit distribution for $m(t)$. However, because B is block diagonal, we have

$$B^x = \left(\begin{array}{c|c} 1 & \mathbf{0}^T \\ \hline \mathbf{0} & C^x \end{array} \right).$$

So, if we ensured that the probability of the chain initially being in state 0 is 0 (that is $\Pr\{m(0) = 0\} = 0$), the chain could never visit such a state at any future time. Because of this property, and because, objectively, state 0 is totally uninteresting (of course we already know that if we are interested in no individual at generation G, we do not need to know any individual at previous generations!) we can declare such a state of the iterated coupon-collection chain as invalid, and reduce the state set to $\{1, 2, \ldots, M\}$. In this situation C is the state transition matrix for the chain, and to model the sampling behaviour of tournament selection over multiple generations we just need to concentrate on the properties of C.

The transition matrix C is ergodic if $n\alpha > 1$ as can be easily seen by the following argument. If $n\alpha > 1$ then each old coupon gives us the right to draw more than one new coupon in the iterated coupon-collection problem. So, if the state of the chain is k $(k < M)$, it is always possible to reach state $k + 1$ in one stage of the game with non-zero probability. From there it is then, of course, possible to reach state $k + 2$ and so on up to M. So, from any lower state it is always possible to reach any higher state in repeated iterations of the game. But, of course, the converse is always true: irrespective of the value of $n\alpha$ there is always a chance of getting fewer coupons than we had before in an iteration of the game, due to resampling. So, from any higher state we can also reach any lower state (in fact, unlike the reverse, we can achieve this in just one iteration of the game).

Since, $\alpha \geq 1$ and $n \geq 2$ for any practical applications, the condition $n\alpha > 1$ is virtually always satisfied and C is ergodic. So, the Perron-Frobenius theorem guarantees that the probability over the states of the chain converges towards a limit distribution which is independent from the initial conditions (see (Nix and Vose, 1992; Davis and Principe, 1993; De Jong et al., 1995; Rudolph, 1994; Poli et al., 2001) for other applications of this result to genetic algorithms and genetic programming). This distribution is given by the (normalised) eigenvector corresponding to the largest eigenvalue of C $(\lambda_1 = 1)$, while the speed at which the chain converges towards such a distribution is determined by the magnitude of the second largest eigenvalue λ_2 (the relaxation time of an ergodic Markov chain is $1/(1 - |\lambda_2|)$). Naturally, this infinite-time limit behaviour of the chain is particularly important if G is sufficiently big that $m(t)$ settles into the limit distribution well before we iterate back to generation 0. Otherwise the transient behaviour is what one needs to focus on. Both are provided by the theory.

Because the transition matrices we are talking about are relatively small $(M \times M)$, they are amenable to numerical manipulation. We can, for example, find the eigenvalues and eigenvectors of C for quite respectable population sizes, certainly well in the range of those used in most applications of evolutionary algorithms, thereby determining the limit distribution and the speed at which this is approached.

If $x(t)$ is a probability vector representing the probability distribution over $m(t)$, then the expected value of $m(t)$ is

$$E[m(t)] = \left(1\ 2 \cdots M \right) \cdot x(t) = \left(1\ 2 \cdots M \right) \cdot C^t x(0). \tag{2}$$

Typically $m(0)$ will be deterministic and so $x(0) = e_{m(0)}$ (where e_l is a base vector containing all zeros except for element l which is 1).

If x^* denotes the limit distribution for $x(t)$, then for large enough G, the average number γ of individuals (in generations 0 through to $G-1$) that have no effect whatsoever on a designated set of $m(0)$ individuals of interest at generation G is approximately $\gamma = M - (1 \ 2 \ \cdots \ M) \cdot x^*$. So, a question that naturally springs to mind is whether we could run an evolutionary algorithm without creating and evaluating these unnecessary individuals.

If we could do this, because of the ergodicity of the selection process, for large enough G, it is almost irrelevant whether $m(0)$ is as small as 1 or as large as M, and so we might want to know the entire makeup of generation G and still have a saving of approximately $G \times \gamma$ individual creations and evaluations.

In the next two sections we will consider two different ways in which we could modify our evolutionary algorithm to achieve this kind of saving.

4   Running Evolutionary Algorithms Efficiently

Normally, in each generation of an evolutionary algorithm we iterate the following phases:

a) the choice of genetic operator to use to create a new individual,
b) the creation of a random pool of individuals for the application of tournament selection,
c) the identification of the winner of the tournament (parent) based on fitness,
d) the execution of the chosen genetic operator,
e) the evaluation of the fitness of the resulting offspring.

Naturally, phases (b) and (c) are iterated as many times as the arity of the genetic operator chosen in phase (a), and the whole process needs to be repeated as many times as there are individuals in the new population.

Interestingly, the genetic makeup of the individuals involved in these operations is of interest only in phase (d) (we need to know the parents in order to produce offspring) and phases (c) and (e) (we must know the genetic makeup of individuals in order to evaluate their fitness). However, phases (a) and (b) do not require any knowledge about the actual individuals involved in the creation of a new individual. In most implementations these phases are just performed by properly manipulating numbers drawn from a pseudo-random number generator.

So, there is really no reason why we could not first iterate phases (a) and (b) as many times as needed to create a full new generation (of course, memorising all the decisions taken), and then iterate phases (c)–(e). This idea was first used in (Teller and Andre, 1997) for the purposed on speeding up genetic programming fitness evaluation[5].

---

[5] The main idea in (Teller and Andre, 1997) was to estimate the fitness of the individuals involved in the tournaments by evaluating them on a subset of the fitness cases available. On the basis of this estimate, for most tournaments it was often possible to determine with a small error probability which individual would win. These tournaments could therefore be decided quickly, while only in a subset of tournaments individuals ended up being evaluated using all fitness cases. This is what produced the speed up

In fact, we could go even further. In many practical applications of evolutionary algorithms, people fix a maximum number of generations they are prepared to run their algorithm for[6]. Let this number be G. So, at the cost of some memory space, we could iterate phases (a) and (b) not just for one generation but for a whole run from the first generation to generation G and then iterate phases (c)–(e) as required (that is, either until generation G or until any other stopping criterion is satisfied).

Because the decisions as to which operator to adopt to create a new individual and which elements of the population to use for a tournament are random, statistically speaking this version of the algorithm is exactly the same as the original. In fact, if the same seed is used for the random number generator in both algorithms, *they are indistinguishable*! However, the version of the algorithm we propose (let us call it an *EA with macro-selection*) makes it possible to avoid wasting the computation involved in generating and evaluating the individuals "neglected" by tournament selection: after macro-selection (the iteration of phases (a) and (b) up until generation G) is completed we analyse the information stored during that phase and identify which population members were not involved in any tournament in each generation, we mark them, and we avoid calculating and evaluating them when iterating phases (c)–(e). We will call this algorithm the *EA with efficient macro-selection (EA-EMS)*.

Irrespective of the problem being solved and the parameter settings used, the behaviours of the standard algorithm and the efficient version proposed above will have to be on average identical. So, what are the differences between the two evolutionary algorithms?

Obviously, the standard algorithm requires more fitness evaluations and creations of individuals while the one proposed above requires more bookkeeping and use of memory. Also, clearly, in any particular run, the plots of average fitness and max fitness in each generation may differ (since in EA-EMS not all individuals are considered in calculating these statistics). However, when averaged over multiple runs the average fitness plots would have to coincide.

A more important difference derives from the fact that most practitioners keep track of the best individual seen so far in a run of an EA and designate that as the result of the run. In EA-EMS we can either return the best individual in generation G or the best individual seen in a run *out of those that have been sampled by tournament selection*. Because the fast algorithm does not create and evaluate individuals that did not get sampled, the end-of-run results may differ in the two algorithms. Of course, quite often the best individual seen in a run is actually a member of the population at the last generation. So, if one creates and evaluates all individuals in generation G (which leads to only a minor inefficiency in the EA with efficient macro-selection), most of the time the two algorithms will behave identically from this point of view too.

One remaining inefficiency in EA-EMS derives from the fact that there may be individuals which were sampled by selection when creating a particular indi-

---

[6] This is a limit that is virtually always present, even if another stopping criterion, e.g., based on fitness, is present

vidual in a new generation which in turn, however, was not sampled by selection. If the first individual was involved only in the creation of the second, further computational savings could be achieved by not creating and evaluating the first individual at all. This is just an example of a more general issue: it is possible that some of the "ancestors" of the individuals neglected by selection were unnecessarily created and evaluated. How could we improve our algorithm to get rid of these individuals too?

Clearly the iteration of phases (a) and (b) over multiple generations induces a graph structure containing $(G + 1)M$ nodes representing all the individuals evolved during a run and where edges connect each individual to the individuals which were involved in the tournaments necessary to select the parents of such an individual. If we are interested in calculating and evaluating all the individuals in the population at generation G, maximum efficiency would be achieved by considering (marking for evaluation) only the individuals which are directly or indirectly connected with the M individuals in generation G. So, the problem can be solved with a trivial connected-component algorithm. Let us call the resulting algorithm an *EA with efficient macro-selection and connected-component detection (EA-EMS-CCD)*.

This would appear to be the best we can get from our EA: it saves on fitness evaluations and it is really as close as we can get to the original in terms of behaviour. However, the recursive nature of connected-component detection and the similarity between the mechanics of EAs and that of rule-based systems give us suggestions on how to make further substantial improvements, as will be discussed in the next section.

5   Backward-Chaining Evolutionary Algorithms

Running evolutionary algorithms from generation 0, to generation 1, to generation 2, and so on is the norm: this is what happens in nature, and this is certainly what has been done for decades in the field of evolutionary computation. This is similar to running a rule-based system in forward-chaining mode (Russell and Norvig, 2003). In these systems, we start with a working memory containing some premises, we apply a set of IF-THEN inference rules which modify the working memory by adding or removing facts, and we iterate this process until a certain condition is satisfied (e.g., a fact which we consider to be a conclusion is asserted). The rules in the knowledge base are a bit like the genetic operators in an evolutionary algorithm, the working memory is a bit like a population and the facts in it are a bit like individuals in an evolutionary algorithm.

This loose analogy between rule-based systems and evolutionary algorithms is not, in itself, terribly useful, except for one thing: it suggests the possibility of running an evolutionary algorithm in backward chaining mode, like one can do with a rule-based system. Broadly speaking, when a rule-based system is run in backward chaining, the system focuses on one particular conclusion that it wants to prove and operates as follows: a) it looks for all the rules which have such a conclusion as a consequent (i.e. a term following the "THEN" part of a rule), b) it analyses the antecedent (the "IF" part) of each such rule, c) if the antecedent is a fact (in other words, it is already in the working memory) then the original conclusion is proven and can be placed in the working memory, otherwise the system saves the state of the inference and recursively restarts the process with the antecedent as a new conclusion to prove (if there is no rule which has this term as a consequent, the recursion is stopped and another way of proving the original conclusion is attempted). If a rule has more than one condition (which is quite common), the system attempts to prove the truth of all the conditions, one at a time, and will assert the conclusion of the rule only if all conditions are satisfied. In this modality only the rules that can contribute to determining the truth or falsity of the target conclusion are ever considered, which of course can lead to major efficiency gains.

So, how would we run an evolutionary algorithm in backward-chaining mode? Let us suppose we are interested in knowing the makeup of the population at generation G and let us start by focusing on the first individual in the population. Let r be such an individual. Effectively r plays the role of a conclusion we want to prove. In order to generate r we only need to know what operator to apply to produce it and what parents to use. In turn, in order to know which parents to use, we need to perform tournaments to select them[7]. In each such tournaments we will need to know the makeup of n (the tournament size) individuals from the previous generation (which of course, at this stage we may still not know). Let us call $S = \{s_1, s_2, \ldots\}$ the set of the individuals that we need to know in generation $G - 1$ in order to determine r. Clearly, $s_1, s_2, \ldots$ are like the premises in a rule which, if applied, would allowed us to work out r (this would require evaluating the fitness of each element of S, deciding the winners of the tournament(s) and applying the chosen genetic operator to generate r). Normally we will not know the makeup of these individuals. However, we can recursively consider each of them as a subgoal. So, we determine which operator should be used to compute s_1, we determine which set of individuals at generation $G - 2$ is needed to do so, and we continue with the recursion. When we emerge from it, we repeat the process for s_2, etc. The recursion can terminate in one of two ways: a) we reach generation 0, in which case we can directly instantiate the individual in question by invoking the initialisation procedure for the particular EA we are considering, or b) the individual for which we need to know the genetic makeup has already been constructed and evaluated. Clearly the individuals in generation 0 have a role similar to that of the initial contents of the working memory in a rule-based system. Once we have finished with r we repeat the process with all the other individuals in the population at generation G, one by one.

Clearly, at its top-level, the algorithm just described is a recursive depth-first traversal of the graph mentioned at the end of the previous section. While we traverse the graph (more precisely, when we re-emerge from the recursion), we are in a position to know the genetic makeup of all the nodes encountered and so we can invoke the fitness evaluation procedure for each of them. Thus, we

---

[7] Decisions regarding operator choice and tournaments are trivial and can be made on the spot by drawing random numbers or can be all made in advance as in the EA with macro-selection

can label each node with the genetic makeup and fitness of the individual represented by such a node. Recursion stops when we reach a node without incoming links (a generation-0 individual, which gets immediately labelled randomly and evaluated) or when we reach a node that has been previously labelled. We will call an EA running in this mode a *Backward-Chaining EA (BC-EA)*.

Statistically a BC-EA is fully equivalent to the EA-EMS-CCD, and so it presents the same level of equivalence to an ordinary EA. In particular, if the same seed is used for the random number generators and all decisions regarding operators and tournaments are performed in a batch before the graph traversal, generations G of a BC-EA and an EA are indistinguishable.

So, if there are no differences why bother with a BC-EA instead of using a simpler "forward-chaining" version of the algorithm? One important difference between the two modes of operation is the order in which individuals in the population are evaluated. Let us consider an example where we have a population of $M = 5$ individuals which we run for $G = 3$ generations using tournament selection with $n = 2$ and we use crossover with 50% probability. Let us further suppose that, in the first instance, we are interested in knowing the first individual in the last generation. The ancestors of this individual might form a graph like the one in Figure 2.

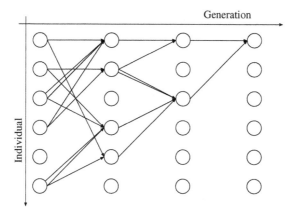

Fig. 2. Ancestors of the first individual in generation 3

Let us denote the nodes in row i (for individual) and column g (for generation) in the graph with the notation r_{ig}. In a forward chaining EA, even if we knew which individuals are unnecessary to define our target individual r_{13} (e.g., individuals r_{50}, r_{31}, r_{61}, and so on), we would evaluate individuals column by column from the left to the right in the following sequence: r_{10}, r_{20}, r_{30}, r_{40}, r_{60}, r_{11}, r_{21}, r_{41}, r_{51}, r_{12}, r_{32}, and finally r_{13}. That is, generation 0 individuals are computed before generation 1 individuals, which in turn are computed before generation 2 individuals, and so on. A backward chaining EA would instead evaluate nodes in a different order. For example, it might do it according to the sequence: r_{10}, r_{30}, r_{40}, r_{11}, r_{20}, r_{21}, r_{12}, r_{60}, r_{41}, r_{51}, r_{32}, and finally r_{13}. So, the algorithm would move back and forth evaluating nodes at different generations.

Why is this important? Typically, in an EA the average fitness of the population and the maximum fitness in each generation grow as the generation number grows. In our forward chaining EA the first 3 individuals evaluated have an expected average fitness equal to the average fitness of the individuals at generation 0, and the same is true for the BC-EA. However, unlike for the forward-chaining EA, the fourth individual created and evaluated by BC-EA belongs to generation 1, so its fitness is expected to be higher than that of the previous individuals. Individuals 5 and 6 have same expected fitness in the two algorithms. However, the seventh individual drawn by BC-EA is a generation 2 individual, while the forward EA draws a generation 1 individual. So, again the BC-EA is expected to produce a higher fitness sample than the other EA. Of course, this process is not going to continue indefinitely, and at some point the individuals evaluated by BC-EA start being on average inferior. This is unavoidable since the sets of individuals sampled by the two algorithms are identical.

This behaviour is typical: for problems where fitness tends to increase generation after generation a BC-EA will converge faster than an ordinary EA in the first part of a run and slower in the second part. So, if one restricts oneself to that first phase, the BC-EA is not just faster than an ordinary EA because it avoids evaluating individuals neglected by tournament selection, *BC-EA is also a faster converging algorithm*. How can we make sure we work in the region where the BC-EA is superior to the corresponding forward EA? Simple: like in any ordinary EA, in a BC-EA one does not need to continue evolution until all the individuals in generation G are known and evaluated, e.g., we can stop the algorithm whenever the best fitness seen so far reaches a suitably high value. In this way we can avoid at least a part of the phase where BC-EA is slower converging than the forward EA.

It is worth noting that this faster convergence behaviour is present in a BC-EA irrespective of the value of the tournament size, although, of course, the differences in behaviour between the two algorithms depend on it.

6   Experimental Results

We have implemented a backward chaining version of genetic algorithm (BC-GA) and run a variety of experiments on the counting ones problem. The choice of algorithm and problem was simply dictated by simplicity, since the notion of BC-EA is completely general.

Let us start by corroborating experimentally the expected faster convergence behaviour of BC-EA. To assess this we performed 100 independent runs of both a backward and a forward chaining version of the algorithm applied to a 100-bit problem. In these runs the maximum number of generations G was set to 99 (i.e, we did 100 generations). The population size M was 100. Only tournament selection and mutation (mutation rate $p_m = 0.01$) were used. To make a comparison between the algorithms possible, in our BC-GA we computed maximum fitness and average fitness every 100 fitness evaluations, and we treated this interval as a generation. In the BC-GA we computed *all* the individuals at generation G.

Figure 3 shows the fitness vs. generation plots for the two algorithms when the tournament size n is 2. It is clear from the figure that BC-GA performs about 20% fewer fitness evaluations than the standard EA, reaching, however, the same average and maximum fitness values. So, as predicted in the previous sections this significant computational saving comes without altering in any substantial way the behaviour of the EA.

Figure 4 shows the fitness vs. generation plots for the two algorithms when the tournament size n is 3. With this tournament size, there is still a saving of about 6% which is definitely worth having, but clearly for higher selection pressures the disadvantages of using a BC-GA in terms of memory use and bookkeeping become quickly preponderant.

Similar results should be expected when using a two-offspring version of crossover, although, of course, higher fitness values would be observed.

These experiments illustrate how a BC-EA typically converges faster than the traditional version of the algorithm. However, the most important question about BC-EA is how the expected number of fitness evaluations changes as a function of M, n, and G. We investigate this in the remainder of this section.

In order to assess the impact of both the transient and the limit-distribution behaviour of BC-EA, we performed a series of experiments where the task was to evaluate just one individual at generation G (that is $m(0) = 1$). In these experiments, we set G to be big enough so that all transients had finished well before generation 0, thereby revealing also the limit-distribution sampling behaviour. In the experiments we used populations of size $M = 10$, $M = 1000$ and $M = 100\,000$. We set $G = 49$ (i.e., we performed exactly 50 generations, 0 through to 49), so forward runs required exactly 500, $50\,000$ and $5\,000\,000$ fitness evaluations to complete, respectively. For each setting we did 100 independent runs.

Figure 5 shows the average proportion of individuals evaluated by BC-EA when *mutation only* is used ($\alpha = 1$) for tournaments sizes $n = 2$ and $n = 3$ as a function of the population size M, while Figure 6 shows the average proportion of individuals evaluated by BC-EA when *one-offspring crossover* with $p_c = 0.5$ is used ($\alpha = 1.5$) for the same tournaments sizes[8].

From these figures we can see that, as expected, the limit-distribution saving is largely independent from the size of the population. E.g., for $n = 2$, after the transient about 80% of the population is an "ancestor" of the individual of interest in generation G if mutation only is used, while this goes up to 94% when $\alpha = 1.5$. For EAs where long runs are used, these percentages provide an approximate estimation of the total proportion of fitness evaluations required by a backward chaining version of the algorithm w.r.t. the standard algorithm.

The figures also show that during most of the transient the number of individuals sampled by tournament selection grows very quickly (backward from generation 49). As clearly shown in Figure 7, the growth is exponential. The reasons for this are quite simple: when only a few samples are drawn from a

---

[8] Plots for crossovers producing two offspring in each application would show the same behaviour as the mutation only case

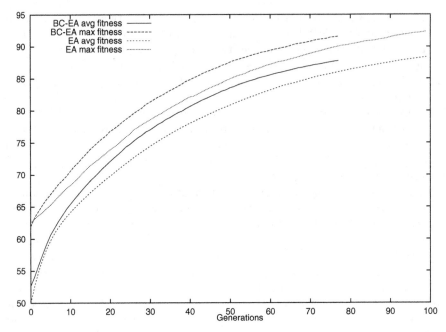

Fig. 3. Comparison between BC-GA and standard GA when a tournament size of 2 is used. Means over 100 independent runs

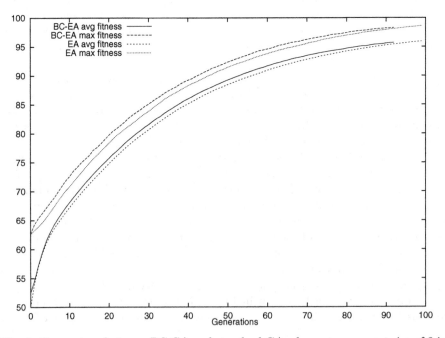

Fig. 4. Comparison between BC-GA and standard GA when a tournament size of 3 is used. Means over 100 independent runs

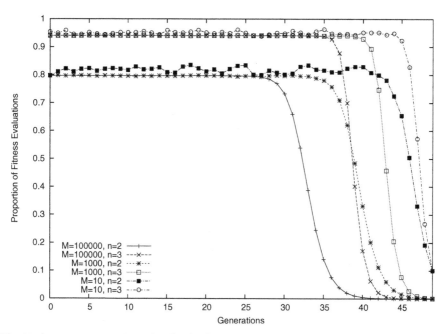

Fig. 5. Average proportion of individuals evaluated by BC-EA when mutation only is used ($\alpha = 1$) for tournaments sizes $n = 2$ and $n = 3$. Means over 100 independent runs

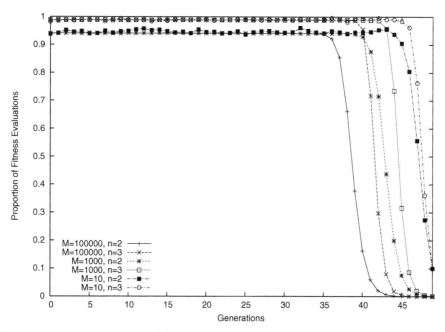

Fig. 6. Average proportion of individuals evaluated by BC-EA when one-offspring crossover with $p_c = 0.5$ is used ($\alpha = 1.5$) for tournaments sizes $n = 2$ and $n = 3$. Means over 100 independent runs

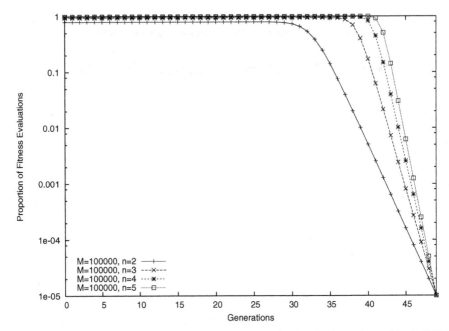

Fig. 7. Logarithmic plot of the average proportion of individuals evaluated by BC-EA when mutation only is used ($\alpha = 1$) for tournaments sizes $n = 2$–5 and a population size $M = 100\,000$

population, resampling is very unlikely, and, so, the ancestors of the individual of interest will tend to form a tree rather than a graph for at least some generations before the last. The branching factor of the tree is $n\alpha$. So, for small enough g, generation $G - g$ will include $(n\alpha)^g$ ancestors of the individual of interest in generation G. Naturally, this exponential growth continues only until $(n\alpha)^g$ becomes comparable with the expected number of individuals processed in the limit distribution case.

For small populations or high selective pressures the transient is short. However, there are cases where the transient lasts for many generations. For example, in a population of $M = 100\,000$ individuals and $\alpha = 1$ (i.e., in the case of mutation only or two-offspring crossover), the transient lasts for almost 20 generations (although it is exponential only for around 16 or 17). This population size may appear very big and these generation numbers may appear quite small. However, big populations and short runs are actually typical of at least one important class of EAs: genetic programming. So, it is worth evaluating the impact of the transient on the total number of fitness evaluations.

Let us assume $G - g_e$ is the last generation in which the transient is effectively exponential. (E.g., for $n = 2$, $\alpha = 1$, a population of size M and, of course, $m(0) = 1$, we have an exponential transient of $g_e \approx \log_2 M$ generations.) Then the number of individuals evaluated during the last g_e generations of a run is

$$F = \frac{(\alpha n)^{g_e+1} - 1}{\alpha n - 1}.$$

Because by definition $(\alpha n)^{g_e} \le M$, we have the following simple upper bound for the number of individuals required in the exponential transient:

$$F < \left(\frac{\alpha n}{\alpha n - 1} \right) M$$

Then, based on the previous equation, if we ran our BC-EA for $G = g_e$ generations, whatever α and n the total number of fitness evaluations F is upper bounded by $2M$ – i.e., F is less than the effort required to initialise a population and create one generation in the standard version of the EA!

Even when runs last for more than g_e generations, the effects of the exponential transient are marked. To illustrate this, Table 1 reports the mean *total* number of fitness evaluations recorded in the experiments shown in Figures 5 and 6. Taking, for example, the case of $n = 2$ and no mutation, where the limit distribution effort would be around 80%, we can see that efforts of as low as 53.4% of those required by a forward EA are achieved.

Table 1. Mean number of fitness evaluations recorded during 50 generations in the experiments shown in Figures 5 and 6. Forward EA fitness evaluations are also reported for reference

M	BC-EA with mutation		BC-EA with two-offspring crossover		Forward EA
	$n = 2$	$n = 3$	$n = 2$	$n = 3$	
10	384	454	452	476	500
1 000	31 980	40 967	40 895	44 615	50 000
100 000	2 671 667	3 702 667	3 692 295	4 160 496	5 000 000

7   Discussion

In the previous sections we have developed a deeper understanding of the sampling behaviour of tournament selection, in particular focusing on a source of inefficiency: the creation and evaluation of individuals that cannot influence future generations. We have then proposed general methods to remove this source of inefficiency and speed up evolutionary algorithms based on tournament selection. One of these methods, the backward chaining evolutionary algorithm, provides the additional benefit of converging faster that a standard (forward) algorithm due to its constructing and evaluating individuals belonging to later generations sooner.

The implementation of a backward chaining evolutionary algorithm is not very complex and the added book keeping required is quite limited. However, there is no doubt that BC-EAs require more memory than their forward counterparts. So, if one did not want to leave the well known and safe terrain of forward chaining evolutionary algorithms, what could one do to mitigate the effects of the potential sampling inefficiency of tournament selection? One possibility is to modify the manner in which the "random samples" are taken for the tournaments: instead of using uniform random samples, which can sometimes fail to select a subset of the total sample domain, we could use "super-uniform" random samples, which mimic the expected behaviour of the uniform random sample.

In the context of tournament selection, and for the case of $\alpha = 1$ (e.g., in a mutation-only algorithm) the following method would capture the spirit of tournament selection while more uniformly distributing a subset of the samples used in its calculation. For the i-th tournament of size n, choose the first sample as the i-th member of the previous population, and choose the remaining $n - 1$ samples uniformly at random. Effectively, this amounts to choosing the first samples with a super-uniform distribution, since exactly M tournaments are required when $\alpha = 1$ [9]. Clearly it would be easy to adapt this scheme for the case $\alpha > 1$. For example, one could use the scheme just described for M of the required tournaments and then use standard tournament selection for the remaining $M(\alpha-1)$. With this revised version of tournament selection all samples would be used in the calculation of the next generation. So, this method is elitist, in that it is guaranteed to keep the best individual. As a result, the method may also present stronger selection pressure than the original. Future research will be needed to fully understand the properties of the proposed method.

If one, however, is prepared to adopt the ideas behind BC-EA, the computational savings can be very big. These are achievable not only when we exploit the transient behaviour of the algorithm (as we illustrated in the previous section), but also in the limit-distribution behaviour, as will be illustrated below.

Maximum savings are achieved when αn is minimum. The smallest value α can take is 1 and with standard tournament selection the minimum for n is 2. So, we already know that the best we can do is saving around 20% fitness evaluations. However, a form of tournament selection exists (e.g., see (Mitchell, 1996)) that we can modify to obtain even more spectacular savings.

In this form of tournament selection, one picks up two individuals at random and then chooses the one with the higher fitness with probability p, the other with probability $1 - p$. For $p = 1$ this form of selection is equivalent to standard tournament selection with $n = 2$, while it is a form of random selection for $p = 0.5$. By acting on p it is possible to vary the selection pressure of the method continuously between these two extremes. An alternative description of the method is that we choose the higher fitness individual with probability q and randomly between the two with probability $1 - q$ (naturally $p = q + (1 - q)/2 = (1 + q)/2$). In this case q can be varied in the interval $[0, 1]$.

This second version of the algorithm can be modified for our purposes. Instead of first choosing a pair of individuals and then deciding whether we select the best or we pick one at random, we first decide which selection strategy we are going to use, and then, based on this, we randomly draw individuals from the population. If we decide to go for the best in the tournament, then we must draw two individuals from the population. However, if we decide to choose randomly between the two members of the tournament, then we can just draw one

---

[9] Effectively, this method stands to standard tournament selection as stochastic universal selection stands to roulette wheel selection

random individual from the population (instead of drawing two individuals and then randomly discarding one).

With this method, the expected number of individuals drawn in each tournament is $n = 2 \times q + 1 \times (1 - q) = q + 1 \leq 2$. So, clearly the smaller q the bigger the saving we should expect in a BC-EA. Just to get a feel for the order of magnitude of these savings, let us assume $\alpha = 1$ and let us use Equation 1 to estimate the expected proportion of individuals not sampled. This is approximately $e^{-(1+q)}$. So, for very low selection pressures saving of over 35% fitness evaluations are possible.

Naturally, much more substantial savings can be obtained when exploiting the transient behaviour of BC-EAs. In the previous section we showed that when running a BC-EA with $m(0) = 1$, the effort is of the order of the population size. However, the reader will probably wonder about the usefulness of evaluating just one individual in the last generation. Normally we would want to have the whole generation. However, we need to remember that the individual provided by a BC-EA (with $m(0) = 1$) at generation G is effectively a random sample drawn from the population at that generation. Although we expect one individual to be insufficient, one important question is whether we really need to have the whole of generation G in order to solve a problem. Elsewhere (Poli and Langdon, 2005) we have experimented with a backward-chaining genetic programming implementation showing that even when run with $m(0) = 1$ BC-GP can solve problems well. So, the answer to the above mentioned question appears to be that, at least is some cases, we do not need the whole population. To get a more complete and satisfactory answer, future work on BC-EAs will need to include a thorough investigation of the best way to choose $m(0)$ and G.

As we already noted, BC-EAs are based on changing the order of various operations on an EA, which requires memorising choices and individuals over multiple generations. Let us evaluate the space complexity of BC-EA and compare it to the space complexity of a standard EA. For simplicity, we consider EAs where the representation of each individual requires a fixed amount of memory: b bytes. The space complexity of a forward generational EA is

$$C_F = 2 \times (b + 4) \times M$$

where we assumed that we store both the current and the new generation and that fitness values are stored in a vector of floats (4 byte each). So, for $b \gg 1$, $C_F \approx 2bM$. In BC-EA, instead, the space complexity is

$$C_B = G \times M \times (b + 4 + \frac{1}{8})$$

since we need to store one array of individuals, one of floats, and one bit array, all of size $G \times M$ [10]. So, for $b \gg 1$, $C_B \approx GbM$. So, the difference in space

---

[10] This calculation is based on an implementation where the graph structure induced by tournament selection on the population is not explicitly stored. Instead, it is created dynamically and recursively. The bit array is used to flag the individuals that have been constructed and evaluated in a previous recursions. The calculation ignores the small amount of memory required in the stack during recursion

complexity between the two algorithms is

$$\Delta C = C_B - C_F = (G - 2) \times M \times (b + 4) + \frac{GM}{8},$$

which indicates that in most conditions the use of BC-EA carries a significant memory overhead. However, this does not prevent the use of BC-EAs. For example, if the representation of an individual requires $b = 100$ bytes, and we run a population of $M = 1000$ individuals for $G = 100$ generations, BC-EA requires only around 9.6MB of memory to run.

8   Conclusions

In this paper we have analysed the sampling behaviour of tournament selection over multiple generations and exploited this analysis to come up with more efficient implementations of evolutionary algorithms based on this selection method. In particular we have proposed a new way of running evolutionary algorithms, the BC-EA, which offers a combination of fast convergence, increased efficiency in terms of fitness evaluations, complete statistical equivalence to a standard EA and broad applicability. Because of these interesting properties we think the class of BC-EAs is an area worthy of further investigation.

Acknowledgements

I would like to thank Bill Langdon, Chris Stephens and the referees for their extremely useful comments, criticisms and suggestions for improvements. Also, I would like to thank Bill for agreeing to give my talk at the workshop when I had to suddenly leave for family reasons.

References

T. Bäck, D. B. Fogel, and T. Michalewicz, editors. *Evolutionary Computation 1: Basic Algorithms and Operators.* Institute of Physics Publishing, 2000.

T. Blickle and L. Thiele. A mathematical analysis of tournament selection. In L. J. Eshelman, editor, *Proceedings of the Sixth International Conference on Genetic Algorithms (ICGA'95)*, pages 9–16, San Francisco, California, 1995. Morgan Kaufmann Publishers.

Tobias Blickle and Lothar Thiele. A comparison of selection schemes used in evolutionary algorithms. *Evolutionary Computation*, 4(4):361–394, 1997.

Thomas E. Davis and Jose C. Principe. A Markov chain framework for the simple genetic algorithm. *Evolutionary Computation*, 1(3):269–288, 1993.

Kenneth A. De Jong, William M. Spears, and Diana F. Gordon. Using Markov chains to analyze GAFOs. In L. Darrell Whitley and Michael D. Vose, editors, *Proceedings of the Third Workshop on Foundations of Genetic Algorithms*, pages 115–138, San Francisco, July 31–August 2 1995. Morgan Kaufmann. ISBN 1-55860-356-5.

W. Feller. *An Introduction to Probability Theory and Its Applications*, volume 2. John Wiley, 1971.

Melanie Mitchell. *An introduction to genetic algorithms*. Cambridge MA: MIT Press, 1996.

Tatsuya Motoki. Calculating the expected loss of diversity of selection schemes. *Evolutionary Computation*, 10(4):397–422, 2002.

Allen E. Nix and Michael D. Vose. Modeling genetic algorithms with Markov chains. *Annals of Mathematics and Artificial Intelligence*, 5:79–88, 1992.

Riccardo Poli and William B. Langdon. Backward-chaining genetic programming. 2005.

Riccardo Poli, Jon E. Rowe, and Nicholas F. McPhee. Markov chain models for GP and variable-length GAs with homologous crossover. In *Proceedings of the Genetic and Evolutionary Computation Conference (GECCO-2001)*, San Francisco, California, USA, 7-11 July 2001. Morgan Kaufmann.

Günter Rudolph. Convergence analysis of canonical genetic algorithm. *IEEE Transactions on Neural Networks*, 5(1):96–101, 1994.

S. J. Russell and P. Norvig. *Artificial Intelligence: A Modern Approach*. Prendice Hall, Englewood Cliffs, New Jersey, second edition, 2003.

Astro Teller and David Andre. Automatically choosing the number of fitness cases: The rational allocation of trials. In John R. Koza, Kalyanmoy Deb, Marco Dorigo, David B. Fogel, Max Garzon, Hitoshi Iba, and Rick L. Riolo, editors, *Genetic Programming 1997: Proceedings of the Second Annual Conference*, pages 321–328, Stanford University, CA, USA, 13-16 July 1997. Morgan Kaufmann. URL http://www.cs.cmu.edu/afs/cs/usr/astro/public/papers/GR.ps.

A Schema-Based Version of Geiringer's Theorem
for Nonlinear Genetic Programming
with Homologous Crossover

Boris Mitavskiy and Jonathan E. Rowe

School of Computer Science, University of Birmingham, Birmingham B15 2TT, UK
{B.S.Mitavskiy,J.E.Rowe}@cs.bham.ac.uk

Abstract. Geiringer's theorem is a statement which tells us something about the limiting frequency of occurrence of a certain individual when a classical genetic algorithm is executed in the absence of selection and mutation. Recently Poli, Stephens, Wright and Rowe extended the original theorem of Geiringer to include the case of variable length genetic algorithms and linear genetic programming. In the current paper a rather powerful version of Geiringer's theorem which has been established recently by Mitavskiy is used to derive a schema-based version of the theorem for nonlinear genetic programming with homologous crossover.

1 Introduction

Geiringer's classical theorem (see [3]) is an important part of GA theory. It has been cited in a number of papers: see, for instance, [9], [10], [16] and [17]. It deals with the limit of the sequence of population vectors obtained by repeatedly applying the crossover operator $\mathcal{C}(p)_k = \sum_{i,j} p_i p_j r_{(i,\,j\to k)}$ where $r_{(i,\,j\to k)}$ denotes the probability of obtaining the individual k from the parents i and j after crossover. In other words, it speaks to the limit of repeated crossover in the case of an infinite population. In [6], a new version of this result was proved for *finite* populations, addressing the limiting distribution of the associated Markov chain, as follows. Let $\Omega = \prod_{i=1}^{n} A_i$ denote the search space of a given genetic algorithm (intuitively A_i is the set of alleles corresponding to the ith gene and n is the chromosome length). Fix a population P consisting of m individuals with m being an even number. P can be thought of as an m by n matrix whose rows are the individuals of the population P. Write

$$P = \begin{pmatrix} a_{11} & a_{12} & \cdots & a_{1n} \\ a_{21} & a_{22} & \cdots & a_{2n} \\ \vdots & \vdots & \ddots & \vdots \\ a_{m1} & a_{m2} & \cdots & a_{mn} \end{pmatrix}.$$

Notice that the elements of the ith column of P are members of A_i. Continuing with the notation used in [9], denote by $\Phi(h, P, i)$ where $h \in A_i$ the proportion of rows, say j, of P for which $a_{ji} = h$. In other words, let $R_h = \{j \mid 1 \leq j \leq m \text{ and } a_{ji} = h\}$. Now simply let $\Phi(h, P, i) = \frac{|R_h|}{m}$. The classical Geiringer theorem (see [3] or, [9] for modern notation) says that if one starts with a population P of individuals and runs a genetic algorithm (GA) in the absence of selection and mutation (crossover being the only operator involved) then, in the "long run", the frequency of occurrence of the individual $(h_1, h_2, \ldots, h_n)$ before time t, call it $\Phi(h_1, h_2, \ldots, h_n, t)$, approaches independence:

$$\lim_{t \to \infty} \Phi(h_1, h_2, \ldots, h_n, t) = \prod_{i=1}^{n} \Phi(h, P, i).$$

Thereby, Geiringer's theorem tells us something about the limiting frequency with which certain elements of the search space are sampled in the long run, provided one uses crossover alone. In [9] this theorem has been generalized to cover the cases of variable-length GA's and homologous linear genetic programming (GP) crossover. The limiting distributions of the frequency of occurrence of individuals belonging to a certain schema under these algorithms have been computed. The special conditions under which such a limiting distribution exists for linear GP under homologous crossover have been established (see theorem 9 and section 4.2.1 of [9]). In [6] a rather powerful extension of the finite population version of Geiringer's theorem has been established. It was also shown how the finite population versions of these results given in [3] and in [9] are special cases of the generalized Geiringer theorem proved in [6]. In the current paper we shall use the recipe described in [6] to derive a version of Geiringer's theorem for nonlinear GP with homologous crossover (see section 3 or [7] for a detailed description of how nonlinear GP with homologous crossover works) which is based on Poli's hyperschemata (see section 3 or [7]). The first step in this procedure is to describe the search space and the appropriate family of reproduction transformations so that the resulting GP algorithm is bijective and self-transient in the sense of definition 5.2 of [6]. Then the generalized Geiringer theorem (theorem 5.2 of [6]) as well as corollaries 6.1 and 6.2 of [6] apply. A simplified version of the necessary details is presented in the next section.

2   General Framework

In this section we introduce the necessary framework needed to state the schema-based version of Geiringer theorem for GP [1]. First, we describe how a general evolutionary search algorithm works. Let Ω denote a set, which we shall refer to as a *search space*. Denote by $\mathcal{F}$ a family of functions on Ω^2 (every $F \in \mathcal{F}$ is simply a function $F : \Omega^2 \to \Omega^2$). Intuitively, $\mathcal{F}$ is the family of *reproduction transformations*. A typical evolutionary search algorithm works by cycling through a sequence of steps such as selection, reproduction (crossover) and asexual reproduction (mutation). Geiringer theorem deals only with the reproduction steps. In the current paper we shall concentrate only on the bisexual reproduction step which is described in a general setting below (for a more detailed description see [6] and [5]):

A given population $P = (x_1, x_2, \ldots, x_m)$ with $x_i \in \Omega$ is taken as an input. The individuals in P are partitioned into pairs according to some probability distribution $\wp$

---

[1] The theorem presented in this section is a special case of the extended Geiringer theorem established in [6], yet it is general enough for most applications

on the set of all partitions of the set $\{1, 2, \ldots, m\}$ into 2-element subsets. (For simplicity of presentation we assume that m is an even integer. This assumption can be safely ignored and all of the results would still hold. See [6] for the details.) We shall assume that $\wp$ assigns a positive probability to every possible partition. Although this assumption can be weakened, we shall not bother in the current presentation. For instance, the couples could be

$$Q_1 = (x_{i_1^1}, x_{i_2^1}), \; Q_2 = (x_{i_1^2}, x_{i_2^2}), \; \ldots, \; Q_j = (x_{i_1^j}, x_{i_2^j}), \; \ldots, \; Q_{\frac{m}{2}} = (x_{i_1^{\frac{m}{2}}}, x_{i_2^{\frac{m}{2}}}).$$

Transformations $T_j \in \mathcal{F}$ are chosen independently according to some probability distribution p on the family $\mathcal{F}$. (The general result holds even under the assumption that the probability distributions are different for every j. However, in practice, this distribution is usually fixed.) Once the choices are made, replace the pairs Q_j with the pairs $T_j(y_{i_1^j}, y_{i_2^j})$. This way a new population $P' = (y_1, y_2, \ldots, y_m)$ is obtained.

Remark 1 Given some pair of individuals $(x, y) \in \Omega^2$, it is quite possible that there are two or more transformations, let's say for the sake of concreteness F_1 and $F_2 \in \mathcal{F}$, with $F_1 \neq F_2$ but $F_1(x, y) = F_2(x, y)$.

We now consider the following Markov chain: The states of this Markov chain are all populations[2]. For two populations $\mathbf{x}$ and $\mathbf{y}$ the transition probability of going from $\mathbf{x}$ to $\mathbf{y}$, $p_{\mathbf{x} \to \mathbf{y}}$ is the probability that population $\mathbf{y}$ is obtained from the population $\mathbf{x}$ after a single reproduction step. (Evidently $p_{\mathbf{x} \to \mathbf{y}}$ depends on $\mathcal{F}$, p, and $\wp$) Denote by $p_{\mathbf{x}, \mathbf{y}}^n$ the probability that population $\mathbf{y}$ is obtained from $\mathbf{x}$ upon the completion of n reproduction steps (this is the nth power of the Markov transition matrix defined above).

Definition 2 We say that a given algorithm is bijective and self-transient if the following conditions hold:
  1. Every transformation $T \in \mathcal{F}$ is bijective (i. e. one-to-one and onto).
  2. $1 \in \mathcal{F}$ and $p(1) > 0$ [3] (Here $1 : \Omega^2 \to \Omega^2$ denotes the identity map.)

We shall consider the following relation on the set of all populations:

Definition 3 Fix an evolutionary algorithm $\mathcal{A}$ with a reproduction step as described above. Fix populations $\mathbf{x}$ and $\mathbf{y}$. We shall write $\mathbf{x} \xrightarrow{\mathcal{A}} \mathbf{y}$ if $p_{\mathbf{x}, \mathbf{y}}^n > 0$ for some n.

The following facts have been established in [6]. Appendix A of [6] reveals the mathematics behind all of the facts listed in the remainder of this section.

Proposition 4 *If a given algorithm $\mathcal{A}$ is bijective and self-transient then $\xrightarrow{\mathcal{A}}$ is an equivalence relation.*

Definition 5 Given a population $P \in \Omega^m$ denote by $[P]_{\mathcal{A}}$ the equivalence class of the population P under the equivalence relation $\xrightarrow{\mathcal{A}}$.

---

[2] In the current presentation a population is an ordered m-tuple, i. e. an element of Ω^m. Lothar Schmitt used this representation in some of his work (see [14] and [15])

[3] This condition may be weakened but we want to make the presentation as simple as possible to follow

When a given algorithm $\mathcal{A}$ starts running with the initial population P and reproduction is the only step performed, thanks to proposition 4, only the populations in $[P]_{\mathcal{A}}$ may occur with nonzero probability. It makes sense, therefore, to restrict the state space of our Markov chain to include only the elements of the equivalence class $[P]_{\mathcal{A}}$. We shall call such a Markov process "the Markov chain initiated at P". The generalized Geiringer theorem of [6] tells us something nice about this Markov chain:

Theorem 6 *Let $\mathcal{A}$ denote a bijective and self-transient algorithm. Then the Markov chain initiated at some population $P \in \Omega^m$ is irreducible and its unique stationary distribution is the uniform distribution on $[P]_{\mathcal{A}}$.*

The classical versions of Geiringer theorem, such as the ones established in [3] and in [9] are stated in terms of the "limiting frequency of occurrence" of a certain element of the search space. The following definitions, which also appear in [6], make these notions precise in the general setting:

Definition 7 We define the characteristic function $\mathcal{X} : \Omega^m \times \mathcal{P}(\Omega) \to \mathbb{N} \cup \{0\}$ as follows: $\mathcal{X}(P, S) = $ the number of individuals of P which are the elements of S. (Recall that $P \in \Omega^m$ is a population consisting of m individuals and $S \in \mathcal{P}(\Omega)$ simply means that $S \subseteq \Omega$.)

Example 8 For instance, suppose $\Omega = \{0, 1\}^n$, $P = \begin{pmatrix} 0\,1\,0\,1\,0 \\ 0\,1\,0\,1\,0 \\ 1\,0\,1\,0\,1 \\ 0\,0\,1\,0\,1 \\ 0\,1\,0\,1\,0 \\ 1\,0\,1\,0\,1 \end{pmatrix}$ and $S \subseteq \Omega = \{0, 1\}^n$ is determined by the Holland schema $(*, 1, *, 1, *)$. Then $\mathcal{X}(P, S) = 3$ because exactly three rows of P, the 1st, the 2nd, and the 5th are in S.

Definition 9 Fix an evolutionary algorithm $\mathcal{A}$ and an initial population $P \in \Omega^m$. Let $P(t)$ denote the population obtained upon the completion of t reproduction steps of the algorithm $\mathcal{A}$ in the absence of selection and mutation. For instance, $P(0) = P$. Denote by $\Phi(S, P, t)$ the proportion of individuals from the set S which occur before time t. That is, $\Phi(S, P, t) = \frac{\sum_{s=1}^{t} \mathcal{X}(P(s), S)}{tm}$. (Notice that tm is simply the total number of individuals encountered before time t. The same individual may be repeated more than once and the multiplicity contributes to Φ.) Denote by $\mathcal{X}(\square, S) : \Omega^m \to \mathbb{N}$ the restriction of the function $\mathcal{X}$ when the set S is fixed (the notation suggests that one plugs a population P into the box).

Intuitively, $\Phi(S, P, t)$ is the frequency of encountering the individuals in S before time t when we run the algorithm starting with the initial population P.

3   Nonlinear Genetic Programming (GP) with Homologous Crossover

In genetic programming, the search space, Ω, consists of the parse trees which usually represent various computer programs.

Example 10 A typical parse tree representing the program $(+(\sin(x), *(x, y)))$ is drawn below:

Since computers have only a finite amount of memory, it is reasonable to assume that there are finitely many basic operations which can be used to construct programs and that every program tree has depth less than or equal to some integer L. Under these assumptions Ω is a finite set. We may then define the search space as follows:

Definition 11 Fix a signature $\Sigma = (\Sigma_0, \Sigma_1, \Sigma_2, \ldots, \Sigma_N)$ where Σ_i's are finite sets[4]. We assume that $\Sigma_i \neq \emptyset$ for some i and $|\Sigma_j| \neq 1 \forall j$ [5]. The search space Ω consists of all parse trees having depth at most L. Interior nodes having i children are labelled by the elements of Σ_i. The leaf nodes are labelled by the elements of Σ_0.

In order to study the appropriate family of reproduction (crossover) transformations with the aim of applying the generalized Geiringer theorem, it is most convenient to exploit Poli's hyperschemata ([7] for a more detailed description).

Definition 12 A Poli's hyperschema is a rooted parse tree which may have two additional labels for the nodes, namely $\#$ and $=$ signs (it is assumed, of course, that neither one of these denotes an operation). $=$ sign may label any interior node v of the tree. Since v does occur in the tree, we must have $|\Sigma_i| > 0$.) The $\#$ sign can only label a leaf node. A given Poli's hyperschema represents the set of all programs whose parse tree can be obtained by replacing the $=$ signs with any operation of the appropriate arities and attaching any program trees in place of the $\#$ signs. Different occurrences of $\#$ or $=$ may be replaced differently. We shall denote by S_t the set of programs represented by a hyperschema t.

A couple of programs fitting the hyperschema $(+(= (\#, x), *(\sin(y), \#))$ are shown below:

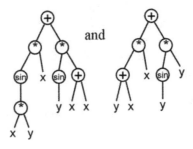

---

[4] Intuitively Σ_i is the set consisting of i-ary operations and Σ_0 consists of the input variables. Formally this does not have to be the case though

[5] The assumption that $|\Sigma_j| \neq 1 \forall j$ does not cause any problems since we are free to select any elements from the search space that we want. On the other hand, this assumption helps us to avoid unnecessary complications when dealing with the poset of Poli's hyperschemata later

In order to model the family of reproduction (crossover) transformation in a way which makes it obvious that GP is a bijective and self-transient algorithm, we shall introduce a partial order on the set of all Poli's hyperschema which will make it into a complete lattice (every two elements have a least upper bound). The notion of the least upper bound will be also used to define the *common region* (see [8] for an alternative description of the notion of a common region).

Definition 13 Denote by $\mathcal{O}$ the set of all basic operations which can be used to construct the programs and by $\mathcal{V}$ the set of all variables. Put the following partial order, $\preceq$, on the set $\mathcal{O} \cup \mathcal{V} \cup \{=, \#\}$:
    1. $\forall\, a,\, b \in \mathcal{O} \cup \mathcal{V}$ we have $a \preceq b \Longleftrightarrow a = b$.
    2. $\forall\, a \in \mathcal{O}$ we have $a \preceq =$.
    3. $\forall\, a \in \mathcal{O} \cup \mathcal{V}$ we have $a \preceq \#$.
    4. $=\preceq=$, $\# \preceq \#$ and $=\preceq \#$.

It is easy to see that $\preceq$ is, indeed a partial order. Moreover, every collection of elements of $\mathcal{O} \cup \mathcal{V} \cup \{=, \#\}$ has the least upper bound under $\preceq$. We are now ready to define the partial order relation on the set of all Poli's hyperschemata:

Definition 14 Let t_1 and t_2 denote two Poli's hyperschemata. We say that $t_1 \geq t_2$ if and only if the following two conditions are satisfied:
    1. the tree corresponding to t_1 when all of the labels are deleted is a subtree of the tree corresponding to t_2 with all of the labels deleted.
    2. Every one of the labels (which represents an operation or a variable) of t_1 is $\succeq$ the label of the node in the corresponding position of t_2.

Example 15 For instance, the hyperschema $t_1 = (+(= (\#, x)), *(\sin(y), \#)) \geq t_2 = (+(+(*(\sin(x), y), x)), *(\sin(y), = (\#)))$. Indeed, the parse trees of t_1 and t_2 appear on the picture below:

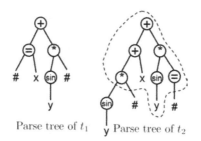

Parse tree of t_1 Parse tree of t_2

When all the labels in the dashed subtree of the parse tree of t_2 are deleted one gets the tree isomorphic to that obtained from t_1 by deleting all the labels. Thus condition 1 of definition 14 is satisfied. To see that condition 2 is fulfilled as well, we notice that the labels of t_1 are $\preceq$ to the corresponding labels of the dashed subtree of t_2: Indeed, we have $+ \succeq +$, $= \succeq +$, $* \succeq *$, $\# \succeq *$, $x \succeq x$, $\sin \succeq \sin$, $\# \succeq =$ and $y \succeq y$.

Again it is easy to check that $\geq$ is, indeed, a partial order relation on the collection of Poli's hyperschemata. Proposition 16 below tells us even more:

Proposition 16 *Any given collection of Poli's hyperschemata has the least upper bound under $\geq$.*

Proof. Denote by $\mathcal{S}$ a given collection of Poli's hyperschemata. We provide an algorithm to construct the least upper bound of $\mathcal{S}$ as follows: Copies of all the trees in $\mathcal{S}$ are recursively jointly traversed starting from the root nodes to identify the parts with the same shape, i. e. the same arity in the nodes visited. Recursion is stopped as soon as an arity mismatch between corresponding nodes in some two trees from $\mathcal{S}$ is present. All the nodes and links encountered are stored. This way we obtain a tree. It remains to stick in the labels. Each one of the interior nodes is labeled by the least upper bound of the corresponding labels of the trees in $\mathcal{S}$. The label of a leaf node is a variable, say x, if all the labels of the corresponding nodes of the trees in $\mathcal{S}$ are x (which implies that they are leaf nodes themselves). In all other cases the label of the leaf node is the # sign. It is not hard to see that this produces the least upper bound of the collection $\mathcal{S}$ of parse trees.

It was pointed out before, that programs themselves are Poli's hyperschemata. The following fact is almost immediate from the explicit construction of the least upper bound carried out in the proof of proposition 16:

Proposition 17 *A given Poli's hyperschema t is the least upper bound of the set S_t of programs determined by t.*

From proposition 17 it follows easily that $\geq$ is order isomorphic to the collection of subsets determined by the Poli's hyperschemata:

Proposition 18 *Let t and s denote Poli's hyperschemata. Denote by S_t and S_s the subsets of the search space determined by the hyperschemata t and s respectively. Then $t \geq s \iff S_t \supseteq S_s$.*

There is another type of schemata which is useful to introduce in order to define the family of reproduction (crossover) transformations:

Definition 19 A shape schema is just a rooted ordered tree. If $\tilde{t}$ is a given shape schema then $S_{\tilde{t}}$ is just the set of all programs whose underlying tree when all the labels are deleted is precisely $\tilde{t}$. Given a Poli's hyperschema s, we shall denote by $\tilde{s}$ the underlying shape schema of s, i. e. the tree obtained by deleting all the labels in s.

The notion of a common region which is equivalent to the one defined below also appears in [8]:

Definition 20 Given two Poli's hyperschemata t and s we define their common region to be the underlying shape schema of the least upper bound of t and s.

Definition 21 Fix a shape schema $\tilde{t}$. We shall say that the set $C_{\tilde{t}} = \{(a, b) \mid a, b$ are program trees and $\tilde{t}$ is the common region of a and $b\}$ is a component corresponding to the shape $\tilde{t}$.

Notice that sets determined by the shape schemata partition the search space:

Remark 22 Notice that $\Omega^2 = \bigcup_{\tilde{t} \text{ is a shape}} C_{\tilde{t}}$. Moreover, $C_{\tilde{t}} \cap C_{\tilde{s}} = \emptyset$ for $t \neq s$. (This is so because least upper bounds in a poset are uniquely determined and so the function sending $(a, b) \to \sup(a, b) \to$ the underlying shape of $\sup(a, b)$ is well defined. But then the sets $C_{\tilde{t}}$ are simply the pre-images under a function of singleton subsets of the set of all shapes and, hence, form a partition of Ω^2.)

We now proceed to define the family of reproduction transformations. Our goal is to introduce a family of functions on Ω^2 in such a way that each one of them is easily seen to be bijective (see theorem 6 and definition 2). The idea is to define these transformations on each of the components first:

Definition 23 Fix a shape schema $\tilde{t}$. Fix a node, v of $\tilde{t}$. A one-point partial homologous crossover transformation $T_v : C_{\tilde{t}} \to C_{\tilde{t}}$ is defined as follows: For given $(a, b) \in C_{\tilde{t}}$ let $T_v(a, b) = (c, d)$ where c and d are obtained from the program trees of a and b as follows: First identify the node v in the parse trees of a and b respectively. Now obtain the pair (c, d) by swapping the subtrees of a and b rooted at v. (This procedure is described in detail in [8] and it is also illustrated in the example below). Let $\mathcal{F}_{\tilde{t}} = \{T_v \mid v \text{ is a node of } \tilde{t}\}$ denote the family of all partial homologous one-point crossover transformations associated to the shape $\tilde{t}$.

The following example illustrates the concepts in definitions 19, 20 and 23:

Example 24 In the upper left part of the picture parse trees of the two sample programs a and b are shown. Then on the upper right one can see the least upper bound of a and b. On the lower right the underlying tree of the least upper bound of a and b is drawn. According to definition 20, this tree is precisely the common region of the programs a and b. The isomorphic subtrees inside both, a and b, are emphasized inside the dashed areas:

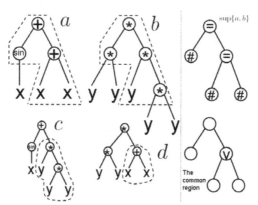

A node v is selected inside the common region. The pair of children $(c, d) = T_v(a, b)$ appear on the lower left of the picture above. The subtrees rooted at v which are swapped during crossover are emphasized inside the dashed area.

Remark 25 One does need to show that for $(a, b) \in C_{\tilde{t}}$ we have $T_v(a, b) \in C_{\tilde{t}}$. A rigorous argument can be given as follows: Clearly $T_v : C_{\tilde{t}} \to \bigcup_{\tilde{t} \text{ is a shape}} C_{\tilde{t}}$ is a well-defined map. Moreover, since v is a node of the least upper bound of a and b and the pair (c, d) is obtained simply by swapping the corresponding subtrees rooted at v, we get $s = \sup\{c, d\} \leq \sup\{a, b\}$. Now consider the transformation $F_v : C_{\tilde{s}} \rightarrow \bigcup_{\tilde{t} \text{ is a shape}} C_{\tilde{t}}$ and notice that, by definition, we have $F_v(c, d) = (a, b)$. But then, according to the reasoning above, we have $\sup\{c, d\} \leq \sup\{a, b\}$. Thereby, we get $\sup\{c, d\} \leq \sup\{a, b\} \leq \sup\{c, d\} \implies \sup\{c, d\} = \sup\{a, b\} \implies \tilde{t} = \tilde{s}$. This shows that T_v does, indeed, map into $C_{\tilde{v}}$. Moreover, in the process, we have also observed a couple of very important facts:

1. $T_v \circ T_v = 1_{C_{\tilde{t}}}$ where $1_{C_{\tilde{t}}}$ denotes the identity map on $C_{\tilde{t}}$. This shows, in particular, that T_v is a bijection.

2. T_v preserves the least upper bounds: $\sup\{a, b\} = \sup T_v(a, b)$.

We are finally ready to define the family of reproduction transformations on the search space Ω of all programs:

Definition 26 For every shape schema $\tilde{t}$ fix a node $v_{\tilde{t}}$ of $\tilde{t}$. Define a one point crossover transformation $T_{\{v_{\tilde{t}}\}_{\tilde{t} \text{ is a shape schema}}} : \Omega^2 \rightarrow \Omega^2$ to be the set-theoretic union of all partial crossover transformations of the form $T_{v_{\tilde{t}}}$. More explicitly, this means that whenever a given pair $(a, b) \in \Omega^2$ we must have $(a, b) \in C_{\tilde{s}}$ for a unique shape schema $\tilde{s}$ (since, according to remark 22, Ω^2 is a disjoint union of components corresponding to various shapes). But then $T_{\{v_{\tilde{t}}\}_{\tilde{t} \text{ is a shape schema}}}(a, b) = T_{v_{\tilde{s}}}(a, b)$. Denote by $\mathcal{F}$ the family of all crossover transformations together with the identity map on Ω^2. For simplicity of notation we shall denote the transformations in $\mathcal{F}$ by plain English letters: T, F etc., keeping in mind that every such transformation is determined by making choices of partial crossover transformations on every one of the components.

Remark 27 Thanks to remark 25, everyone of the crossover transformations in the family $\mathcal{F}$ is bijective (since it is a union of bijections on the pieces of a partition). It follows now that the generalized Geiringer theorem (theorem 6) applies to the case of homologous GP.

Remark 28 It is also possible to model uniform GP crossover (this type of crossover is examined in detail in [8]) in the analogous manner. All of the results established in the current paper apply to this case without any modification.

4 The Statement of the Schema-Based Version of Geiringer's Theorem for Non-linear GP Under Homologous Crossover

As mentioned before, the schema-based version of Geiringer's theorem for non-linear GP is stated in terms of Poli's hyperschemata.

Definition 29 A Poli's hyperschema of order i is a Poli's hyperschema which has exactly i nodes whose label is not a # or an = sign.

A configuration schema is a 0-order Poli's hyperschema (i.e a hyperschema which has only the equal signs in the interior nodes and # signs in the leaf nodes.)

An operation schema is a Poli's hyperschema of order 1 (i. e. a hyperschema which has exactly one node whose label is not a # or an = sign).

Fix an individual (a parse tree) $\mathbf{u} \in \Omega$. Let v denote any node of $\mathbf{u}$. Let $B(v)$ denote the branch of the shape schema of $\mathbf{u}$ from the root down to the node v. Let $B^+(v) = B(v) \cup \{w \mid w$ is a child of some node z of B with $z \neq v\}$. Now define $cs(v)$ to be the configuration schema whose underlying shape schema is $B^+(v)$. Let o denote an operation or a variable (an element of Σ_i for some i between 0 and N). Now obtain the operation schema os_o from $cs(v)$ by attaching the node labelled by o in place of the $\#$ sign at the node corresponding to v of $cs(v)$. Unless v is the leaf node of $\mathbf{u}$, all the children of this new node are the leaf nodes of os_o labelled by the $\#$ sign. When o is the operation (or the variable) labelling the node v of $\mathbf{u}$, we shall write $os(v)$ instead of os_o.

Notice that if v is a root node then $cs(v)$ is just the schema which determines the entire search space, i. e. the parse tree consisting of a single node labelled by the $\#$ sign. Example 30 illustrates definition 29.

Example 30 Below we list all of the configuration schemata and operation schemata for the individual of example 10:

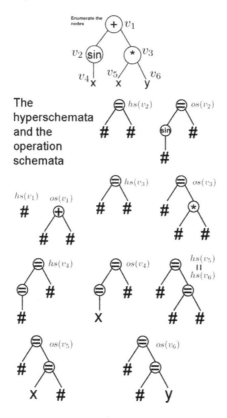

Recall from definition 7 that $\mathcal{X}(P, S)$ denotes the number of individuals in the population P which are the elements of $S \subseteq \Omega$. The following definition makes it more convenient to state the schema-based version of Geiringer's theorem:

Definition 31 Given a Poli's hyperschema H, we shall write $|H(P)|$ instead of $\mathcal{X}(P, S_H)$ (see definition 12) to denote the number of individuals (counting repetitions) in the population P fitting the hyperschema H.

We can now finally state the Geiringer's theorem for non-linear GP under homologous crossover:

Theorem 32 *Fix an initial population $P \in \Omega^m$ and an individual $\mathbf{u} \in \Omega$. Suppose every pair of individuals has a positive probability to be paired up for crossover and every transformation in $\mathcal{F}$ has a positive probability of being chosen[6]. Then the limiting frequency of occurrence of a given individual $\mathbf{u}$,*

$$\lim_{t\to\infty} \Phi(\mathbf{u},\, P,\, t) = \prod_{v \text{ is a node of } \mathbf{u}} \frac{|os(v)(P)|}{|cs(v)(P)|}.$$

Example 33 To illustrate how theorem 32 can be applied in practice, suppose we are interested in computing the frequency of encountering the individual $\mathbf{u}$ from examples 10 and 30 when the initial population of 6 individuals pictured below is chosen:

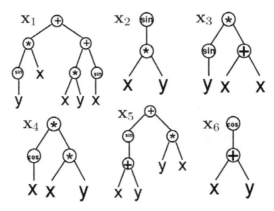

The number of individuals in P fitting the operation schema $os(v_1)$ is 2 (these are $\mathbf{x}_1$ and $\mathbf{x}_5$) while every individual fits the configuration schema $cs(v_1)$. Therefore $\frac{|os(v_1)(P)|}{|cs(v_1)(P)|} = \frac{2}{6} = \frac{1}{3}$. 4 individuals, namely $\mathbf{x}_1$, $\mathbf{x}_3$, $\mathbf{x}_4$ and $\mathbf{x}_5$ fit $cs(v_2) = cs(v_3)$, among these only 2 individuals, namely $\mathbf{x}_3$ and $\mathbf{x}_5$, fit $os(v_2)$ and 2 individuals, $\mathbf{x}_4$ and $\mathbf{x}_5$ fit $os(v_3)$ so that $\frac{|os(v_2)(P)|}{|cs(v_2)(P)|} = \frac{|os(v_3)(P)|}{|cs(v_3)(P)|} = \frac{2}{4} = \frac{1}{2}$. Individuals $\mathbf{x}_3$, $\mathbf{x}_4$ and $\mathbf{x}_5$ fit the configuration schema $cs(v_4)$ while only $\mathbf{x}_4$ fits the operation schema $os(v_4)$ so that $\frac{|os(v_4)(P)|}{|cs(v_4)(P)|} = \frac{1}{3}$. $\mathbf{x}_1$, $\mathbf{x}_3$, $\mathbf{x}_4$ and $\mathbf{x}_5$ fit $cs(v_5) = cs(v_6)$. Among these only $\mathbf{x}_3$ and $\mathbf{x}_4$ fit $os(v_5)$ while only $\mathbf{x}_4$ fits $os(v_6)$ so that $\frac{|os(v_5)(P)|}{|cs(v_5)(P)|} = \frac{2}{4} = \frac{1}{2}$ and $\frac{|os(v_6)(P)|}{|cs(v_6)(P)|} = \frac{1}{4}$. Thereby, according to theorem 32, we obtain:

$$\lim_{t\to\infty} \Phi(\mathbf{u},\, P,\, t) = \prod_{i=1}^{6} \frac{|os(v_i)(P)|}{|cs(v_i)(P)|} =$$

---

[6] These conditions can be slightly relaxed, but we try to present the main idea only

$$= \frac{1}{3} \cdot \frac{1}{2} \cdot \frac{1}{2} \cdot \frac{1}{3} \cdot \frac{1}{2} \cdot \frac{1}{4} = \frac{1}{288}.$$

Roughly speaking, this means that if we run GP starting with the population P pictured above, in the absence of mutation and selection (crossover being the only step) for an infinitely long time, the individual $\mathbf{u}$ will be encountered on average 1 out of 288 times.

Example 34 Notice that linear GP (or, equivalently, variable length GA) as described in [9] is a special case of nonlinear GP when $\forall i > 1$ $\Sigma_i = \emptyset$ and Σ_0 and $\Sigma_1 \neq \emptyset$. Indeed, the elements of such a search space are parse trees such that every interior node has exactly one child and the depth of the tree is bounded by some integer N. One can think of such a tree as a sequence of labels $(a_1, a_2, \ldots, a_n)$, the first label affiliated with the root node, second label with the child of the root node and so on. The label a_n is affiliated with the leaf node. This gives us a one-to-one correspondence, call it ϕ between the search space for nonlinear GP in our specific case when $\forall i > 1$ $\Sigma_i = \emptyset$ while Σ_0 and $\Sigma_1 \neq \emptyset$ and the search space for linear GP which preserves crossover. The following types of schemata have been introduced in [9]:

Definition 35 The schema $H = (*^{i-1}, h_i, \#)$ represents the subset $S_H = \{\mathbf{x} = (x_1, x_2, \ldots, x_l) \mid l > i \text{ and } x_i = h_i\}$. In words, S_H is simply the set of all individuals whose length is at least $i + 1$ and whose i^{th} allele is h_i.

Definition 36 The schema $H = (*^i, \#)$ represents the subset

$$S_H = \{\mathbf{x} = (x_1, x_2, \ldots, x_l) \mid l > i\}.$$

In words, S_H is simply the subset of all individuals whose length is at least $i + 1$.

Definition 37 The schema $H = (*^{i-1}, h_i)$ represents the subset

$$S_H = \{\mathbf{x} = (x_1, x_2, \ldots, x_i) \mid x_i = h_i\}$$

of the search space which is simply the set of all individuals of length exactly equal to i whose i^{th} (last) allele is h_i.

The reader may check that under the correspondence ϕ the configuration schemata correspond to the schemata $H_i = (*^i, \#)$ for $i \geq 1$, operation schemata correspond to the schemata of the form $H = (*^{i-1}, h_i, \#)$ and of the form $H = (*^{i-1}, h_i)$ for $i > 1$. Finally, the hyperschema $t_{(1, 1)}$ corresponds to the schema $H = (h_1, \#)$. Fix a population $P \in \Omega^m$. Recall that we denote by $|H|$ the number of individuals in P which fit the schema H counting repetitions. Also recall from definition 9 that $\Phi(S_H, P, 1) = \frac{|H|}{m}$ denotes the fraction of the number of individuals of P which fit the schema H. To abbreviate the notation we shall write $\Phi(H, P, 1)$ instead of $\Phi(S_H, P, 1)$. Fix an individual $\mathbf{u} = (h_1, h_2, \ldots, h_n) \in \Omega$. Theorem 32 tells us that

$$\lim_{t \to \infty} \Phi(\mathbf{u}, P, t) = \frac{|(h_1, \#)|}{m} \cdot \left(\prod_{i=1}^{n-2} \frac{|(*^i, h_{i+1}, \#)|}{|(*^i, \#)|} \right) \cdot \frac{|(*^{n-1}, h_n)|}{|(*^{n-1}, \#)|} =$$

$$= \frac{|(h_1, \#)|}{m} \cdot \left(\prod_{i=1}^{n-2} \frac{\frac{|(*^i, h_{i+1}, \#)|}{m}}{\frac{|(*^i, \#)|}{m}} \right) \cdot \frac{\frac{|(*^{n-1}, h_n)|}{m}}{\frac{|(*^{n-1}, \#)|}{m}} =$$

$$\Phi(h_1, \#) \cdot \left(\prod_{i=1}^{n-2} \frac{\Phi(*^i, h_{i+1}, \#)}{\Phi(*^i, \#)} \right) \cdot \frac{\Phi(*^{n-1}, h_n)}{\Phi(*^{n-1}, \#)} =$$

$$= \Phi(*^{n-1}, h_n) \cdot \frac{\prod_{i=n-2}^{0} \Phi(*^i, h_{i+1}, \#)}{\prod_{i=n-1}^{1} \Phi(*^i, \#)} = \Phi(*^{n-1}, h_n) \cdot \prod_{i=n-1}^{i=1} \frac{\Phi(*^{i-1}, h_i, \#)}{\Phi(*^i, \#)}$$

which is precisely the formula obtained in [9].

5   How Do We Obtain Theorem 32 from Theorem 6?

The following couple of corollaries from [6] are useful in obtaining the schema-based versions of Geiringer theorem for various evolutionary algorithms. Throughout, we shall denote by $\varrho_{[P]_\mathcal{A}}$ the uniform probability distribution on the set $[P]_\mathcal{A}$ (see definition 5).

Corollary 38   *Fix a bijective and self-transient algorithm $\mathcal{A}$ and an initial population $P \in \Omega^m$. Fix a set S of individuals in Ω ($S \subseteq \Omega$). Then $\lim_{t \to \infty} \Phi(S, P, t) = \frac{1}{m} E_{\varrho_{[P]_\mathcal{A}}}(\mathcal{X}(\square, S))$ (here $E_{\varrho_{[P]_\mathcal{A}}}(f)$ denotes the expectation of the random variable f with respect to the uniform distribution on the set $[P]_\mathcal{A})$ [7].*

To state the next corollary which brings us one step closer to deriving results similar in flavor to Geiringer's original theorem we need one more, purely formal, assumption about the algorithm:

Definition 39   We say that a given algorithm $\mathcal{A}$ is regular if the following is true: for every population $P = (x_1, x_2, \ldots, x_m) \in \Omega^m$ and for every permutation $\pi \in \mathcal{S}_m$, the population obtained by permuting the elements of P by π, namely $\pi(P) = (x_{\pi(1)}, x_{\pi(2)}, \ldots, x_{\pi(m)}) \in [P]_\mathcal{A}$. In words this says that the equivalence classes $[P]_\mathcal{A}$ are permutation invariant.

Remark 40   Definition 39 is only needed because our description of an evolutionary search algorithm uses the ordered multi-set model. This makes the generalized Geiringer theorem (theorem 6) look nice (the stationary distribution is uniform on $[P]_\mathcal{A}$). A disadvantage of the multi-set model is that it allows algorithms which are not regular. If we were to use the model of [17] where the order of elements in a population is not taken into account (a reasonable assumption since most evolutionary algorithms used in practice are, indeed, regular) then the Generalized Geiringer theorem would have to be modified accordingly since the stationary distribution of the corresponding Markov process would be different from uniform (it is not difficult to compute it though since the corresponding Markov chain is just a "projection" of the one used in the current paper).

Corollary 41   *Fix a regular bijective and self-transient algorithm $\mathcal{A}$ and an initial population $P \in \Omega^m$. Denote by $\varrho_{[P]_\mathcal{A}}$ the uniform probability distribution on $[P]_\mathcal{A}$ (see*

---

[7] Throughout the paper, whenever a limit is involved, the equality is meant to hold for almost every infinite sequence of trials

definition 5). Fix a set S of individuals in Ω ($S \subseteq \Omega$). Then $\lim_{t\to\infty} \Phi(S, P, t) = \varrho_{[P]_A}(\mathcal{V}_S)$ where

$$\mathcal{V}_S = \{P \mid P \in [P]_A \text{ and the } 1^{st} \text{ individual of } P \text{ is an element of } S\}.$$

Corollaries 38 and 41 are proved in section 6 of [6]. When deriving schema-based versions of Geiringer theorem for a specific algorithm the following strategy may be implemented: Continuing with the notation in corollaries 38 and 41, suppose we are given a nested sequence of subsets of the search space: $S_1 \supseteq S_2 \supseteq \ldots \supseteq S_n$. According to corollary 41,

$$\lim_{t\to\infty} \Phi(S_n, P, t) = \varrho_{[P]_A}(\mathcal{V}_{S_n}) = \frac{|\mathcal{V}_{S_n}|}{|[P]_A|} = \frac{|\mathcal{V}_{S_n}|}{|\mathcal{V}_{S_{n-1}}|} \cdot \frac{|\mathcal{V}_{S_{n-1}}|}{|[P]_A|} =$$

$$= \frac{|\mathcal{V}_{S_n}|}{|\mathcal{V}_{S_{n-1}}|} \cdot \frac{|\mathcal{V}_{S_{n-1}}|}{|\mathcal{V}_{S_{n-2}}|} \cdot \ldots \cdot \frac{|\mathcal{V}_{S_2}|}{|\mathcal{V}_{S_1}|} \cdot \frac{|\mathcal{V}_{S_1}|}{|[P]_A|} =$$

$$= \varrho_{[P]_A}(\mathcal{V}_{S_1}) \cdot \prod_{j=0}^{n-2} \frac{|\mathcal{V}_{S_{n-j}}|}{|\mathcal{V}_{S_{n-j-1}}|} = \frac{1}{m} E_{\varrho_{[P]_A}}(\mathcal{X}(\Box, S)) \cdot \prod_{j=0}^{n-2} \frac{|\mathcal{V}_{S_{n-j}}|}{|\mathcal{V}_{S_{n-j-1}}|}$$

Notice that $\frac{|\mathcal{V}_{S_j}|}{|\mathcal{V}_{S_{j-1}}|}$ is just the proportion of populations in $[P]_A$ whose first individual is a member of S_j inside the set of populations in $[P]_A$ whose first individual is a member of S_{j-1}.

Corollary 42 *Fix a regular, bijective and self-transient algorithm A and an initial population $P \in \Omega^m$. Fix a nested sequence of subsets $S_1 \supseteq S_2 \supseteq \ldots \supseteq S_n$ of individuals in Ω ($S_1 \subseteq \Omega$). Then $\lim_{t\to\infty} \Phi(S_n, P, t) = \frac{1}{m} E_{\varrho_{[P]_A}}(\mathcal{X}(\Box, S)) \cdot \prod_{j=0}^{n-2} \frac{|\mathcal{V}_{S_{n-j}}|}{|\mathcal{V}_{S_{n-j-1}}|}$ where, as before, $\mathcal{V}_S$ denotes the set of all populations whose first individual is a member of S for a given subset $S \subseteq \Omega$.*

Denote by A a given GP algorithm. Fix an individual $x \in \Omega$. In order to apply corollary 42, we may choose a descending chain of Poli's hyperschemata $t_1 \geq t_2 \geq \ldots \geq t_n = x$. Fix an initial population P. To avoid putting many subscripts, we shall write $\mathcal{V}_t$ instead of $\mathcal{V}_{S_t}$ for the set of all populations in $[P]_A$ (see definition 3) whose 1st individual is a member of S_t (the set of individuals determined by the hyperschema t). Now fix an individual $x \in \Omega$. In order to construct the desired sequence of nested hyperschemata, we assign the following numerical labelling to the nodes of the parse tree of **u**: The nodes are labelled by the pairs of integer coordinates. The first coordinate shows the depth of the tree and the second coordinate shows how far to the right a given node at the depth specified by the first coordinate is located. Notice, for instance, that the root node is labelled by the coordinates $(1, 1)$. We also introduce the following lexicographic linear ordering on the set of coordinate pairs:

Definition 43 $(a, b) \leq (c, d)$ if and only if either $a \leq c$ or ($a = c$ and $b \leq d$).

It is well known and easy to verify that this defines a linear ordering.

Definition 44 Given a pair of coordinates (i, j), denote by $\uparrow (i, j)$ the immediate successor of (i, j) under the lexicographic ordering defined above. Explicitly,

$$\uparrow (i, j) = \begin{cases} (i+1, 1) \text{ if } (i,j) \text{ labels the rightmost node of } \mathbf{u} \text{ at depth } i \\ (i, j+1) \text{ otherwise} \end{cases}$$

We obtain the desired nested sequence of hyperschemata for the given individual $\mathbf{u}$ recursively in the following manner:

Definition 45 Define $t_{(1,1)}$ to be the hyperschema whose root node has the same label (operation) and arity as that of the root node of $\mathbf{u}$. All children of the root node are the leaf nodes labelled by the # sign. Once the hyperschema $t_{(i,j)}$ has been constructed, we obtain the hyperschema $t_{\uparrow(i,j)}$ by attaching the node of $\mathbf{u}$ with coordinate $\uparrow (i, j)$ in place of the # sign at coordinate $\uparrow (i, j)$ to the parse tree of $t_{(i,j)}$. Unless this node, call it v, is a leaf node of $\mathbf{u}$, all children of this new node are the leaf nodes of $t_{\uparrow(i,j)}$ labelled by the # sign.

We illustrate the construction with an explicit example:

Example 46 Below, the nested sequence $t_{(1, 1)} \geq t_{(2, 1)} \geq t_{(2, 2)} \geq t_{(3, 1)}, \geq t_{(3, 2)} \geq t_{(3, 3)}$ corresponding to the program of example 10 is drawn explicitly:

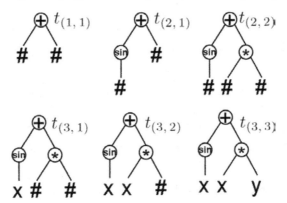

The formula for the limiting frequency of occurrence of a given program u in corollary 42 involves the ratios of the form $\frac{\mathcal{V}_{t_{\uparrow(i,j)}}}{\mathcal{V}_{t_{(i,j)}}}$. It turns out that these ratios can be expressed nicely in terms of the presence of certain configuration and operation schemata in the initial population P:

Definition 47 Given a program tree $\mathbf{u}$ and the corresponding nested sequence $t_{(1, 1)} \geq t_{(2, 1)} \geq \dots \geq t_{(i, j)} \geq t_{\uparrow(i, j)}, \geq \dots \geq t_{(l, k)} = \mathbf{u}$ of hyperschemata as in definition 45, for every $(i, j) \neq (l, k)$, denote by $cs_{(i, j)}$ $(os_{(i, j)})$ the configuration schema $cs(v)$ (operation schema $os(v)$) where v is the node of $\mathbf{u}$ with coordinate $\uparrow (i, j)$.

Example 48 Continuing with examples 10 and 30 notice that for the individual in these examples we have $cs_{(1,1)} = cs_{(2,1)} = cs(v_2) = cs(v_3)$ while $os_{(1,1)} = os(v_2)$ and $os_{(2,1)} = os(v_3)$ (see example 30), $cs_{(2,2)} = cs(v_4)$ while $os_{(2,2)} = os(v_4)$ and $cs_{(3,1)} = cs_{(3,2)} = cs(v_5) = cs(v_6)$ while $os_{(3,1)} = os(v_5)$ and $os_{(3,2)} = os(v_6)$.

The following "orbit description" lemma is the reason for introducing configuration and operation schemata: We prove the lemma under the following special assumption:

Definition 49 We say that a population P is special with respect to the individual $\mathbf{u}$ if for every node v of $\mathbf{u}$ and for every operation (or variable) o we have $|os_o(P)| \leq 1$ where os_o is obtained from $cs(v)$ by means of attaching the operation o at the leaf node of $cs(v)$ corresponding to v as described in definition 29.

Definition 49 basically requires that no 2 operations (or variables) occurring in P at the specified location are the same. It turns out that the orbit description lemma stated below is a lot more convenient to prove under this special assumption. The general case will then follow by introducing enough extra labels for the operations and variables involved and then deleting the extra labels.

Lemma 50 *Fix an initial population P and a program $\mathbf{u} \in \Omega$ Assume that the population P is special with respect to the individual $\mathbf{u}$. Suppose every pair of individuals has a positive probability to be paired up for crossover and every transformation in $\mathcal{F}$ has a positive probability of being chosen[8]. Consider the sequences of hyperschemata*
$$t_{(1,1)} \geq t_{(2,1)} \geq \cdots \geq t_{(i,j)} \geq t_{\uparrow(i,j)}, \geq \cdots \geq t_{(l,k)} = \mathbf{u}, \{cs_{(i,j)} \mid (i,j) \text{ is a}$$
coordinate of $\mathbf{u}$, (i,j) is not the maximal coordinate $\}$ and $\{os_{(i,j)} \mid (i,j)$ is a coordinate of $\mathbf{u}$, (i,j) is not the maximal coordinate $\}$ corresponding to the individual $\mathbf{u}$. For a given hyperschema t, denote by $|t(P)|$ the number of individuals in P which fit the hyperschema t counting repetitions. Suppose $\forall$ non-maximal pairs of coordinates (i,j) we have $|os_{(i,j)}(P)| \neq 0$ and $|t_{(1,1)}(P)| \neq 0$. Then it is true that
$$\forall (i,j) \ \frac{\mathcal{V}_{t_{\uparrow(i,j)}}}{\mathcal{V}_{t_{(i,j)}}} = \frac{1}{|cs_{(i,j)}(P)|}.$$

Proof. The key idea is to observe the following fact:

Claim: Fix a coordinate (i,j). Fix any two operation schemata os_1 and os_2 which are obtained from $cs_{(i,j)}$ by attaching either a variable or an operation at the node (i,j). Suppose $\exists$ individuals in P fitting both, os_1 and os_2. Then $|\mathcal{V}_{t_{(i,j)}} \cap \mathcal{V}_{os_1}| = |\mathcal{V}_{t_{(i,j)}} \cap \mathcal{V}_{os_2}|$.

Proof. Consider the map $F : [P]_{\mathcal{A}} \to [P]_{\mathcal{A}}$ defined as follows: Given a population, say $Q \in [P]_{\mathcal{A}}$, notice that $\exists$ an individual, say $\mathbf{x}_1$, in Q fitting the operation schema os_1 (due to the way crossover is defined, the number of individuals fitting the operation schema $os_1(Q)$ is the same in every population $Q \in [P]_{\mathcal{A}}$). Moreover, such an individual is unique since we assumed that all operations appearing in the individuals of P are distinct. Likewise, $\exists$ unique individual in Q, say $\mathbf{x}_2$ fitting the operation schema os_2. Pair up individuals $\mathbf{x}_1$ and $\mathbf{x}_2$ and pair up the rest of the individuals arbitrarily for crossover. Select the crossover transformation T_v where v is the node with coordinate (i,j) for the pair $(\mathbf{x}_1, \mathbf{x}_2)$ and choose the identity transformation for the rest of the pairs. Now let $F(Q)$ be the population obtained upon the completion of the reproduction step described above (notice that $F(Q) \in [P]_{\mathcal{A}}$ by definition of $[P]_{\mathcal{A}}$). Notice also that F is its own inverse (i. e. $F \circ F = 1_{[P]_{\mathcal{A}}}$). This tells us, in particular, that F is bijective. Moreover, it is clear from the definitions that $F(\mathcal{V}_{t_{(i,j)}} \cap \mathcal{V}_{os_1}) \subseteq \mathcal{V}_{t_{(i,j)}} \cap \mathcal{V}_{os_2}$

---

[8] These conditions can be slightly relaxed, but we try to present the main idea only

and, likewise, $F(\mathcal{V}_{t_{(i,j)}} \cap \mathcal{V}_{os_2}) \subseteq \mathcal{V}_{t_{(i,j)}} \cap \mathcal{V}_{os_1}$. The desired conclusion follows at once.

Now observe that $t_{\uparrow(i,j)} = t_{(i,j)} \cap os_{(i,j)}$ so that $\mathcal{V}_{t_{\uparrow(i,j)}} = \mathcal{V}_{t_{(i,j)}} \cap \mathcal{V}_{os_{(i,j)}}$ and $t_{(i,j)} = \bigcup_{o \text{ is an operation or a variable}} (t_{(i,j)} \cap os_o)$ where os_o is obtained from $cs_{(i,j)}$ by attaching the operation (or variable) o at the node $\uparrow (i,j)$. Therefore we also have $\mathcal{V}_{t_{(i,j)}} = \bigcup_{o \text{ is an operation or a variable,}} (\mathcal{V}_{t_{(i,j)}} \cap \mathcal{V}_{os_o})$. Since operations can not appear or disappear from a population during crossover, $\mathcal{V}_{os_o} \neq \emptyset \implies \exists$ an individual in P fitting the operation schema os_o. Thus the only sets of the form $\mathcal{V}_{t_{(i,j)}} \cap \mathcal{V}_{os_o}$ which may possibly contribute to the union above are these for which $\exists$ an individual in P fitting the operation schema os_o. According to the claim above, all such sets contribute exactly the same amount. Moreover, by assumption $os_{(i,j)}(P) \neq \emptyset$, and so we have

$$|\mathcal{V}_{t_{(i,j)}}| = n \cdot |\mathcal{V}_{t_{(i,j)}} \cap \mathcal{V}_{os_{(i,j)}}| = n \cdot |\mathcal{V}_{t_{(i,j)} \cap os_{(i,j)}}| = n \cdot |\mathcal{V}_{t_{\uparrow(i,j)}}| \implies \frac{|\mathcal{V}_{t_{\uparrow(i,j)}}|}{|\mathcal{V}_{t_{(i,j)}}|} = \frac{1}{n}$$

where n is the number of operation schemata of the form os_o for which $\exists$ an individual in P fitting the operation schema os_o and the last implication holds under the condition that $|\mathcal{V}_{t_{(i,j)}}| \neq 0$. This condition is, indeed satisfied. (Suppose not. Let (a,b) denote the smallest coordinate such that $|\mathcal{V}_{t_{(a,b)}}| = 0$. Notice that $(a,b) \neq (1,1)$ since $|\mathcal{V}_{t_{(1,1)}}| \neq 0$. (By assumption $\exists$ an individual, say $\mathbf{x}$, in P fitting the hyperschema $t_{(1,1)}$. Even if $\mathbf{x}$ is not the 1st individual of P, by performing crossover of $\mathbf{x}$ with the 1st individual of P at the root node one gets a population $Q \in \mathcal{V}_{t_{(1,1)}}$.) But then $(a,b) = \uparrow (i,j)$ for some coordinate (i,j) and according to the equation above we have $|\mathcal{V}_{t_{(i,j)}}| = n \cdot |\mathcal{V}_{t_{\uparrow(i,j)}}| = n \cdot |\mathcal{V}_{t_{(a,b)}}| = 0$ which contradicts the minimality of the coordinate (a,b). So we conclude that $|\mathcal{V}_{t_{(i,j)}}| \neq 0$) Thereby we have $\frac{|\mathcal{V}_{t_{\uparrow(i,j)}}|}{|\mathcal{V}_{t_{(i,j)}}|} = \frac{1}{n}$. But $cs_{(i,j)} = \bigcup_{o \text{ is an operation or a variable}} os_o \implies cs_{(i,j)}(P) = \bigcup_{o \text{ is an operation or a variable}} os_o(P)$. Since we assumed that all of the operations and variables are distinct, $\exists$ at most one individual in P fitting the operation schema os_o and it now follows that $|cs_{(i,j)}(P)| =$ the number of operation schemata of the form os_o such that $os_o(P) \neq \emptyset$ which is precisely the number n. We finally obtain $\frac{|\mathcal{V}_{t_{\uparrow(i,j)}}|}{|\mathcal{V}_{t_{(i,j)}}|} = \frac{1}{|cs_{(i,j)}|}$ which is precisely the conclusion of the lemma.

Remark 51 Given an individual $\mathbf{u}$ and a population P consisting of m individuals, observe that the number of individuals fitting the hyperschema $t_{(1,1)}$ is the same in every population from $[P]_{\mathcal{A}}$, i. e. $\forall Q \in [P]_{\mathcal{A}}$ we have $|t_{(1,1)}(Q)| = |t_{(1,1)}(P)| = 1$. It follows immediately now that $\frac{1}{m} E_{\varrho[P]_{\mathcal{A}}}(\mathcal{X}(\square, S_{t_{(1,1)}})) = \frac{1}{m}$.

We now combine corollary 42, remark 51 and lemma 50 to obtain the following special case of Geiringer theorem for nonlinear GP under homologous crossover in case when all of the operations appearing in the individuals of the initial population P are distinct:

$$\lim_{t \to \infty} \Phi(\mathbf{u}, P, t) = \frac{1}{m} \cdot \prod_{(i,j) \text{ is not the maximal coordinate of } \mathbf{u}} \frac{1}{|cs_{(i,j)}(P)|} =$$

$$= \prod_{v \text{ is a node of } \mathbf{u}} \frac{1}{|cs(v)(P)|}$$

(recall that when v is the root node of $\mathbf{u}$, $cs(v)$ determines the entire search space, and so $\frac{1}{|cs(v)(P)|} = \frac{1}{m}$.) To obtain the general case, suppose we are given an initial population P. Fix a node v of $\mathbf{u}$ and consider the set of operations $\mathcal{O}(v) = \{o \,|\, |os_o(P)| \geq 1$ where os_o is obtained from $cs(v)$ as in definition 29$\}$. For every node v of $\mathbf{u}$ and for every operation (or variable) $o \in \mathcal{O}(v)$ fix an enumeration $\mathbf{x}_1^o, \mathbf{x}_2^o, \ldots, \mathbf{x}_{|os_o(P)|}^o$ of the individuals in P fitting the operation schema $os_o(P)$. Relabel the operation o occurring in the node v of $\mathbf{x}_i^o$ by the formally different operation (o, i) (i. e. by the ordered pair (o, i) whose first element is the operation o itself and the second element is the index telling us in which individual of P the operation o labels the node v). After all of the relabelling is complete we obtain a new population P' which is special with respect to the individual $\mathbf{u}$ in the sense of definition 49. Formally speaking, we expand our signature $\Sigma = (\Sigma_1, \Sigma_2, \ldots, \Sigma_N)$ as in definition 11 by adding the operations (variables) (o, i) into Σ_j where j is the arity of the operation o. This gives us a new signature $\Sigma^* = (\Sigma_1^*, \Sigma_2^*, \ldots, \Sigma_N^*)$ where

$$\Sigma_j^* = \{o \,|\, o \in \Sigma_j \text{ and } o \notin \bigcup_{v \text{ is a node of } \mathbf{u}} \mathcal{O}(v)\} \cup$$

$$\cup \{(o, i) \,|\, o \in \mathcal{O}(v) \text{ for some } v \text{ and } 1 \leq i \leq |os_o(P)|\}.$$

Denote by Ω^* the search space induced by the signature Σ^*. The natural projection maps $p_j : \Sigma_j^* \to \Sigma_j$ sending $0 \to o$ when $o \notin \bigcup_{v \text{ is a node of } \mathbf{u}} \mathcal{O}(v)$ and $(o, i) \to o$ when $o \in \mathcal{O}(v)$ for some node v of $\mathbf{u}$, induce the natural "deletion of the extra labels" projection of the search spaces $\varphi : \Omega^* \to \Omega$ where the individual $\varphi(\mathbf{w}) \in \Omega$ is obtained from the individual $\mathbf{w} \in \Omega^*$ by replacing the label of every node w of $\mathbf{w}$ with $p_j(w)$ where j is the arity of the node w. It is easily seen that the natural projection φ commutes with the crossover transformations in the sense that for any individuals $\mathbf{x}, \mathbf{y} \in \Omega^*$ and for any crossover transformation $T \in \mathcal{F}$ (see definition 26) we have $\varphi(T(\mathbf{x}, \mathbf{y})) = T(\varphi(\mathbf{x}), \varphi(\mathbf{y}))$ [9]. Notice also that the population P can be obtained from the population P' by applying the natural projection φ to every individual of P'. Therefore, running the algorithm with the initial population P is the same thing as running the algorithm with the initial population P' and reading the output by applying the natural projection φ. The special case does apply to the population P', as mentioned above, and so we have

$$\lim_{t \to \infty} \Phi(\mathbf{u}, P, t) = \sum_{\mathbf{w} \in \varphi^{-1}(\mathbf{u})} \lim_{t \to \infty} \Phi(\mathbf{w}, P, t) = \sum_{\mathbf{w} \in \varphi^{-1}(\mathbf{u})} \prod_{v \text{ is a node of } \mathbf{w}} \frac{1}{|cs(v)(P)|}.$$

Notice that $\mathbf{w} \in \varphi^{-1}(\mathbf{u})$ precisely when the underlying shape schema of $\mathbf{w}$ is the same as that of $\mathbf{u}$, call this shape schema $t_{\mathbf{u}}$, and the label of every node v of $\mathbf{w}$ is (o, i) where o is the label of the node v of $\mathbf{u}$. According to the way the population P' was introduced, there are precisely $|os(v)(P)|$ such labels (see also definition 29). We can

---

[9] Of course, formally speaking, the two transformations T involved in the equation above are distinct, as they have different domains (Ω^* and Ω respectively), but they are determined by the same set of shape schemata and the same choice of nodes for crossover so we denote them by the same symbol

then identify the preimage $\varphi^{-1}(\mathbf{u})$ with the set $\prod_{j=1}^{K}\{i \mid 1 \leq i \leq |os(v_j)|\}$ of ordered K-tuples of integers where K is the number of nodes in the parse tree of $\mathbf{u}$ and v_1, $v_2, \ldots, v_K$ is any fixed enumeration of the nodes of $\mathbf{u}$, in the following manner: The identification map $\imath : \prod_{j=1}^{K}\{i \mid 1 \leq i \leq |os(v_j)(P)|\} \to \varphi^{-1}(\mathbf{u})$ sends a given ordered K-tuple $(i_1, i_2, \ldots, i_K)$ into the tree $\mathbf{w} = \imath((i_1, i_2, \ldots, i_K))$ whose underlying shape schema is $t_\mathbf{u}$ and the label of a node v_j of $\mathbf{w}$ is (o_j, i_j) where o_j is the label of the node v_j in the parse tree of $\mathbf{u}$. We finally obtain:

$$\lim_{t \to \infty} \Phi(\mathbf{u}, P, t) = \sum_{\mathbf{w} \in \varphi^{-1}(\mathbf{u})} \prod_{v \text{ is a node of } \mathbf{w}} \frac{1}{|cs(v)(P)|} =$$

$$= \sum_{(i_1, i_2, \ldots i_K) \in \prod_{j=1}^{K}\{i \mid 1 \leq i \leq |os(v_j)|\}} \prod_{v \text{ is a node of } \mathbf{u}} \frac{1}{|cs(v)(P)|} =$$

$$= \sum_{i_1=1}^{|os(v_1)(P)|} \sum_{i_2=1}^{|os(v_2)(P)|} \cdots \sum_{i_K=1}^{|os(v_K)(P)|} \prod_{v \text{ is a node of } \mathbf{u}} \frac{1}{|cs(v)(P)|}$$

$$= \prod_{j=1}^{K} \sum_{i_j=1}^{|os(v_j)(P)|} \frac{1}{|cs(v_j)(P)|} = \prod_{v \text{ is a node of } \mathbf{u}} \frac{|os(v)(P)|}{|cs(v)(P)|}$$

which is precisely the assertion of theorem 32.

6   Conclusions

In the current paper we applied the methods developed in [6] to obtain a schema-based version of Geiringer's theorem for non-linear GP with homologous crossover. The result enables us to calculate exactly the limiting distribution of the Markov chain associated with the evolution of a finite (fixed size) population under the action of repeated crossover. This is an extension of the results for fixed and variable length strings given in [6] for finite populations. The infinite population versions are given by the classical Geiringer theorem (in the case of fixed length strings) and the generalization given in [9] (for variable length strings). The corresponding infinite population result for non-linear GP is not yet established, although it seems to follow from the embedding theorems of [5] together with the corresponding result of [9] for linear GP. We are currently working on this issue. What is not known, is under which general conditions does the finite population result imply a corresponding limit in the infinite population case. This remains an open question.

References

1. Booker, L. (1993). Recombination distributions for genetic algorithms. In L. Darrell Whitley, editor, *Foundations of Genetic Algorithms 2*, pages 29-44, Morgan Kaufmann.
2. Coffey, S. (1999) An Applied Probabilist's Guide to Genetic Algorithms. *A Thesis Submitted to The University of Dublin for the degree of Master in Science.*

3. Geiringer, H. (1944). On the probability of linkage in Mendelian heredity. *Annals of Mathematical Statistics*, 15:25-57.
4. Mitavskiy B. (2004). Crossover Invariant Subsets of the Search Space for Evolutionary Algorithms. *Evolutionary Computation*, 12(1): 19-46.
5. Mitavskiy B. (2003). Comparing Evolutionary Computation Techniques via Their Representation. In Eric Cantú-Paz *et al.* editor, *Proceedings of the Genetic and Evolutionary Computation (GECCO) Conference*, Vol. 1, pages 1196-1209, Springer-Verlag.
6. Mitavskiy B. (recently submitted). A Generalization of Geiringer Theorem for a wide class of evolutionary search algorithms. *Evolutionary Computation*.
   http://www.math.lsa.umich.edu/∽bmitavsk/
7. Poli, R. (2000). Hyperschema Theory for GP with One-Point Crossover, Building Blocks, and Some New Results in GA Theory. In R. Poli, W. Banzhaf, and *et al.*, editors, *Genetic Programming, Proceedings of EuroGP'2000*, Springer-Verlag
8. Poli, R. and Langdon, W. (1998). On the search Properties of Different Crossover Operators in Genetic Programming.
9. Poli, R., Stephens, C., Wright, A., Rowe, J. (2002). A Schema-Theory-Based Extension of Geiringer's Theorem for Linear GP and variable-length GAs under Homologous Crossover. *FOGA 2002*
10. Rabani, Y., Rabinovich, Y., Sinclair, A. (1995) A Computational View of Population Genetics. In *Annual ACM Symposium on the Theory of Computing* pages 83-92.
11. Rosenthal, Jeffrey (1995). Convergence Rates for Markov Chains. *SIAM Review* 37(3): 387-405.
12. Rowe, J., Vose, M., and Wright, A. (2002). Group properties of crossover and mutation. *Evolutionary Computation*, 10(2): 151-184.
13. Rowe, J., Vose, M., and Wright, A. (to appear) Structural Search Spaces and Genetic Operators. *Evolutionary Computation*.
14. Schmitt, L. (2001) "Theory of Genetic Algorithms." *Theoretical Computer Science*, 259: 1-61.
15. Schmitt, L. (2004) "Theory of Genetic Algorithms II: Models for Genetic Operators over the String-Tensor representation of Populations and Convergence to Global Optima for Arbitrary Fitness Function under Scaling." *Theoretical Computer Science*, 310: 181-231.
16. Spears, W. (2000) "The Equilibrium and Transient Behavior of Mutation and Recombination." *Foundations of Genetic Algorithms 6 (FOGA-6)*.
17. Vose, M. (1999). The simple genetic algorithm: foundations and theory. *MIT Press, Cambridge, Massachusetts*.

Coarse Graining Selection and Mutation

Jonathan E. Rowe[1], Michael D. Vose[2], and Alden H. Wright[3]

[1] School of Computer Science, University of Birmingham
J.E.Rowe@cs.bham.ac.uk
[2] Computer Science Department, University of Tennessee
vose@cs.utk.edu
[3] Dept. of Computer Science, University of Montana
alden.wright@umontana.edu

Abstract. Coarse graining is defined in terms of a commutative diagram. Necessary and sufficient conditions are given in the continuously differentiable case. The theory is applied to linear coarse grainings arising from partitioning the population space of a simple Genetic Algorithm (GA). Cases considered include proportional selection, binary tournament selection, and mutation. A nonlinear coarse graining for ranking selection is also presented. Within the context of GAs, the primary contribution made is the introduction and illustration of a technique by which the possibility for coarse grainings may be analyzed. A secondary contribution is that a number of new coarse graining results are obtained.

1 Introduction

Managing complexity involves quotients (or some generalization thereof) if by "managing complexity" one intends to reduce complexity while simultaneously maintaining important aspects of fidelity. The following diagram is an abstraction of the general scheme being considered. In that illustration, $x \in X$ represents state and $h : X \to X$ transforms state. Complexity is managed by Ξ, which maps state into a simpler form, and by $\tilde{h}$ which has reduced complexity by virtue of transforming simplified state.

$$
\begin{array}{ccc}
x & \xrightarrow{\;h\;} & h(x) \\
\Xi \downarrow & & \downarrow \Xi \\
\Xi x & \xrightarrow{\;\tilde{h}\;} & \Xi h(x)
\end{array}
$$

Maintaining important aspects of fidelity is interpreted to mean the diagram commutes; both paths from x to $\Xi h(x)$ yield identical results. Thus Ξ can be regarded as defining what aspects of fidelity *are maintained* – if leeway exists in choosing it – or what aspects of fidelity are *capable of preservation* – if there is virtually no leeway. The reduced complexity model $\tilde{h}$ is the *quotient* of h corresponding to the *coarse graining* Ξ.

Whereas modeling h in an approximate fashion (by relaxing commutativity of the diagram) is interesting, the central question this paper is concerned with

A.H. Wright et al. (Eds.): FOGA 2005, LNCS 3469, pp. 176–191, 2005.

is whether one can do better than approximation, and if so, then how? Moreover, knowledge of what it is that can be exact may identify a useful starting point for what it is that later will be approximated or perturbed from.

This abstract framework may provide a useful context in which to consider systems comprised of large collections of components interacting with each other (and with possibly some background environment). Assuming practical limitations to exact computation of the dynamics $x, h(x), h \circ h(x), \ldots$, approximation may be the best one can do. One would like to know, however, if that was the case or whether useful quotients did exist. It is natural to ask whether the underlying components could somehow be partitioned into a collection of disjoint subsets which could be considered as units in their own right. If obtaining a description of the dynamics of the subsets – *in terms of the subsets alone* – is possible, then the original system might be coarse grained into higher level units (the subsets) having dynamics compatible with the dynamics of the original system.

This scenario will be made concrete by taking the system to be a Genetic Algorithm (GA). In that case the underlying components comprise the search space, the environment is modeled by the fitness function (which determines competition between population members), and the state space is the set of possible populations. The primary contribution made by this paper is to introduce and illustrate a technique by which the possibility for coarse grainings may be analyzed. We are concerned with analytical tools rather than establishing particular results about any specific fitness function. The potential utility of those tools is demonstrated by obtaining a number of new coarse graining results.

This paper is organized as follows. First, some conceptual examples of quotients are discussed. Second, a necessary and sufficient condition characterizing quotients is described (assuming h is continuously differentiable, X is an open subset of a finite dimensional Euclidean space, ...), followed by a reduction to special cases. Third, aspects of the theory of the Simple Genetic Algorithm (Vose, 1999) are reviewed, followed by an application of the necessary and sufficient condition (characterizing quotients) to investigate coarse grainings of selection and mutation. The paper concludes with a summary of results.

2   Conceptual Overview

A few real world examples are briefly mentioned to make the framework introduced above less abstract and to illustrate that in practice complex systems are frequently managed and understood with the aid of coarse grainings. It should be kept in mind that we must necessarily coarse grain some *model* of the real world, because the state space X and the transformation h are mathematical abstractions.

1. Modeling the motion of a body by assuming it is rigid leads to the typical coarse-graining (of that rigid model) where $\Xi(x)$ is the center of gravity.

   Examples of this sort employ coarse grainings to *transfer* the domain of analysis to a simplified setting (namely, $\tilde{h}$ acting on ΞX).

2. Conservation laws assert that the dynamics h (of some model of a physical system) is compatible with a coarse graining under which the quotient $\tilde{h}$ is the identity map. For instance, $E = mc^2$ corresponds to the coarse graining $\Xi(x) = E(x) - m(x)c^2$.

   Examples of this sort show the existence of coarse grainings may be used to *constrain* the analysis (in the original setting X) by invariants.

3. The quantum mechanics describing the hardware of a computer is usually modeled by digital logic. A familiar coarse-graining (of that gate-level digital model) is the high level gnu/linux interface seen by the C programmer.

   Examples of this sort suggest that the quotient $\tilde{h}$ may be the primary object of concern; commutativity of the coarse graining ($\Xi \circ h = \tilde{h} \circ \Xi$) may serve as a proof of correctness for the implementation h.

The quotient in the last example above is obtained only if the state transition $x \mapsto h(x)$ corresponds to a number of micro cycles which depends on x (namely, that number required for completion of the high level service/command corresponding to x). This point is made to clarify the general phenomenon that even though a desirable quotient of a system's trajectory

$$x \mapsto h(x) \mapsto h^2(x) \mapsto \cdots \tag{1}$$

might not exist (think of h as being analogous to a single micro cycle), it nevertheless could be the case that the trajectory

$$x \mapsto h^{p(x)}(x) \mapsto h^{p(h^{p(x)}(x))}(x) \mapsto \cdots \tag{2}$$

does admit useful quotients. The applications to genetic algorithms presented in sections 4 through 6, however, are limited to scenario (1) – where h corresponds to a single generation – rather than the more general situation (2).

Because models are coarse grained, an exact coarse graining (of a model) can be an *approximation* (to reality) if the model itself is an approximate one. This points to another reason why quotients are significant; they may aid in identifying tractable approximate models (i.e., models which have useful quotients). The applications to genetic algorithms presented in sections 4 through 7, however, are not concerned with approximation since the models being coarse grained are themselves exact.

3 Differentiable Coarse Graining

The following expands upon the account given at Dagstuhl (Vose, 2004). Rather than immediately beginning with Ξ, a coarse graining is instead obtained as a byproduct of a continuously differentiable map Ψ. A reason for this is to constrain the context of the general framework for coarse graining to a more specific setting wherein differential calculus may be brought to bear (most coarse graings appearing in the Evolutionary Computation literature correspond to equivalence relations obtainable as a byproduct of linear – and thus trivially differentiable – maps). The hope is that this may facilitate the *computation* of coarse grainings in some circumstances. That possibility is in fact achieved and is demonstrated in subsequent sections.

Let $\Psi : V \longrightarrow W$ be a continuously differentiable function between open subsets of finite dimensional Euclidean spaces. A *path* with respect to Ψ (called simply a path, when Ψ is understood) is a smooth function[1] $\rho : [0,1] \longrightarrow V$ such that $\Psi \circ \rho$ is constant. The path ρ is said to be *from u to v* provided $\rho(0) = u$ and $\rho(1) = v$. Let the equivalence relation $\equiv$ on V be defined by

$$u \equiv v \iff \text{there exists a path } \rho \text{ from } u \text{ to } v$$

and let $\Xi : V \longrightarrow V/\equiv$ map element v to its equivalence class $\tilde{v}$. Equivalence classes are path connected components of level sets of Ψ. It follows that the image of any path is contained in some equivalence class.

A continuously differentiable function $h : V \longrightarrow V$ is said to be *compatible with* $\equiv$ provided there exists a function $\tilde{h}$ for which the following diagram commutes,

$$
\begin{array}{ccc}
V & \xrightarrow{\ h\ } & V \\
{\scriptstyle \Xi}\downarrow & & \downarrow{\scriptstyle \Xi} \\
V/\equiv & \xrightarrow{\ \tilde{h}\ } & V/\equiv
\end{array}
$$

In that case $\tilde{h}$ is called the quotient of h (with respect to Ξ),

$$\Xi(u) = \Xi(v) \implies \Xi \circ h(u) = \Xi \circ h(v) \tag{3}$$

and $\tilde{h}$ is well defined by $\tilde{h} \circ \Xi(u) = \Xi \circ h(u)$.

By restricting h to the equivalence class $V_w = \Xi^{-1}(\tilde{w})$, compatibility of h with $\equiv$ implies that for each $w \in V$,

$$V_w \xrightarrow{\ h\ } h(V_w) \xrightarrow{\ \Psi\ } \{\Psi(h(w))\}$$

since $\Xi(u) = \Xi(v) = \tilde{w}$ for all $u, v \in V_w$, and so by (3)

$$\Xi \circ h(u) = \Xi \circ h(v) \implies h(u) \equiv h(v) \implies \Psi \circ h(u) = \Psi \circ h(v)$$

Thus on V_w, the composition of Ψ with h is the constant function

$$\widehat{h} = \Psi \circ h \tag{4}$$

Let T_w be the *tangent space* of the equivalence class V_w at w defined by

$$T_w = \mathcal{L}\{ d\rho_0(1) : \rho \text{ is a path from } w \text{ to } v, \text{ for some } v\}$$

---

[1] We call a continuous function *smooth* if it's differential (over the interior of the domain) has a continuous extension to the entire domain

where $\mathcal{L}\{\cdots\}$ denotes the linear span of $\{\cdots\}$ and, for any function f differentiable at x, the differential df_x of f at x is the linear function of z

$$df_x(z) = \lim_{t \to 0} \frac{f(x+tz) - f(x)}{t}$$

Note that $d\rho_0(1)$ occurring in the definition of T_w is the tangent at w of the path ρ from w to v,

$$d\rho_0(1) = \lim_{t \downarrow 0} \frac{\rho(t) - w}{t}$$

Assuming compatibility (of h with $\equiv$), $\widehat{h}$ is constant (on equivalence classes) and so by equation (4)

$$\Psi \circ h \circ \rho$$

is also constant, for every path ρ (since the image of a path is comprised of equivalent elements). Moreover, it follows from the chain rule that

$$d\Psi_{h \circ \rho(t)} dh_{\rho(t)} d\rho_t(1) = 0 \tag{5}$$

for $0 < t < 1$.

Therefore, choosing $x = \rho(0)$ and $t \downarrow 0$ in (5),

$$dh_x : T_x \longrightarrow K_{d\Psi_{h(x)}} \tag{6}$$

where, for any linear function L, the kernel of L is denoted by K_L.

Condition (6) is therefore *necessary* (for all x) in order that h be compatible with $\equiv$. It will also be shown sufficient. Note first, however, that because $\Psi \circ \rho$ is constant (for any path ρ), the chain rule yields

$$d\Psi_{\rho(t)} d\rho_t(1) = 0$$

Hence T_x is a subspace of $K_{d\Psi_x}$ [2]. A condition which implies (6) is therefore

$$dh_x : K_{d\Psi_x} \longrightarrow K_{d\Psi_{h(x)}}$$

Conversely, suppose (6) holds for all $x \in V$. Since the compatibility of h with $\equiv$ is implied by condition (3), suppose $u, v \in V_w$ and let ρ be a path from u to v (i.e., assume $\Xi(u) = \Xi(v)$). Define the function f by

$$f(t) = \Psi \circ h \circ \rho(t)$$

and note that

$$\Psi \circ h \circ \rho(y) - \Psi \circ h(u) = \int_{0+}^{y^-} df_t(1) \, dt = \int_{0+}^{y^-} d\Psi_{h \circ \rho(t)} dh_{\rho(t)} d\rho_t(1) \, dt$$

---

[2] They need not coincide; at an extrema or saddle point (for instance) there may exist dimensions orthogonal to the level set along which $d\Psi_x$ vanishes

If (5) held (for $0 < t < 1$), then the integral would be zero, establishing that $h \circ \rho$ is a path from $h(u)$ to $h(v)$, hence $\Xi \circ h(u) = \Xi \circ h(v)$, and thus condition (3) would hold, implying compatibility of h with $\equiv$. Choosing $x = \rho(t)$ shows that condition (5) is a consequence of (6), provided $d\rho_t(1) \in T_x$. Let ϱ be the path from x to v defined by

$$\varrho(s) = \rho(t(1-s)+s)$$

Note that $d\varrho_0(1) \in T_x$ (by definition of tangent space), and

$$
\begin{aligned}
d\rho_t(1-t) &= \lim_{s \downarrow 0} \frac{\rho(t+(1-t)s) - \rho(t)}{s} \\
&= \lim_{s \downarrow 0} \frac{\rho(t(1-s)+s) - x}{s} \\
&= \lim_{s \downarrow 0} \frac{\varrho(s) - x}{s} \\
&= d\varrho_0(1)
\end{aligned}
$$

Hence $d\rho_t(1) = d\varrho_0(1)/(1-t) \in T_x$. The following theorem has been established.

Theorem 1. *A necessary and sufficient condition for h to be compatible with $\equiv$ is that for all $x \in V$,*

$$dh_x : T_x \longrightarrow K_{d\Psi_{h(x)}}$$

Moreover, T_x is a subspace of $K_{d\Psi_x}$.

In the special case where Ψ is linear, the necessary and sufficient condition reduces to

$$dh_x : K_\Psi \longrightarrow K_\Psi$$

since a linear function is its own differential, and for $x \in K_\Psi$ the path $\rho(t) = x(1-t) + 2tx$ shows $x \in T_x$ (the tangent to ρ is x), hence $K_\Psi \subset T_x \subset K_\Psi$.

If both h and Ψ are linear, then the situation reduces to the case considered in Rowe, Vose, Wright (2004); the kernel of Ψ is an invariant subspace of h.

4   Proportional Selection + Mutation

A brief summary of relevant background (Vose, 1999) is given, followed by an application of theorem 1 to proportional selection + mutation.

Let τ denote the stochastic transition function for a finite population GA over the search space $\Omega = \{0, \ldots, n-1\}$, and let $\mathcal{G}$ be the corresponding infinite population model. The transition matrix Q of the GA's Markov chain is defined by the probability that $\tau(p) = q$ and satisfies

$$Q_{p,q} = r! \prod \frac{(\mathcal{G}(p)_j)^{rq_j}}{(rq_j)!}$$

where r is the population size, and where the the population represented by the n dimensional vector p contains $r\,p_j$ instances of j.

Let $\equiv$ be an arbitrary equivalence relation over Ω, and let $\{0^*,\ldots,(k-1)^*\}$ be equivalence class representatives. The linear operator with $k \times n$ matrix Ξ defined by

$$\Xi_{i,j} = [i^* \equiv j]$$

(where $[expression]$ denotes 1 if $expression$ is true, and 0 otherwise) lifts $\equiv$ to populations by

$$p \equiv p' \iff \Xi p = \Xi p' \tag{7}$$

Compatibility in the stochastic case generalizes the definition given in the previous section; τ is said to be $compatible$ $with$ $\equiv$ iff

$$p \equiv p' \implies \forall q \,.\, \mathrm{Prob}\,\{\tau(p) \equiv q\} = \mathrm{Prob}\,\{\tau(p') \equiv q\}$$

In that case, $\tilde{\tau}$ defined by $\tilde{\tau}(\Xi x) = \Xi \tau(x)$ is referred to as the $quotient$ of τ. It is known that $\tilde{\tau}$ exists if and only if a corresponding quotient $\tilde{\mathcal{G}}$ of $\mathcal{G}$ exists. Moreover, the transition matrix for $\tilde{\tau}$ is obtained from the formula for $Q_{p,q}$ above by replacing $\mathcal{G}$ by $\tilde{\mathcal{G}}$, p by Ξp, and q by Ξq (Vose, 1999).

The "selection + mutation" case refers to the simple GA with (proportional) fitness and mutation, but no crossover. There the infinite population model takes the form

$$\mathcal{G}(p) = \frac{Gp}{\mathbf{1}^T G p}$$

where $G = MF$ is a $n \times n$ matrix and $\mathbf{1}$ is the vector of all 1 s. Here M is a column stochastic mutation matrix ($\mathbf{1}^T M = \mathbf{1}^T$) where $M_{i,j} = \mathrm{Prob}\,\{j \text{ mutates to } i\}$, and F is a diagonal fitness matrix where $F_{i,i} = f_i$ is the fitness of i (the vector f is referred to as the fitness function). In particular, $\mathbf{1}^T G p = f^T p$. The domain of immediate interest is

$$p \in \Lambda = \{\langle x_0, \ldots, x_{n-1}\rangle \,:\, x_i \geq 0,\, \mathbf{1}^T x = 1\}$$

since that is the completion of the population representation space. Note that $\mathbf{1}^T G p$ does not vanish on Λ provided fitness is positive. Positive fitness will be assumed throughout the remainder of this paper. The results of the previous section will be applied with $h = \mathcal{G}$ and V a neighborhood of Λ.

The situation is particularly simple, since choosing $\Psi = \Xi$ yields the equivalence relation above; it follows from (7) that $p \equiv p'$ iff p and p' are contained in a level set of Ψ (i.e., a translate of K_Ξ). Moreover, the coarse graining (as defined in sections 1 and 3) is also Ξ (which is a fortunate happenstance for notation), since Ψp can be regarded as representing the equivalence class $\tilde{p}$ of p. As noted after theorem 1, compatibility reduces to

$$d\mathcal{G}_x : K_\Xi \longrightarrow K_\Xi$$

The differential of $\mathcal{G}$ at x is

$$d\mathcal{G}_x = \frac{\mathbf{1}^T G x G - G x \mathbf{1}^T G}{(\mathbf{1}^T G x)^2}$$

Compatibility is therefore the condition that for all $x \in V$, and for all $v \in K_\Xi$,

$$\Xi \mathbf{1}^T GxGv = \Xi Gx \mathbf{1}^T Gv \tag{8}$$

Assume compatibility, and let $v \in K_\Xi$. Choosing $x = e_i$, where e_i is the ith column of the $n \times n$ identity matrix (indices begin with zero) and solving (8) for ΞGv yields

$$\Xi Gv = \mathbf{1}^T Gv \Xi \frac{Ge_i}{\mathbf{1}^T Ge_i} \tag{9}$$

Replacing e_i with e_j in the right hand side of (9) – the left hand side is invariant under that replacement – and equating the right hand side before replacement with after yields

$$\mathbf{1}^T Gv \Xi \left(\frac{Ge_i}{\mathbf{1}^T Ge_i} - \frac{Ge_j}{\mathbf{1}^T Ge_j} \right) = 0 \tag{10}$$

If $\mathbf{1}^T Gv = 0$, then it follows from (9) that $\Xi Gv = 0$. Otherwise, it follows from (10) that

$$\frac{Ge_i}{\mathbf{1}^T Ge_i} \equiv \frac{Ge_j}{\mathbf{1}^T Ge_j}$$

Taking into account $G = MF$, these alternatives simplify to

$$f^T v = 0 \tag{11}$$

$$Me_i \equiv Me_j \tag{12}$$

First suppose there exist $y, z \in \Omega$ such that $y \equiv z$ and $f_y \neq f_z$. Note that $v = e_y - e_z \in K_\Xi$ and $f^T v \neq 0$. Since condition (11) does not hold, condition (12) must.

Next suppose no such y, z exist. Then all equivalent population members have identical fitness. Note that $v \in K_\Xi$ is equivalent to the condition that for all equivalence class representatives c^*,

$$\sum_{i \equiv c^*} v_i = 0 \tag{13}$$

(in particular, $v \in K_\Xi \implies \mathbf{1}^T v = 0$, i.e., $v \in \mathbf{1}^\perp$). Since fitness is constant over equivalence classes,

$$\sum_{i \equiv c^*} f_i v_i = 0$$

It follows that $F : K_\Xi \to K_\Xi$ (F is compatible with $\equiv$) and $\mathbf{1}^T Gv = f^T v = 0$ for all $v \in K_\Xi$. As observed after (10),

$$\Xi Gv = 0 \tag{14}$$

Moreover, since F is invertible (fitness is positive) quantification over $v \in K_\Xi$ is equivalent to quantification over $v \in \{F^{-1}w : w \in K_\Xi\}$ (an injective linear map on a finite dimensional space is surjective). Replacing v by $F^{-1}w$ in (14) yields

$$w \in K_\Xi \implies \Xi Mw = 0$$

Hence M is compatible with $\equiv$.

Theorem 2. *Suppose positive fitness and zero crossover. Equivalent population members have identical fitness if and only if F is compatible with Ξ. When F is compatible with Ξ, a necessary and sufficient condition for τ to be compatible with $\equiv$ is that M is. If F is not compatible with Ξ, then a necessary and sufficient condition for τ to be compatible with $\equiv$ is that the columns of M are equivalent.*

Proof. It was shown above that F is compatible (with Ξ) when all equivalent population members have identical fitness. Conversely,

$$\Xi F v = 0 \Longrightarrow 1^T \Xi F v = 1^T F v = f^T v = 0$$

so F cannot be compatible if there exists $v \in K_\Xi$ such that $f^T v \neq 0$. But if $y \equiv z$ and $f_y \neq f_z$, then $v = e_y - e_z \in K_\Xi$ and $f^T v \neq 0$.

It was shown above that if all equivalent population members have identical fitness (i.e., F is compatible), then compatibility of $\mathcal{G}$ – which is equivalent to the compatibility of τ (Vose, 1999) – implies M is compatible. Conversely, if K_Ξ is invariant under both M and F, then it is invariant under $G = MF$, and $1^T G v = 1^T \Xi G v = 0$ for all $v \in K_\Xi$. Therefore, the compatibility condition (8) reduces to the identity

$$1^T G x \Xi G v = 0$$

It was shown above that if F is not compatible with Ξ (all equivalent population members do not have identical fitness), then compatibility of τ – which is equivalent to the compatibility of τ (Vose, 1999) – implies the columns of M are equivalent (12). Conversely, if ΞM has the form $c 1^T$ (columns of M are equivalent), then $\Xi G = c f^T$ and the compatibility condition (8) reduces to the identity

$$f^T x c f^T v = c f^T x f^T v \qquad \qquad \square$$

Theorem 2 is put into sharper focus by the following result (recall that e_i is the ith column of the $n \times n$ identity matrix, indices begin with zero).

Theorem 3. *A necessary and sufficient condition for M to be compatible with $\equiv$ is that for all i, j,*

$$i \equiv j \implies M e_i \equiv M e_j$$

Proof. If $i \equiv j$, then $v = e_i - e_j \in K_\Xi$. Therefore if M is compatible, then

$$\Xi M (e_i - e_j) = 0$$

Hence $M e_i \equiv M e_j$. Conversely, if $i \equiv j \Longrightarrow M e_i \equiv M e_j$, then ΞM has the form

$$\sum_{h=0}^{k-1} C_h \sum_{l \equiv h^*} e_l^T$$

where $C_h = \Xi M e_i$ for $i \equiv h^*$ (the choice of i does not matter; $\Xi M e_i = \Xi M e_j$ when $i \equiv j$). If $v \in K_\Xi$, then

$$\Xi M v = \sum_{h=0}^{k-1} C_h \sum_{l \equiv h^*} e_l^T v = \sum_{h=0}^{k-1} C_h \sum_{l \equiv h^*} v_l = 0 \qquad \qquad \square$$

Theorem 3 provides a method by which a mutation operator can be constructed compatible with a given equivalence relation; whenever $i \equiv j$, choose columns i and j of M to differ by an element of K_Ξ. Moreover, since $K_\Xi \subset 1^\perp$, obtaining column i by adding an element $v \in K_\Xi$ to the jth column will not disturb the column stochasticity of M, provided $v + Me_j$ is non negative.

5 Binary Tournament Selection

A zero mutation, zero crossover, tournament selection GA with tournament size t and fitness function f has corresponding infinite population model (Vose, 1999)

$$\mathcal{G}(p)_i \;=\; t! \sum_{v \in X_n^t} \int_{\sum [f_j < f_i](v/t)_j}^{\sum [f_j \le f_i](v/t)_j} \varrho(y)\, dy \prod_{j<n} \frac{p_j^{v_j}}{v_j!}$$

where

$$X_n^t \;=\; \{\langle x_0, \ldots x_{n-1}\rangle : x_i \in \mathcal{Z}^{\ge 0},\, 1^T x = t\}$$

and ϱ is any continuous increasing probability density over $[0,1]$. *Binary tournament selection* refers to the result of choosing $t = 2$ and taking the limit as ϱ tends to point mass at 1. Assuming injective fitness (which will be assumed for the remainder of this paper), the result is

$$\mathcal{G}(p)_i \;=\; p_i^2 + 2p_i \sum_j p_j [f_j < f_i]$$

It follows that

$$(d\mathcal{G}_x v)_i \;=\; 2v_i x_i + 2 \sum_l [f_l < f_i](v_i x_l + x_i v_l) \tag{15}$$

Note that (15) is a symmetric expression in x and v, and therefore $d\mathcal{G}_x v = d\mathcal{G}_v x$ is linear in both x and v. In view of this, the compatibility condition is that for all $x \in V$, and for all $v \in K_\Xi$,

$$d\mathcal{G}_x v \;=\; \sum_h x_h d\mathcal{G}_{e_h} v \in K_\Xi$$

Since K_Ξ is a subspace, compatibility is therefore equivalent to the condition that for all h,

$$v \in K_\Xi \implies d\mathcal{G}_{e_h} v \in K_\Xi$$

Moreover, the ith component of the differential above simplifies (from 15) to

$$(d\mathcal{G}_{e_h} v)_i \;=\; 2v_i [f_h < f_i] + 2[h = i] \sum_l [f_l \le f_i] v_l \tag{16}$$

This equality constrains what equivalence relations are possible. Consider the case where there exist nonequivalent elements a and b. Choosing $h = a$ in (16) and applying condition (13) for membership in K_Ξ, compatibility requires the following implication

$$0 = \sum_{i \equiv b} v_i$$

$$\implies 0 = \sum_{i \equiv b} \left(v_i [f_a < f_i] + [a = i] \sum_l [f_l \le f_i] v_l \right) = \sum_{i \equiv b} v_i [f_a < f_i]$$

Therefore (by suitable choice of v), either every i equivalent to b must satisfy $[f_a < f_i]$ or else no i equivalent to b can satisfy $[f_a < f_i]$. In other words, equivalence classes are "fitness-contiguous" as defined below.

Let θ be a permutation of $\{0, \ldots, n-1\}$ such that $i < j \iff f_{\theta(i)} < f_{\theta(j)}$ and let $\equiv$ be any equivalence relation on Ω for which the equivalence classes are fitness-contiguous, meaning they are

$$\{\theta(0), \ldots, \theta(z_0)\}, \; \{\theta(z_0 + 1), \ldots, \theta(z_1)\}, \; \ldots, \; \{\theta(z_{k-2} + 1), \ldots, \theta(z_{k-1})\}$$

for some $0 \le z_0 < \cdots < z_{k-1} = n - 1$. Let the equivalence class representative of the cth class be $c^* = \theta(z_c)$. It follows that if $b < c$ then everything equivalent to b^* has fitness less than everything equivalent to c^*.

Lemma 1. *If the equivalence classes of $\equiv$ are fitness-contiguous and $v \in K_\Xi$, then for all l,*

$$\sum_{i \equiv j} [f_l \le f_i] v_i = [l \equiv j] \sum_i [f_l \le f_i] v_i$$

Proof. If $l \not\equiv j$, then f_l is either less than everything equivalent to j or else it is greater than everything equivalent to j. In the latter case, both sides above are zero. In the former case, both sides are also zero since then the left hand size vanishes due to (13). By what has been shown so far,

$$\sum_i [f_l \le f_i] v_i = \sum_{c=1}^{k-1} \sum_{i \equiv c^*} [f_l \le f_i] v_i = \sum_{c=1}^{k-1} [l \equiv c^*] \sum_{i \equiv c^*} [f_l \le f_i] v_i = \sum_{i \equiv l} [f_l \le f_i] v_i$$

which completes the proof for the remaining case $l \equiv j$.   □

Theorem 4. *Binary tournament selection is compatible with $\equiv$ if and only if the equivalence relation is fitness-contiguous.*

Proof. The "only if" part has already been established. Let $v \in K_{\equiv}$. By (16) and what has been established above, the "if" part follows from

$$\frac{1}{2}\sum_{i\equiv j}(d\mathcal{G}_{e_h}v)_i = \sum_{i\equiv j}\left(v_i[f_h < f_i] + [h = i]\sum_l[f_l \le f_i]v_l\right)$$

$$= \sum_{i\equiv j}v_i[f_h < f_i] + [h \equiv j]\sum_l[f_l \le f_h]v_l$$

$$= [h \equiv j]\left(\sum_i v_i[f_h < f_i] + \sum_l[f_l \le f_h]v_l\right)$$

$$= [h \equiv j]\sum_i v_i$$

$$= 0$$

$\square$

For "binary tournament selection + mutation" to be made compatible with a fitness-contiguous equivalence relation, mutation may be chosen as in theorem 3.

6   Ranking Selection

A zero mutation, zero crossover, ranking selection GA with parameter ϱ and fitness function f has corresponding infinite population model

$$\mathcal{G}(x)_i = \int_{\sum[f_j < f_i]x_j}^{\sum[f_j \le f_i]x_j} \varrho(y)\, dy$$

where ϱ is any continuous increasing probability density over $[0, 1]$ (Vose, 1999). Define η by

$$\eta_{\theta(0)} = 0$$

$$\eta_{\theta(i+1)} = \eta_{\theta(i)} + x_{\theta(i)}$$

(recall that $i < j \iff f_{\theta(i)} < f_{\theta(j)}$). It follows that

$$\mathcal{G}(x)_i = \varphi(x_i + \eta_i) - \varphi(\eta_i) \tag{17}$$

$$d\mathcal{G}_x v = \sum_i e_i \sum_k (\varrho(x_i + \eta_i)[f_k \le f_i] - \varrho(\eta_i)[f_k < f_i])\, v_k \tag{18}$$

where φ is an anti-derivative of ϱ (Vose, 1999). Choosing $x = e_h$, the last expression above simplifies to yield

$$(d\mathcal{G}_{e_h}v)_i = \varrho([f_i \ge f_h])v_i + [i = h](\varrho(1) - \varrho(0))\sum_k[f_k < f_i]v_k$$

Compatibility requires that for all c^*, and all v satisfying (13),

$$0 = \sum_{i\equiv c^*}\left(\varrho([f_i \ge f_h])v_i + [i = h](\varrho(1) - \varrho(0))\sum_k[f_k < f_i]v_k\right)$$

$$= \sum_{i\equiv c^*}\varrho([f_i \ge f_h])v_i + [h \equiv c^*](\varrho(1) - \varrho(0))\sum_k[f_k < f_h]v_k \tag{19}$$

Assuming the equivalence relation is nontrivial, choose $h \not\equiv c^*$ to obtain

$$0 = \sum_{i \equiv c^*} \varrho([f_i \geq f_h])v_i = \varrho(0)\sum_{i \equiv c^*}[f_i < f_h]v_i + \varrho(1)\sum_{j \equiv c^*}[f_j \geq f_h]v_j \quad (20)$$

As seen in the previous section, this implies $\equiv$ must be fitness-contiguous (it follows from $\varrho(1) > \varrho(0)$ and choosing v to have exactly two nonzero components; either every j equivalent to c must satisfy $[f_j < f_h]$ or else every i equivalent to c must satisfy $[f_i \geq f_h]$ since otherwise $v_i = -v_j \neq 0$ contradicts 20).

Theorem 5. *Ranking selection is compatible with $\equiv$ if and only if the equivalence relation is fitness-contiguous.*

Proof. The "only if" part has already been established. Let $v \in K_{\equiv}$. Appealing to (18) and the fact that $\equiv$ is fitness-contiguous, the "if" part follows from

$$\sum_{i \equiv c^*}\sum_{k}(\varrho(x_i + \eta_i)[f_k \leq f_i] - \varrho(\eta_i)[f_k < f_i])\,v_k$$

$$= \sum_{k \equiv i \equiv c^*}(\varrho(x_i + \eta_i)[f_k \leq f_i] - \varrho(\eta_i)[f_k < f_i])v_k + \sum_{k \not\equiv c^*}v_k\sum_{i \equiv c^*}\varrho(x_i + \eta_i) - \varrho(\eta_i)$$

$$= \sum_{z_{c-1} < u, v \leq z_c}(\varrho(x_{\theta(v)} + \eta_{\theta(v)})[f_{\theta(u)} \leq f_{\theta(v)}] - \varrho(\eta_{\theta(v)})[f_{\theta(u)} < f_{\theta(v)}])v_{\theta(u)}$$

$$= \sum_{z_{c-1} < u \leq z_c}v_{\theta(u)}\sum_{z_{c-1} < v \leq z_c}\varrho(\eta_{\theta(v+1)})[u \leq v] - \varrho(\eta_{\theta(v)})[u < v]$$

$$= \sum_{z_{c-1} < u \leq z_c}v_{\theta(u)}\Big(\sum_{u \leq v \leq z_c}\varrho(\eta_{\theta(v+1)}) - \sum_{u < v \leq z_c}\varrho(\eta_{\theta(v)})\Big)$$

$$= \varrho(\eta_{\theta(z_c+1)})\sum_{z_{c-1} < u \leq z_c}v_{\theta(u)}$$

$$= 0 \qquad\qquad\qquad \square$$

7 Nonlinear Coarse Graining

Applications have so far involved linear coarse grainings corresponding to an equivalence relation over Ω. A nonlinear coarse graining is derived below for ranking selection. To simplify analysis, let $\varphi(x) = x^\gamma$ (where γ is a parameter), and let m and M denote the minimal fitness and maximal fitness elements of Ω, respectively. We seek a coarse graining where Ψ is real valued, independent of γ, and depends on x_m and x_M.

The derivation of Ψ is simplified by exploiting the invariant $1 = x_M + \eta_M$, so we choose to work with $\Psi(x) = \psi(x_m, \eta_M)$ for some function ψ. Let ψ_1 and ψ_2 denote the partial derivative of ψ with respect to its first and second argument, respectively. It follows that

$$\frac{\partial \Psi}{\partial x_j} = \begin{cases} \psi_1 + \psi_2 & \text{if } j = m \\ \psi_2 & \text{if } j \neq m \text{ and } j \neq M \\ 0 & \text{if } j = M \end{cases}$$

The condition $v \in K_{d\Psi_x}$ can therefore be expressed as

$$0 = \sum_j v_j \frac{\partial \Psi}{\partial x_j} = v_m \psi_1(x_m, \eta_M) + \psi_2(x_m, \eta_M) \sum_{j \neq M} v_j \qquad (21)$$

Hence

$$\sum_{j \neq M} v_j = -v_m \psi_1(x_m, \eta_M)/\psi_2(x_m, \eta_M) \qquad (22)$$

In view of (17), and using the form of (21) with $v \leftarrow d\mathcal{G}_x v$ and $x \leftarrow \mathcal{G}(x)$, the sufficient condition for compatibility (namely, $v \in K_{d\Psi_x} \implies d\mathcal{G}_x v \in K_{d\Psi_{\mathcal{G}(x)}}$) requires

$$0 = (d\mathcal{G}_x v)_m \psi_1(z_0, z_1) + \psi_2(z_0, z_1) \sum_{j \neq M} (d\mathcal{G}_x v)_j \qquad (23)$$

where

$$z_0 = x_m^\gamma, \quad z_1 = \sum_{i \neq M} (x_i + \eta_i)^\gamma - \eta_i^\gamma$$

According to (18),

$$\frac{1}{\gamma} \sum_{j \neq M} (d\mathcal{G}_x v)_j$$

$$= \sum_{j \neq M} \sum_k ((x_j + \eta_j)^{\gamma-1}[f_k \leq f_j] - \eta_j^{\gamma-1}[f_k < f_j]) v_k$$

$$= \sum_{j \neq M} \left((x_j + \eta_j)^{\gamma-1} \sum_k [f_k \leq f_j] v_k - \eta_j^{\gamma-1} \sum_k [f_k < f_j] v_k \right)$$

$$= \eta_M^{\gamma-1} \sum_{k \neq M} v_k$$

since the sum telescopes. Combining the last expression above with (22), the compatibility condition (23) becomes

$$0 = x_m^{\gamma-1} v_m \psi_1(z_0, z_1) - \psi_2(z_0, z_1) \eta_M^{\gamma-1} v_m \psi_1(x_m, \eta_M)/\psi_2(x_m, \eta_M)$$

Since Ψ is to be independent of γ, let $\gamma \downarrow 0$ and note that $z_0 \to 1$ and $z_1 \to 1$. After simplifying and rearranging the equation above, the result as $\gamma \downarrow 0$ is

$$\frac{\psi_2(x_m, \eta_M)}{\psi_1(x_m, \eta_M)} = \frac{x_m}{\eta_M} \frac{\psi_2(1,1)}{\psi_1(1,1)} \qquad (24)$$

Focusing attention on an equivalence class – which makes x_m a function of η_M – consider the relation

$$\psi(x_m, \eta_M) = c$$

(for some constant c). Applying the implicit function theorem,

$$\frac{\psi_2}{\psi_1} = -\frac{d}{d\eta_M} x_m$$

and therefore (24) becomes the differential equation

$$\frac{d}{d\eta_M} x_m = \beta \frac{x_m}{\eta_M}$$

(for some constant β). Solving the differential equation yields

$$x_m = c\eta_M^{\beta}$$

and

$$\psi(x_m, \eta_M) = \frac{x_m}{\eta_M^{\beta}}$$

Using the invariant $1 = x_M + \eta_M$, this may be rephrased in terms of x_M by redefining ψ as

$$\psi(x_m, x_M) = \frac{x_m}{(1 - x_M)^{\beta}}$$

8   Conclusion

Coarse graining is a pervasive concept in science, but has so far not been systematically investigated within the field of Genetic Algorithms. Whereas the phrase "coarse graining" has previously been used by other researchers in connection with GAs (most notably by Chris Stephens) that use typically ascribes a different meaning to the phrase than considered here.

Previous examples of coarse grainings (in the sense used here) include the papers by Rabinovich and Wigderson, and by Muhlenbein and Voigt. Rather than considering specific fitness functions or operators (as they do), our intent is to develop methods which may discover, characterize, and elucidate general invariants of the mathematical objects by which genetic search is formalized.

The principal contribution made by this paper is the introduction and illustration of techniques which facilitate the analysis of coarse graining within the context of GAs. Most remarkable is the manner in which coarse gainings are dealt with. They are not guessed or noticed, to be pointed out and subsequently verified. Instead, they are *derived*.

The potential utility of the methods presented has been demonstrated by obtaining a number of new coarse graining results. In several cases, the coarse grainings derived were characterized as being the only ones possible (within the class of linear coarse grainings corresponding to partitions of the search space). In one case (section 7), a non linear coarse graining was computed by solving a differential equation.

Acknowledgment

We would like to thank Adam Prügel-Bennett for making comments leading toward a deeper understanding of the role played by paths in the general definition of equivalence (section 3).

References

H. Muhlenbein and H. M. Voigt: "Gene pool recombination in genetic algorithms" In *Proceedings of the Metaheuristics Inter. Conf., Norwell*, I. H. Osman and J. P. Kelly (eds). Kluwer Academic Publishers, c1995.

Y. Rabinovich and A. Wigderson: "An analysis of a simple generic algorithm" In *Proceedings of the 4th International Conference on Genetic Algorithms*, p 215–221. R. K. Belew and L. B. Booker (eds.). Morgan Kaufmann Publishers, c1991.

J. E. Rowe, M. D. Vose, A. H. Wright: "State aggregation and population dynamics in linear systems" *Artificial Life*, in press
http://www.cs.umt.edu/u/wright/pubs.htm

C. R. Stephens and H. Waelbroeck: "Schemata Evolution and Building Blocks". In *Evolutionary Computation Vol. 7 n. 2*, p 109–124. 1999

M. D. Vose: "Coarse Graining". Dagstuhl Seminar No 04081, 15.02.-20.02.04
http://www.dagstuhl.de/04081/Materials/

M. D. Vose: *The Simple Genetic Algorithm: Foundations and Theory*. Cambridge, Mass. MIT Press, c1999. ISBN: 026222058X

Perturbation Theory and the Renormalization Group in Genetic Dynamics

C.R. Stephens[1,2], A. Zamora[1], and Alden H. Wright[3]

[1] Instituto de Ciencias Nucleares, UNAM
Circuito Exterior, A. Postal 70-543, México D.F. 04510
[2] Dublin Institute for Advanced Studies
10 Burlington Road, Dublin 4, Ireland
[3] Dept. of Computer Science, University of Montana
Missoula, MT 59812, USA

Abstract. Although much progress has been made in recent years in the theory of GAs and GP, there is still a conspicuous lack of tools with which to derive systematic, approximate solutions to their dynamics. In this article we propose and study perturbation theory as a potential tool to fill this gap. We concentrate mainly on selection-mutation systems, showing different implementations of the perturbative framework, developing, for example, perturbative expansions for the eigenvalues and eigenvectors of the transition matrix. The main focus however, is on diagrammatic methods, taken from physics, where we show how approximations can be built up using a pictorial representation generated by a simple set of rules, and how the renormalization group can be used to systematically improve the perturbation theory.

1 Introduction

Although much progress has been made in recent years in furthering our theoretical understanding of Genetic Algorithms (GAs) and Genetic Programming (GP) using coarse-grained formulations (see, for instance, [1–5]), most of this progress has been either at the formal level, for instance in the derivation of exact Schema Theorems, or at the qualitative level where, for example, a deeper understanding of the role of recombination has been gained. Such coarse-grained formulations have also led to a unified theoretical framework for both GAs and GP. However, there remains a conspicuous absence of tools by which the dynamics of evolutionary algorithms (EAs) may be systematically approximated.

The Statistical Mechanics approach [6] offers one possibility but, as emphasised in [7] – *"...it is not a mechanical, procedural method. Some insight about what is important and what is inessential is required"*. Instead of passing directly to a "macroscopic" view, as is done in the statistical mechanics approach, one may wonder if any progress can be made at a more microscopic level? Common wisdom is almost uniformly pessimistic as to whether microscopic formulations can offer a way forward. In this paper we try to argue that perhaps the situation is not as bleak as it first seems, proposing perturbative methods in their various guises as a potential way forward. Of course, perturbation theory appears

A.H. Wright et al. (Eds.): FOGA 2005, LNCS 3469, pp. 192–214, 2005.

ubiquitously throughout the physical sciences, as well as in pure and applied mathematics and engineering.

In biology it has been used, for instance, by Eigen and collaborators [8] to analyse the quasi-species model. It has frequently been utilised at a formal level (see, for example [9]) to determine the stability or convergence properties of fixed points of the dynamics. More recently [10], it was used to consider the evolution of the cumulants of a mutation-selection system. The analysis there however, was restricted to a single elementary landscape. As far as we are aware, perturbative techniques have not been considered in the context of Evolutionary Computation (EC). Furthermore, they have not been considered in conjunction with the renormalization group in either biology or EC.

In this paper, as perturbative methods can be implemented in a myriad of different ways, we will give only a simple introduction to a few aspects of the general methodology. Sticking mainly to mutation-selection systems, we briefly discuss the perturbative construction of the eigenvalues and eigenvectors of the transition matrix. However, we concentrate most of our attention on generating perturbative expansions diagrammatically. This has the advantage of being transparent and intuitive, as it concentrates on constructing the different routes by which a given physical process may be realized. Although we use a one-bit system to make a concrete illustration we also consider multi-locus systems on a range of fitness landscapes in order to show that the methodology is not restricted purely to the standard "toy" models. Note that, in standard fashion, we will consider the population dynamics in the infinite population limit. However, as our main interest is in the transition matrix that determines the Markov chain that describes the dynamics, the results herein can be straightforwardly taken over to the finite population model, where a sampling of the multinomial distribution based on this transition matrix is carried out (see [11] Chapters 5 and 6 for a nice introduction to this).

2   An Introduction to Genetic Dynamics

We begin with the fundamental equations that describe the dynamics of a large class of EAs. We consider the three basic genetic operators – mutation, $\mathcal{M}$, selection, $\mathcal{F}$, and recombination, $\mathcal{R}$ – and, without loss of generality, will consider them acting in the causal order $\mathcal{MRF}$ on a population vector $\mathbf{P}(t)$, whose covariant components, $P_I(t)$, represent the probability to find an object – string, tree etc. – I at time t. For fixed length strings of length N and alleles of cardinality n, $I = i_1 \ldots i_N$ is a multi-index with $i_1 \ldots i_N \in [0, n-1]$. $\mathcal{M}$ and $\mathcal{F}$ in their turn are naturally represented as matrices, $M_I{}^J$ and $F_I{}^J$, with the latter generally being a diagonal matrix with elements proportional to $\delta_I{}^J$, where $\delta_I{}^J = 1$ for $I = J$ and 0 otherwise. In the case of proportional selection for instance, $F_I{}^J = (f_I/\bar{f}(t))\delta_I{}^J$, where f_I is the fitness of string I and $\bar{f}(t)$ the average population fitness. $M_I{}^J$ is the probability that string J mutates to string I, the matrix elements being given by $p^{d_{IJ}}(1-p)^{N-d_{IJ}}$, with d_{IJ} being the Hamming distance between strings I and J and p the mutation probability.

Mathematically, as matrices, M and F are linear machines[1] which take as input (co-/contra)-variant vectors (row/column vectors in the case of matrix algebra) and give as output (contra-/co)-variant vectors. The intuitive interpretation is that each element of these matrices acts on a component of $\mathbf{P}$ as input to give another, possibly the same, component, as output. Recombination, on the other hand, naturally takes as input a pair of strings and gives as output a string[2]. Mathematically, it is therefore represented naturally as a mixed tensor, $R_I{}^{JK}$, with two contra- and one co-variant indices, which is a linear machine that takes as input two co-variant (row) vectors and gives as output a single contra-variant (column) vector. The dynamics can then be written in the covariant form (covariant here meaning that it is written such that its transformation properties under a coordinate transformation are manifest and follow from the simple linear rule of equation (3) below)

$$P_I(t+1) = \sum_{JK} H_I{}^{JK} P_J(t) P_K(t) \tag{1}$$

where $H_I{}^{JK} = \sum_{LMN} M_I{}^L R_L{}^{MN} F_M{}^J F_N{}^K$.

The reader may wonder: Why this particular interpretation of the mathematical nature of the dynamics? The answer is that using tensors is the most natural way to represent the geometrical properties of the fundamental objects in a theory under coordinate transformations. The next question is: why are the coordinate transformation properties of interest? The answer is that the dynamics can be greatly simplified when written in the most appropriate coordinate system providing greater insight and facilitating quantitative analysis. Additionally, writing equations in covariant form ensures that any statement valid in one coordinate system will be valid in any. A coordinate transformation is understood here as a linear map between bases and is explicitly realized by a matrix Λ. One may then enquire as to what is the most appropriate basis [13]? For instance, for binary strings the standard basis in the configuration space is the δ-basis, B_δ. The δ-basis is the set of 2^N characteristic functions defined on the hypercube, C_N, embedded in $\mathbb{R}^N$ – N-dimensional Euclidean space – one function for each of the 2^N vertices of C_N. Each characteristic functions is "delta-like", having non-zero values only at the corresponding vertex of the cube. For example, the basis function at the origin is $\bar{x}_1\bar{x}_2\ldots\bar{x}_N$ and so

$$B_\delta = \{\bar{x}_1\bar{x}_2\ldots\bar{x}_N,\ \bar{x}_1\bar{x}_2\ldots x_N,\ \ldots,\ x_1x_2\ldots x_N\}. \tag{2}$$

If we restrict all the basis functions to the vertices of C_N, each x_i, $1 \le i \le N$, takes the values 0 and 1 and $\bar{x}_i \equiv e - x_i$, where e takes the value 1 at each corner of the unit cube. Arranging the basis elements in columns to form the vector $\mathbf{x}_\delta = (\bar{x}_1\ldots\bar{x}_1,\ldots,x_1\ldots x_N)^T$, one implements a transformation to a

---

[1] We will here use the language of tensor analysis. Readers unfamiliar with this may consult an introductory text such as [12]

[2] Although the output is really a pair of strings, determination of the first child completely fixes the second

new basis $\mathbf{x}_{\delta'}$ via $\mathbf{x}_{\delta'} = \Lambda \mathbf{x}_{\delta}$. A tensor $T_{I_1 \ldots I_r}{}^{J_1 \ldots J_s}$ transforms under a basis transformation between δ and δ' as

$$T_{I'_1 \ldots I'_r}{}^{J'_1 \ldots J'_s} = \sum_{\substack{I_1 \ldots I_r \\ J_1 \ldots J_s}} \Lambda_{I'_1}{}^{I_1} \ldots \Lambda_{I'_r}{}^{I_r} T_{I_1 \ldots I_r}{}^{J_1 \ldots J_s} (\Lambda^{-1})_{J_1}{}^{J'_1} \ldots (\Lambda^{-1})_{J_s}{}^{J'_s} \tag{3}$$

Further insight into the dynamics can be obtained by explicitly subtracting out the linear "cloning" term from the recombination operator to obtain

$$P_I(t+1) = \sum_J M_I{}^J \left((1 - p_c)P'_J + p_c \sum_{KLm} \frac{1}{2}(p(m) + p(\bar{m}))\lambda_J{}^{KL}(m)P'_K P'_L \right) \tag{4}$$

where $P'_I(t) = \sum_J F_I{}^J P_J(t)$ is the probability to select string I, p_c is the probability that recombination takes place and $p(m)$ is the probability to implement the recombination mask m, $\bar{m}$ denoting the conjugate mask. Finally, $\lambda_I{}^{JK}(m)$ is an interaction term between strings I, J and K and represents the conditional probability that, given the selection of parent strings J and K, a child string of type I is produced when recombination is implemented using a mask m. It takes values 0 and 1. Equation (4) has a straightforward intuitive interpretation. The first term in brackets represents the probability that a string is "cloned", while the second term represents the probability that a string is created via recombination. An analogous functional form also holds for the case of GP [2, 14].

Despite the covariance of (1), the facility of its analysis and its physical interpretation are basis-dependent. The dynamics is governed by the mutation matrix $M_I{}^J$, the tensor $\lambda_I{}^{JK}(m)$, the mask probability distribution $p(m)$ and the fitness values f_I, hidden inside P'_I or $F_I{}^J$. In this sense the EA is a "black box" whose output depends on a large set of parameters. It therefore behoves us to look for symmetries and regularities that may be exploited in order to effect a coarse graining which makes manifest the effective degrees of freedom of the dynamics in terms of which the dynamics looks simplest. However, this in its turn depends on choosing an appropriate coordinate system wherein a particular regularity is more clearly seen. For instance, in a selection dominated regime, the string basis is the most appropriate one, as the selection matrix F is diagonal in this basis, i.e. the strings themselves are the appropriate effective degrees of freedom. However, when mutation is the dominant operator, a basis transformation to the Walsh basis, $\hat{\mathbf{x}}$, using the transformation matrix

$$\Lambda^w \equiv 2^{-N/2} \begin{pmatrix} 1 & 1 \\ 1 & -1 \end{pmatrix}^{\otimes N} \tag{5}$$

is useful, where $\otimes N$ is the Nth tensor power of the matrix. The power of the Walsh transform is that it diagonalizes the mutation matrix M so that its matrix elements are $(1 - 2p)^{|I|}\delta_I{}^J$, $|I|$ being the order of the Walsh mode I. Similarly, when recombination is the dominant operator a basis transformation to the Building Block or monomial basis [13, 15], $\mathbf{x}_{BB}$, is appropriate using

$$\Lambda^{BB} \equiv \begin{pmatrix} 1 & 1 \\ 0 & 1 \end{pmatrix}^{\otimes N} \tag{6}$$

The advantage of this transformation is that the tensor $\lambda_I{}^{JK}$ becomes skew-diagonal on the indices J and K for any string I [13, 15], thus showing that recombination builds strings by explicitly combining the Building Blocks of that string.

3   Exact Solutions of the Dynamics

As the key element of a perturbative approach is the development of a power series expansion around a known exact limit, it is important to have a good understanding of the different limits in which an exact solution for the dynamics may be found. There are no known exact solutions of equation (4) in the presence of all three genetic operators[3]. However, solutions in the absence of one or more of the operators may be found[4].

3.1   Explicit Solutions

Selection Only. In the case of selection only the evolution equations (4) are un-coupled and essentially linear in the variables $P_I(t)$, the apparent non-linearity in $\bar{f}(t)$ having as its origin nothing more than the normalisation of the probabilities, $\sum_I P_I(t) = 1$. Passing to unnormalised variables $x_I(t)$, defined via

$$P_I(t) = \frac{x_I(t)}{\sum_I x_I(t)} \tag{7}$$

leads to an explicit solution

$$P_I(t) = \frac{f_I^t P_I(0)}{\sum_I f_I^t P_I(0)} \tag{8}$$

In general, an exponential number of fitness values must be specified. However, in many cases the map will be many-to-one and the phenotypic dynamics may simplify accordingly. The fixed point of (8) is $P_I^* = \lim_{t\to\infty} P_I(t) \to 1 \iff f_I > f_J \ \forall \ J$ such that $P_J(0) \neq 0$. Note that this fixed point depends on the initial conditions and hence is not universal. In the case where all strings are represented however, the fixed point is the global maximum of the fitness landscape in the case where this maximum is unique.

Mutation Only. In the case of mutation only, the equations (4) remain linear, but are coupled in the string basis. Passing to the Walsh basis using the basis transformation (5) one finds the solution

$$\hat{P}_I(t) = (1 - 2p)^{|I|t} \hat{P}_I(0) \tag{9}$$

each eigenvalue denoted by $|I|$ being associated with ${}^N C_{|I|}$ degenerate eigenvectors. Thus, just as the exact solution for selection only is diagonal in the string

---

[3] Note however, that exact solutions may be found [16] for the case of modified recombination operators, such as genepool recombination, and certain specific fitness landscapes, such as functions of unitation

[4] A more leisurely derivation of many of the results in this section can be found in [11]

basis the exact solution for mutation only is diagonal in the Walsh basis. The fixed point of the dynamics is given by $\lim_{t\to\infty} \hat{P}_I(t) = 2^{-N/2}\delta_I{}^0$ and corresponds to the centre of the simplex, i.e. equal proportions of every genotype. The bigger is $|I|$, the faster the decay of the associated transient to the fixed point.

Recombination Only. Finally, in the case of recombination only, although an exact solution is not known for discrete time and arbitrary crossover, a solution is known in the continuous time limit for one-point crossover [3]. The solution is

$$P_I(t) = \sum_{n=0}^{N-1} e^{\frac{-np_c}{N-1}}(1 - e^{\frac{-p_c}{N-1}})^{N-n}\mathcal{P}(n+1) \tag{10}$$

where $\mathcal{P}(n+1) = \sum_i \Pi_{n_i=1}^{N-n} P_{n_i}(0)$. Each $P_{n_i}(0)$ is the initial probability for the Building Block n_i which crossover could combine to give genotype I. The product is over the different numbers, n_i, of Building Blocks and the sum is over the different possible permutations for a given number. For example, for $N = 3$, for $I = 111$ $\mathcal{P}(1) = P_{1**}(0)P_{*1*}(0)P_{**1}(0)$, $\mathcal{P}(2) = P_{11*}(0)P_{**1}(0) + P_{1**}(0)P_{*11}(0)$ (two permutations) and $\mathcal{P}(3) = P_{111}(0)$.

3.2   Formal Solutions

Above we considered explicit exact solutions. One can also get useful information by considering formal, or implicit, exact solutions. An example of this is the case of mutation and selection where the problem is linear and so the trick of passing to unnormalised variables, $x_I(t)$, remains valid. In this setting the equation $x_I(t+1) = \sum_K W_I{}^K x_K(t)$, where the matrix W has elements $W_I{}^K = \sum_J M_I{}^J f_J \delta_J{}^K$, can be simply iterated to obtain the formal solution

$$P_I(t) = \frac{\sum_J (W^t)_I{}^J P_J(0)}{\sum_{IJ}(W^t)_I{}^J P_J(0)} \tag{11}$$

The solution is formal in that (W^t) is the t-th power of an exponentially large matrix. If W can be diagonalized via a similarity transformation, which we assume, then we may interpret this as a basis transformation $\tilde{\mathbf{x}} = \tilde{A}\mathbf{x}_\delta$, where the $\tilde{\mathbf{x}}$ are the normalised eigenvectors of W. Under this transformation $P_I(t) \to \tilde{P}_I(t) = \sum_J \tilde{A}_I{}^J P_J(t)$ and $W \to \tilde{W}$, where $\tilde{W}$ is diagonal with elements $\lambda_I \delta_I{}^J$ and λ_I is the eigenvalue corresponding to eigenvector I. One thus finds

$$\tilde{P}_I(t) = \frac{\lambda_I^t \tilde{P}_I(0)}{\sum_I \lambda_I^t \tilde{P}_I(0)} \tag{12}$$

The general solution in the original string basis can be found by inverting the basis transformation using $\tilde{A}$ to find

$$P_I(t) = \frac{\sum_{JK} \tilde{A}_I{}^J \lambda_J^t (\tilde{A}^{-1})_J{}^K P_K(0)}{\sum_{IK} \lambda_I^t (\tilde{A}^{-1})_I{}^K P_K(0)} \tag{13}$$

Note the functional form as a sum of exponentials, where, as W has only positive entries, at least the biggest eigenvalue is positive. For example, for one-bit

$$P_0(t) = N\left(A_{00}\lambda_0^t + A_{01}\lambda_1^t\right) \tag{14}$$

where $A_{00} = (\tilde{A}_0{}^0(\tilde{A}^{-1})_0{}^0 P_0(0) + \tilde{A}_0{}^0(\tilde{A}^{-1})_0{}^1 P_1(0))$ and $A_{01} = (\tilde{A}_0{}^1(\tilde{A}^{-1})_1{}^0 P_0(0) + \tilde{A}_0{}^1(\tilde{A}^{-1})_1{}^1 P_1(0))$ are the amplitudes of the different exponents and where $N = \sum_{IK} \lambda_I^t(\tilde{A}^{-1})_I{}^K P_K(0)$ is a normalisation constant. The asymptotic behaviour is dominated by the largest eigenvalue, $\lambda_{\max}$, associated with a corresponding eigenvector $\tilde{x}_{\max}$. The corresponding component of $\mathbf{P}$ in this basis is $\tilde{P}_{\max}^*$. In terms of the original string basis the fixed point is $P_I^* = \tilde{A}_I{}^{\max}$, independent of the initial population.

4 Perturbation Theory

Perturbation theory is an ubiquitous tool in the physical sciences. However, in all its guises its conceptual basis is the same – finding approximate solutions as power series expansions with respect to a "small" parameter, ϵ, around a known solution. Conceptually, the methodology is simple. In the context at hand one writes P_I (or the unnormalised variable x_I) as a power series in ϵ

$$P_I(t) = \sum_{n=0}^{\infty} \epsilon^n P_I^{(n)}(t) \tag{15}$$

where the expansion coefficients $P_I^{(n)}$ are to be determined. One assumes that the operator $H_I{}^{JK}$ can be written in the form $H_I{}^{JK} = D_I{}^{JK} + \epsilon O_I{}^{JK}$, where $O_I{}^{JK}$ is the perturbation operator and the solution of $P_I(t+1) = \sum_{JK} D_I{}^{JK} P_J(t) P_K(t)$ is known. One subsequently substitutes the ansatz (15) into equation (1) and matches powers of ϵ^n from both sides of the equation. For instance, to $O(1)$ and $O(\epsilon)$ one finds

$$P_I^{(0)}(t+1) = \sum_{JK} D_I{}^{JK} P_J^{(0)}(t) P_K^{(0)}(t) \tag{16}$$

$$P_I^{(1)}(t+1) = \sum_{JK} \left(D_I{}^{JK} P_J^{(1)}(t) P_K^{(0)}(t) + \right.$$
$$\left. D_I{}^{JK} P_J^{(1)}(t) P_K^{(0)}(t) + O_I{}^{JK} P_J^{(0)}(t) P_K^{(0)}(t)\right) \tag{17}$$

The solution of (16) is assumed known. Once $P_I^{(0)}$ has been determined then equation (17) is a linear inhomogeneous difference equation for $P_I^{(1)}$ where the inhomogeneity is a known function of $P_I^{(0)}$. This equation can be solved using as initial condition $P_I^{(1)}(0) = 0$ [5]. The solution to $O(\epsilon)$ is thus $P_I(t) = P_I^{(0)}(t) + \epsilon P_I^{(1)}(t) + O(\epsilon^2)$. The formal expansion parameter ϵ can now be put to one[6].

---

[5] We can naturally set $P_I^{(n)}(0) = 0 \ \forall \ n \neq 0$. This is intuitive, in that ϵ gauges the effect of the perturbation which perturbs the initial population *after* $t = 0$

[6] ϵ is only taken to be small in a formal sense here in order to generate systematic power series expansions. Physically, the relevant small parameter for mutation-selection systems is the mutation rate, or the deviation from a flat fitness landscape, and it is these parameters that will govern the accuracy of the approximation

5 Perturbation Theory for Mutation-Selection Systems

To illustrate the general methodology we restrict attention to the case of mutation and selection. This problem is, in principle, straightforward, requiring only the eigenvalues and eigenvectors of the matrix MF. However, computationally this is extremely difficult for large matrices.

First, we transform to the unnormalised variables defined in (7), $x_I(t)$, remembering that we can consider them in either the string or the Walsh basis. The equation to be solved is

$$x_I(t+1) = \sum_{JK} M_I{}^J F_J{}^K x_K(t) \tag{18}$$

where, without change of notation, we now take F to have elements $F_I{}^J = f_I \delta_I{}^J$, the scalar $\bar{f}(t)$ having been removed by the change to unnormalised variables. The idea now is to solve this approximately by some perturbative expansion around some known exact limit. From section 3, two natural limits are the limits $M_I{}^J \to \delta_I{}^J$ and $F_I{}^J \to \delta_I{}^J$, associated with zero mutation and zero selection gradient respectively. In this case one writes

$$M_I{}^J = (\delta_I{}^J + \epsilon dM_I{}^J) \tag{19}$$

$$F_I{}^J = (\delta_I{}^J + \epsilon dF_I{}^J) \tag{20}$$

where dM and dF are the perturbation operators and contain the deviations of M and F from the unit matrix. Thus, in the case of selection we are using dF to measure deviations from a constant fitness value, which we take to be one. For example, for one bit, in the string basis the deviations are given by

$$dM = \begin{pmatrix} -p & p \\ p & -p \end{pmatrix} \qquad \text{and} \qquad dF = \begin{pmatrix} f_0 & 0 \\ 0 & f_1 \end{pmatrix} \tag{21}$$

with f_0 and f_1 measuring deviations from flat fitness, while in the Walsh basis

$$\widehat{dM} = \begin{pmatrix} 0 & 0 \\ 0 & -2p \end{pmatrix} \qquad \text{and} \qquad \widehat{dF} = \frac{1}{2}\begin{pmatrix} (f_0 + f_1) & (f_0 - f_1) \\ (f_0 - f_1) & (f_0 + f_1) \end{pmatrix} \tag{22}$$

Alternatively, given that (18) is exactly solvable when $W = MF$ is any diagonal matrix, we could divide W into a diagonal part, D, and an off-diagonal part, O, and write $W = D + \epsilon O$.

5.1 Perturbative Construction of Eigenvalues and Eigenvectors

There are several alternatives for constructing a perturbation theory depending on what quantities one wishes to construct. In EC the string proportions, $P_I(t)$, are of direct interest. Hence, it is natural to implement a formalism that focuses directly on them. However, there is another implementation that focuses more on the perturbative construction of the eigenvalues and eigenvectors of W.

We assume, as in section 3.2, that W can be diagonalized via a basis transformation $W \to \tilde{W} = \tilde{\Lambda} W \tilde{\Lambda}^{-1}$. In distinction to section 3.2 though, where it was assumed that $\tilde{\Lambda}$ could be determined exactly, we will here construct the transformation perturbatively. As the eigenfunctions of the unperturbed problem form a complete set of basis functions – string or Walsh basis functions – one may consider the basis transformation $\tilde{\mathbf{x}} = \tilde{\Lambda}\mathbf{x}$ as an expansion of the exact eigenfunctions of W in terms of the unperturbed ones (i.e. of M or F alone), where $\mathbf{x}$ will refer to the unperturbed eigenfunctions. One now seeks perturbative solutions by writing power series expansions for the eigenvalues, λ_i and the expansion coefficients, $\tilde{\Lambda}_i^{\ j}$

$$\lambda_i = \sum_{n=0}^{\infty} \epsilon^n \lambda_i^{(n)}; \qquad \tilde{\Lambda}_i^{\ i} = \sum_{n=0}^{\infty} \epsilon^n \tilde{\Lambda}_i^{\ i(n)}; \qquad \tilde{\Lambda}_i^{\ j} = \sum_{n=1}^{\infty} \epsilon^n \tilde{\Lambda}_i^{\ j(n)} \qquad (23)$$

where we are using lower case letters i and j to index the eigenvectors and eigenvalues. Note that the expansion of the non-diagonal elements of $\tilde{\Lambda}$ starts at $O(\epsilon)$, in distinction to the diagonal ones. This recognises the fact that only the presence of the perturbation can induce such non-diagonal terms. In the basis where $\tilde{W}$ is diagonal, an eigenvector $\tilde{\mathbf{x}}^i$ with components $\tilde{x}_I^i$ is a solution of

$$\sum_J \tilde{W}_I^{\ J} \tilde{x}_J^i = \lambda_i \tilde{x}_I^i \qquad (24)$$

Substituting the ansatz (23) into (24), matching coefficients of ϵ^n and using the fact that the unperturbed eigenfunctions are orthogonal, i.e. $\sum_J x_i^J x^j_{\ J} = 0$ for $i \neq j$, one finds to $O(\epsilon)$

$$\lambda_i = \lambda_i^{(0)} + \epsilon \sum_{JK} x_i^J O_J^{\ K} x_K^i \qquad (25)$$

where O is the perturbation operator. To be more concrete, consider the example of one-bit with perturbation operator $O_I^{\ J} = dF_I^{\ J}$. In this case it is appropriate to work in the Walsh basis using equation (22). In this basis, as $\hat{M}$ is diagonal, the unperturbed eigenvalues, $\lambda_+ = 1$ and $\lambda_- = (1 - 2p)$, can be read off directly from it. The corresponding eigenvectors are $x_+ = (1\ 0)^T$ and $x_- = (0\ 1)^T$. The $O(\epsilon)$ contribution to λ_+, $\lambda_+^{(1)}$, is

$$\lambda_+^{(1)} = \frac{1}{2}(1\ 0) \begin{pmatrix} (f_0 + f_1) & (f_0 - f_1) \\ (1 - 2p)(f_0 + f_1) & (1 - 2p)(f_0 - f_1) \end{pmatrix} \begin{pmatrix} 1 \\ 0 \end{pmatrix} = \frac{(f_0 + f_1)}{2} \qquad (26)$$

where, once again, f_0 and f_1 refer to deviations from flat fitness. The analogous expression for λ_- is found by substituting $(1\ 0)$ for $(0\ 1)$ in (26) to find $\lambda_-^{(1)} = (1 - 2p)(f_0 + f_1)/2$. Thus, to $O(\epsilon)$ the two eigenvalues are

$$\lambda_+ = \left(1 + \frac{\epsilon(f_0 + f_1)}{2}\right) \qquad (27)$$

$$\lambda_- = (1 - 2p)\left(1 + \frac{\epsilon(f_0 + f_1)}{2}\right) \qquad (28)$$

In order to construct a solution of the form (13), as well as the eigenvalues we also need the basis transformation matrix $\tilde{\Lambda}$ that relates the exact basis to the string or Walsh basis. The columns of the transformation matrix are, in fact, just the eigenvectors of W. Hence, a perturbative calculation of the eigenvectors is equivalent to an expansion of the elements of $\tilde{\Lambda}$. For our example one-bit case, as we are working in the Walsh basis, it is the eigenvectors of $\hat{W}$. Explicitly, for the coefficients of the transformation between unperturbed and perturbed eigenstates, to $O(\epsilon)$ one finds $\tilde{\Lambda}_i^{i(1)} = 0$ and for $j \neq i$

$$\tilde{\Lambda}_i^{j(1)} = \frac{\sum_{IJ} x_i^I O_I^J x_J^j}{(\lambda_i^{(0)} - \lambda_j^{(0)})} \tag{29}$$

For one bit, for the case $\hat{O}_I^J = \widehat{dF}_I^J$

$$\tilde{\Lambda}_+^{-(1)} = \frac{1}{4p}(1\ 0)\begin{pmatrix} (f_0 + f_1) & (f_0 - f_1) \\ (1-2p)(f_0 + f_1) & (1-2p)(f_0 - f_1) \end{pmatrix}\begin{pmatrix} 0 \\ 1 \end{pmatrix} = \frac{(f_0 - f_1)}{4p} \tag{30}$$

With the expansion coefficients in hand the exact eigenvectors $\tilde{x}$ may be calculated, which are then used to compute the basis transformation matrix $\tilde{\Lambda}$. As seen in section 3.2 it is in fact this matrix which provides important information, such as the fixed points of the dynamics, the eigenvalues merely governing the approach to the fixed point.

5.2   Diagrammatic Perturbative Construction of P_I

Although conceptually straightforward and well known, the above methodology for calculating eigenvalues and eigenvectors is complicated to implement beyond leading order, especially in terms of calculating the expansions of the eigenvectors, and these are essential if one wishes to construct expressions for the $P_I(t)$ and, in particular, if the asymptotic behaviour in the vicinity of any fixed point is required. Additionally, when there are several eigenvectors that correspond to the same eigenvalue, orthogonal combinations of the associated eigenvectors must be found. We thus consider now how to calculate the $x_I(t)$ directly. Initially, we will consider a general fitness landscape and arbitrary string length and population, as a great deal of useful information can be gleaned from the general case without having to specialise to a particular problem.

We will illustrate the methodology in the context of an expansion around the no selection limit (the corresponding expansion around zero mutation is very similar). In this case it is appropriate to first do a coordinate transformation to the Walsh basis. In the Walsh basis, the solution of the unperturbed (i.e. no selection and no crossover) system is

$$\hat{x}_I(t) = (1 - 2p)^{|I|t}\hat{x}_I(0) \tag{31}$$

One can interpret (31) and, in particular, a factor $(1 - 2p)^{|I|(t-t')}$ as describing the propagation in time, between t' and t, of an elementary "excitation" of

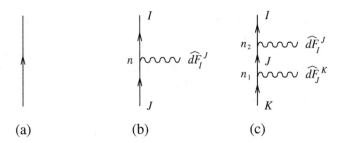

Fig. 1. Diagrammatic representation of the $O(1)$, $O(\epsilon)$ and $O(\epsilon^2)$ perturbative terms [(a), (b) and (c) respectively]

type I^7 which can be represented diagrammatically as a straight line, as shown in Figure 1(a). As $(1 - 2p) < 1$ the excitation decays exponentially, the rate of decay depending on the order of the Walsh coefficient. The only excitation that does not decay is the zeroth order one which corresponds to the uniform population limit in the string basis. The presence of the perturbation, in this case selection, can be interpreted as an interaction between the excitation and some external operator and can be represented diagrammatically by a wavy line as shown in Figure 1(b). These diagrams are a simple, intuitive mnemonic for the algebraic expression

$$\hat{x}_I(t) = (1 - 2p)^{|I|t}\hat{x}_I(0) + \epsilon \sum_{J} \sum_{n=0}^{t-1}(1 - 2p)^{|I|(t-n)}\widehat{dF}_I^{\,J}(1 - 2p)^{|J|n}\hat{x}_J(0), \quad (32)$$

The two terms represent the different physical processes that can contribute to $O(\epsilon)$ to the appearance of a Walsh mode I at time t. The first term, corresponding to Figure1(a), represents the process where the mode I was present at $t = 0$ and propagates forward to t. The second term, corresponding to Figure 1(b) however, represents the probability that it is produced by first starting with a mode J at $t = 0$, which then propagates to time $t = n$. At $t = n$ it interacts with the Walsh-transformed perturbation selection operator, $\widehat{dF}_I^{\,J}$ to produce the mode I, which then propagates from $t = n$ to t. $\sum_J$ represents the fact that one must consider all possible initial starting states as potential contributions, while $\sum_{n=0}^{t-1}$ represents the possibility that the interaction may take place at any one of the t time steps of the evolution.

One may sum the second term (32) to find

$$\hat{x}_I(t) = (1-2p)^{|I|t}\hat{x}_I(0)+\epsilon(1-2p)^{|I|t}\sum_{J}\frac{\left(1 - (1 - 2p)^{(|J|-|I|)t}\right)}{1 - (1 - 2p)^{(|J|-|I|)}}\widehat{dF}_I^{\,J}\hat{x}_J(0) \quad (33)$$

There are three distinct cases to take into account: $|I| < |J|$, $|I| > |J|$ and $|I| = |J|$. In the first case the contribution from the corresponding interactions

---

[7] In the Walsh basis this excitation is analogous to a normal mode, while in the string basis these elementary excitations are obviously the strings themselves

only modifies the amplitude of the zeroth-order transient behaviour, associated with $(1 - 2p)^{|I|t}$, from 1 to $1 + \sum_{|J|>|I|} \widehat{dF}_I^{\ J} \hat{x}_J(0)/(1 - (1 - 2p)^{(|J|-|I|)})$, where the sum is over those Walsh modes for which $|J| > |I|$. When $|I| > |J|$ the contribution from the interaction dominates, leading to a decay for $\hat{x}_I(t)$ of the form $(1 - 2p)^{|J|t}$, which is slower than $(1 - 2p)^{|I|t}$. Finally, in the limit $|J| \to |I|$ there is an apparent singularity in the dynamical factor. However, $\lim_{|J| \to |I|} (1 - (1 - 2p)^{(|J|-|I|)t})/(1 - (1 - 2p)^{(|J|-|I|)}) = t$. This term is the analog of a secular term as found in perturbative solutions of ordinary differential equations [17]. At first glance it invalidates perturbation theory, as it leads to a linearly growing perturbation in time. However, as this term is suppressed by the exponential decay, $(1 - 2p)^{|I|t}$, it does not affect the value found for the fixed point except, at first glance, for the zeroth mode. That this is not a problem can be seen by returning to the normalised variables P_I via equation (7). First we pass to the normalised Walsh variables, $\hat{P}_I$, which are related to the $\hat{x}_I$ via

$$\hat{P}_I(t) = \frac{1}{2^{\frac{N}{2}}} \frac{\hat{x}_I(t)}{\hat{x}_0(t)} \tag{34}$$

with $\hat{P}_0(t) = 1/2^{N/2}$ just the Walsh transformed constraint $\sum_I P_I(t) = 1$. Substituting (33) into (34) and expanding the denominator in ϵ one finds to $O(\epsilon)$

$$\hat{P}_I(t) = (1 - 2p)^{|I|t} \hat{P}_I(0) + \epsilon(1 - 2p)^{|I|t} \sum_J \frac{(1 - (1 - 2p)^{(|J|-|I|)t})}{1 - (1 - 2p)^{(|J|-|I|)}} \widehat{dF}_I^{\ J} \hat{P}_J(0)$$

$$- \epsilon(1 - 2p)^{|I|t} 2^{\frac{N}{2}} \hat{P}_I(0) \sum_J \frac{(1 - (1 - 2p)^{|J|t})}{1 - (1 - 2p)^{|J|}} \widehat{dF}_0^{\ J} \hat{P}_J(0) \tag{35}$$

Taking the limit $t \to \infty$ one finds the fixed point for $|I| \neq 0$

$$\hat{P}_I^* = \frac{\epsilon}{2^{\frac{N}{2}}} \frac{\widehat{dF}_I^{\ 0}}{((1 - 2p)^{-|I|} - 1)} \tag{36}$$

Thus, we see that the fixed point associated with the centre of the simplex is modified by selection and is independent of the initial conditions. This is intuitive given that the non-zero modes are associated with exponentially decaying excitations. We also see that the biggest contribution to the asymptotic behaviour will come from the most important Walsh components of the fitness landscape. For instance, for a unitation type landscape only the $O(1)$ Walsh coefficients of the landscape are non-zero and hence $\hat{P}_I^* = 0$ for $|I| \neq 1$.

To $O(\epsilon^2)$ the corresponding diagram is Figure 1(c) and represents the production of a Walsh mode I by starting with a Walsh mode K which propagates to $t = n_1$, interacts with the perturbation selection operator $\widehat{dF}_J^{\ K}$ to produce a Walsh mode J, which in its turn propagates from $t = n_1$ to $t = n_2$. This Walsh mode then interacts with the perturbation selection operator $\widehat{dF}_I^{\ J}$ at $t = n_2$ to produce a Walsh mode I, which finally propagates from $t = n_2$ to t. The corresponding algebraic expression for this second order process is

$$\epsilon^2 \sum_{JK} \sum_{n_2=1}^{t-1} \sum_{n_1=0}^{n_2-1} (1-2p)^{|I|(t-n_2)} \widehat{dF}_I^{\ J} (1-2p)^{|J|(n_2-n_1)} \widehat{dF}_J^{\ K} (1-2p)^{|K|n_1} \hat{x}_K(0)$$

$$(37)$$

Once again, one must sum over different possible initial and intermediate states, J and K, and sum over the different possibilities for the times at which the excitations interact with the selection operator. Note that causally the second interaction with the selection operator must come *after* the first one, hence the sum over n_2 begins at $t = 1$ not $t = 0$. Evaluating (37) and adding to (32) one obtains to second order

$$\hat{x}_I(t) = (1-2p)^{|I|t} \hat{x}_I(0) + \epsilon(1-2p)^{|I|t} \sum_J \frac{\left(1 - (1-2p)^{(|J|-|I|)t}\right)}{1 - (1-2p)^{(|J|-|I|)}} \widehat{dF}_I^{\ J} \hat{x}_J(0)$$

$$+ \epsilon^2 (1-2p)^{|I|t} \sum_{JK} \left(\frac{(1-2p)^{(|J|-|I|)}(1 - (1-2p)^{(|J|-|I|)(t-1)})}{(1 - (1-2p)^{(|K|-|J|)})(1 - (1-2p)^{(|J|-|I|)})} \right.$$

$$\left. - \frac{(1-2p)^{(|K|-|I|)}(1 - (1-2p)^{(|K|-|I|)(t-1)})}{(1 - (1-2p)^{(|K|-|J|)})(1 - (1-2p)^{(|K|-|I|)})} \right) \widehat{dF}_I^{\ J} \widehat{dF}_J^{\ K} \hat{x}_K(0)$$

$$(38)$$

As at $O(\epsilon)$, one must take care over the limits $|I| = |J|$, $|J| = |K|$, $|I| = |K|$ or $|I| = |J| = |K|$. Note that this expression is valid for an arbitrary fitness landscape as long as the selection pressure is weak. To get back to the probabilities $P_I(t)$ from the $\hat{x}_I(t)$ is straightforward. One first passes to the variables $\hat{P}_I(t)$ using equation (34). One is then faced with a choice – to expand the denominator, $\hat{x}_0$, as a power series in ϵ into the numerator, or to evaluate it numerically without this last expansion. Schematically, it is the difference between writing at $O(\epsilon)$: $\hat{P}_I = (1/2^{N/2})(\hat{x}_I/\hat{x}_0) = (a_I + \epsilon b_I)(a_0 + \epsilon b_0)^{-1}$, where both numerator and denominator are now evaluated numerically for a given landscape, or writing $(a_0 + \epsilon b_0)^{-1} \approx (a_0 - \epsilon b_0)$ and *then* evaluating the expression numerically. The true spirit of perturbation theory is to do the latter and we shall follow that procedure here. However, under certain circumstances it is possible to envision the former. Finally, one passes to the $P_I(t)$ using the inverse Walsh transform from (5).

This diagrammatic formulation gives a powerful pictorial representation of the underlying problem, wherein the different diagrams represent the different ways in which a process may occur – for instance, production of a particular string. The problem then may be turned around to be associated with the specification of the *rules*[8] by which the diagrams that represent the different possibilities may be constructed. In fact, one may take these rules as being a definition of the theory, as their particular form depends on the theory in question, e.g. selection only, selection and mutation etc.. In the case at hand, for mutation and selection with an expansion around the zero selection limit, the rules for constructing a solution to a given perturbative order are:

1. Draw all possible topologically distinct diagrams contributing to the process under study to the desired perturbative order

---

[8] In physics, in quantum field theory, these rules are known as Feynman rules

2. To each internal line attach a propagator $(1 - 2p)^{|I|(t-t')}$

3. At each interaction vertex insert a factor $\widehat{\epsilon dF}_I^{\ J}$

4. Sum over all internal times associated with the interaction vertices

5. Sum over intermediate states on internal lines

These rules are also valid when expanding around the zero mutation limit if the propagator is replaced by $f_I^{(t-t')}$ and the vertex factor by $\widehat{\epsilon dM}_I^{\ J}$. Note that in calculating $x_I(t)$ we summed over all possible initial states as we were interested ultimately in $P_I(t)$ not in the conditional probability $P(I, t|J, t')$. One may, of course, always revert to the algebraic formulation if there is any doubt or ambiguity over constructing the diagrams and their associated algebraic expressions.

In order to give a feeling for the capabilities of the method we present some results for the case of $N = 7$ and with various fitness landscapes. We use a mutation probability $p = 0.1$ which, it is worth noting given that we are formally in the "high" mutation perturbation limit, is not much different from typically used rates. The three fitness landscapes we consider are representative of different classes of landscape – consisting of the Eigen model ("needle-in-a-haystack"), counting ones (unitation models) and a model where fitnesses are assigned to strings randomly (akin to a Kauffman NK-model with $N = K$). Specifically: for the Eigen landscape $f_{1111111} = 1.5$ and $f_I = 1.0$ for $I \neq 1111111$; for counting ones $f_I = 1 + (0.5/7) \sum_i 1_i$; and finally for the random landscape $f_I \in [1, 1+R]$, where R is a random number chosen with uniform probability from the interval $[0, 0.5]$. In all three cases the parameters have been chosen so that the maximum deviation from fitness value 1 is 0.5, corresponding to a 50% difference in fitness between the fittest string and the least fit string, i.e. a 50% "planarity deviation", i.e. the deviation from the no-selection limit. As ϵ is set to one this is a good test of the approximations as the corresponding perturbation is really not then particularly "small" at all. In all cases a random initial population was chosen.

In Figures 2-4 we compare the perturbative approximations of $\bar{f}(t)$ to $O(\epsilon)$ and $O(\epsilon^2)$ with the exact solution, obtained by explicitly integrating equation (18). Notice that the $O(\epsilon^2)$ approximation gives uniformly better results (by a factor of between 2 for the Eigen model and 10 for counting ones) in the asymptotic regime but not necessarily for the transients. This is due to the presence of secular terms, which are also responsible for artefacts like the peak in the second order curve in Figure 4.
Note that even at $O(\epsilon)$ the results are asymptotically very accurate with deviations from the exact answer being less than about 0.1%. The population fitness in this sense is quite robust in terms of approximations.

A more sensitive object is the proportion of optimal strings in the population as a function of time. In Figures 5-7 we see graphs of precisely this quantity. For the Eigen model the optimal sequence is the "master sequence", i.e. the needle – arbitrarily chosen to be the sequence 1111111. For the counting ones landsacpe the optimu is also the string 1111111. For the random landscape the optimal sequence was found by examining all 128 strings. Note that asymptotically the quality of the approximation is quite sensitive to the landscape considered. At

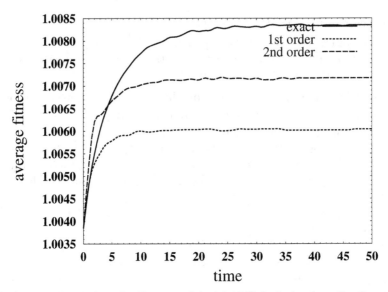

Fig. 2. Average fitness for 7-bit Eigen model with 50% deviation from flat fitness limit

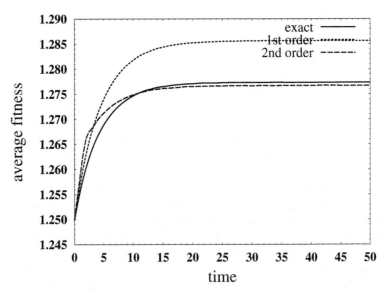

Fig. 3. Average fitness for 7-bit counting ones model with 50% deviation from flat fitness limit

$O(\epsilon)$ the error for the Eigen model is about 28% whereas it is only about 2% for the counting ones landscape. The $O(\epsilon^2)$ results are better than the $O(\epsilon)$ results, as one might expect, except in the case of the counting ones landscape which is both interesting, somewhat counterintuitive and worthy of further investigation.

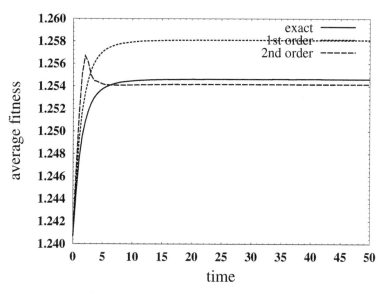

Fig. 4. Average fitness for 7-bit random landscape model with 50% deviation from flat fitness limit

We also repeated the experiments for $N = 3$ with very similar results, the approximation being generally somewhat worse for this shorter string length. This is to be expected as for shorter strings the neglect of mutation events where a higher proportion of bits change is less valid.

6   Perturbation Theory for Mutation-Selection Systems – A Simple Example

This section gives a fairly complete analysis of the case of one bit. Naturally, this is meant only to illustrate the general techniques and the relationship between the different methodologies in a transparent context. In this case a state is represented by the two-component vector $\begin{pmatrix} x_1(t+1) \\ x_0(t+1) \end{pmatrix}$ and $F = \begin{pmatrix} f_0 & 0 \\ 0 & f_1 \end{pmatrix}$.

6.1   The Exact Solution

The exact solution is determined by calculating the eigenvalues and eigenvectors of $W = MF$. The eigenvalues are the solutions of the 2^N-dimensional characteristic equation, which in this case is quadratic with solutions

$$\lambda_\pm = \frac{1}{2}\left[(1-p)(f_1 + f_0) \pm \beta_0\right], \tag{39}$$

where $\beta_0 = [(1-p)^2(f_1+f_0)^2 - 4(1-2p)f_1 f_0]^{1/2}$. The corresponding eigenvectors are

$$\lambda_+ \to \begin{pmatrix} \alpha \\ b\alpha \end{pmatrix}, \qquad \lambda_- \to \begin{pmatrix} \beta \\ a\beta \end{pmatrix}, \tag{40}$$

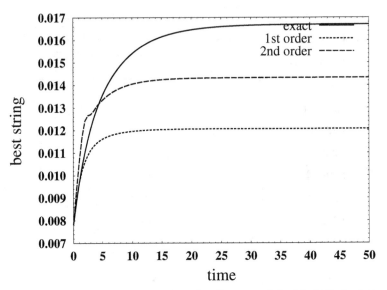

Fig. 5. Proportion of optimal strings for 7-bit Eigen model with 50% deviation from flat fitness limit

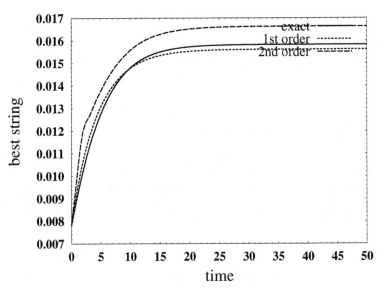

Fig. 6. Proportion of optimal strings for 7-bit counting ones model with 50% deviation from flat fitness limit

where $a = \dfrac{-(1-p)f_1 + \lambda_-}{pf_0}$ and $b = \dfrac{-(1-p)f_1 + \lambda_+}{pf_0}$ and the normalisation factors are $\alpha = (1 + b^2)^{-1/2}$ and $\beta = (1 + a^2)^{-1/2}$. The transformation that diagonalizes W is implemented using the similarity-transformation matrix

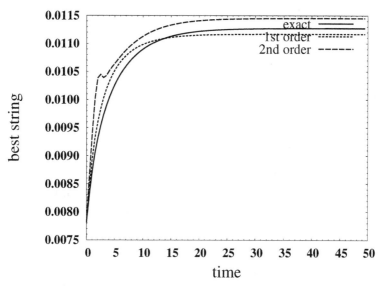

Fig. 7. Proportion of optimal strings for 7-bit random landscape model with 50% deviation from flat fitness limit

$$\tilde{\Lambda} = \frac{1}{\alpha\beta(b-a)} \begin{pmatrix} -\alpha & b\alpha \\ \beta & -a\beta \end{pmatrix}, \tag{41}$$

which relates the eigenvector basis, $\tilde{x}_I$ and the string basis x_I via $\tilde{x}_I = \tilde{\Lambda}_I{}^J x_J$. The solution in the eigenvector basis is $\tilde{x}_I(t) = \lambda_I^t \tilde{x}_I(0)$ which, changing basis back to the string basis, gives the solution

$$x_I(t) = A_{I-}\lambda_-^t + A_{I+}\lambda_+^t \tag{42}$$

which is of the general form posited in equation (14). Explicitly, the four amplitudes, A_{ij}, $i = 0, 1$ and $j = \pm$, are

$$A_{1-} = \left(\frac{(\lambda_- - (1-p)f_0)x_1(0) + pf_0x_0(0)}{\lambda_- - \lambda_+} \right) \tag{43}$$

$$A_{1+} = \left(\frac{(-\lambda_+ + (1-p)f_0)x_1(0) - pf_0x_0(0)}{\lambda_- - \lambda_+} \right) \tag{44}$$

$$A_{0-} = \left(\frac{(\lambda_- - (1-p)f_1)x_0(0) + pf_1x_1(0)}{\lambda_- - \lambda_+} \right) \tag{45}$$

$$A_{0+} = \left(\frac{(-\lambda_+ + (1-p)f_1)x_0(0) - pf_1x_1(0)}{\lambda_- - \lambda_+} \right) \tag{46}$$

If we consider weak selection, i.e. $f_I \to (1 + \epsilon f_I)$, or weak mutation, i.e. $M_I{}^J \to \delta_I{}^J + \epsilon dM_I{}^J$, then the eigenvalues (42) can be perturbatively expanded in ϵ. To $O(\epsilon)$ for weak mutation

$$\lambda_- = f_0(1-p) \qquad \lambda_+ = f_1(1-p) \tag{47}$$

While for weak selection, with $f_I \to 1 + f_I$,

$$\lambda_- = (1 - 2p)\left(1 + \frac{(f_0 + f_1)}{2}\right) \qquad \lambda_+ = \left(1 + \frac{(f_0 + f_1)}{2}\right) \tag{48}$$

where we have put $\epsilon = 1$. The corresponding amplitudes for the case of weak selection can be found by expanding (43-46) to this order.

6.2   Diagrammatic Perturbation Theory

In this case one evaluates explicitly equations (33) or (38), corresponding to the diagrams in Figure 1, depending on the order of perturbation theory required. To $O(\epsilon)$ the solutions for the case of weak selection are

$$\hat{x}_0(t) = \left(1 + \epsilon \widehat{dF}_0{}^0 t\right)\hat{x}_0(0) + \epsilon\left(\frac{1 - (1 - 2p)^t}{2p}\right)\widehat{dF}_0{}^1 \hat{x}_1(0) \tag{49}$$

$$\hat{x}_1(t) = (1 - 2p)^t\left(1 + \epsilon \widehat{dF}_1{}^1 t\right)\hat{x}_1(0) + \epsilon\left(\frac{1 - (1 - 2p)^t}{2p}\right)(1 - 2p)\widehat{dF}_1{}^0 \hat{x}_0(0) \tag{50}$$

which agree with the expressions derived from the exact solution *after* expanding in powers of ϵ *both* the amplitudes (43-46) and the factors λ^t, where the eigenvalues are given by (39), when expanded to $O(\epsilon)$. Notice the secular term linear in t. Schematically, this arises from expanding $\lambda^t = (a + \epsilon b + \epsilon^2 c + ...)^t \approx a^t(1 + \epsilon(b/a)t)$ to $O(\epsilon)$. Passing to the variables $\hat{P}_I(t)$, one finds $\hat{P}_0(t) = 1/2^{1/2}$ and

$$\hat{P}_1(t) = \frac{(1 - 2p)^t}{2^{\frac{1}{2}}}\left(\hat{P}_1(0) + \epsilon\widehat{dF}_1{}^1\left(\frac{1 - (1 - 2p)^{-t}}{1 - (1 - 2p)^{-1}}\right)\right.$$
$$\left. - \epsilon\widehat{dF}_1{}^0 \hat{P}_1^2(0)\left(\frac{1 - (1 - 2p)^t}{1 - (1 - 2p)}\right)\right) \tag{51}$$

Thus, we see that the secular terms cancel out of the $\hat{P}_I$ and hence out of the probabilities P_I. This however, is a property only of the one-bit case. For $N > 1$, generically they will remain. Although they do not destroy the validity of the perturbation expansion entirely, due to the presence of exponential suppression factors $(1 - 2p)^{|I|t}$, they do remain somewhat problematic, as we know that the exact functional form is a sum of pure exponentials. A polynomial times an exponential does not fit this pattern. This can be further understood by considering the contributions from $O(\epsilon^2)$.

$$\hat{x}_0(t) = 1^t\left[\left(1 + \epsilon t \widehat{dF}_0{}^0 + \frac{\epsilon^2}{2}\widehat{dF}_0{}^0 \widehat{dF}_0{}^0 t(t - 1)\right.\right.$$
$$\left. + \epsilon^2 \widehat{dF}_0{}^1 \widehat{dF}_0{}^1 \frac{(1 - 2p)}{2p}(t - 1)\right)\hat{x}_0(0)$$
$$\left. + \left(\frac{\epsilon}{2p}\widehat{dF}_0{}^1 + \frac{\epsilon^2}{2p}\widehat{dF}_0{}^0 \widehat{dF}_0{}^1 (t - 1)\right)\hat{x}_1(0)\right]$$

$$+(1-2p)^t \left[\left(-\frac{\epsilon}{2p}\widehat{dF}_0^{\ 1} - \frac{\epsilon^2}{2p}\widehat{dF}_0^{\ 1}\widehat{dF}_1^{\ 1}(t-1) \right) \hat{x}_1(0) \right.$$

$$\left. +\epsilon^2\widehat{dF}_0^{\ 1}\widehat{dF}_0^{\ 1}\frac{(1-2p)}{4p^2}\hat{x}_0(0) \right] \qquad (52)$$

which agree with the expressions derived from the exact solution when expanded to $O(\epsilon^2)$.

Now, from (13), we know on general grounds the functional form of $\hat{x}_I(t)$, i.e. a sum of exponentials with different time independent amplitudes. In (52) there are exponentials, however, the exponents are what one would expect from the mutation only system as the perturbative expansions contain a part whose origin was the expansion of the eigenvalues of the exact expression. The question arises then – is it possible to restore the general functional form of equation (13)? and thereby improve the approximation, and is it possible to separate out amplitudes from exponents? The answer to both these questions is yes and requires a tool known in physics as the renormalization group.

7   Perturbation Theory and the Renormalization Group

The solution to the two questions just posed begins with the observation that the secular terms invalidate perturbative expressions for $\hat{x}_I(t)$ when, for example, $\epsilon\widehat{dF}_0^{\ 0}t$ is no longer small. Notice for instance though, that the first three coefficients of $\hat{x}_0(0)$ in the amplitude of the exponent of the leading eigenvalue $-1-$ are the first three terms in an expansion of $(1+\epsilon\widehat{dF}_0^{\ 0})^t$, i.e. an exponential. Notice further that from the exact answer (39) and its expansion to $O(\epsilon)$ for weak selection, as given in (27), that this posited exponential is the same as the perturbative expansion of the exact one. Thus, it would seem that a resummation of the perturbative series in (52) is required. However, given that we only have the first two terms in the equations how can we determine what the series should sum to? and how do we determine what should be summed and what shouldn't? At the heart of the problem is the fact that we are trying to be too greedy with the perturbative approximation. In the regime $\epsilon dFt \ll 1$ there is no problem. However, we wish to investigate the dynamics well away from the initial starting point at $t = 0$. To circumvent this difficulty we will introduce a new initial condition, $\hat{x}_I(\tau)$, at some arbitrary time τ and demand that the parameters $\hat{x}_I(\tau)$ are related to the physical initial conditions $\hat{x}_I(0)$ via

$$\hat{x}_I(0) = \sum_J \hat{Z}_I^{\ J}(\tau)\hat{x}_J(\tau) \qquad (53)$$

and posit a perturbative expansion for the coefficients $Z_I^{\ J}$

$$\hat{Z}_I^{\ J}(\tau) = 1 + \sum_{n=1}^{\infty} \epsilon^n \delta_I^{\ J} a_I \qquad (54)$$

The idea now is to "renormalize" (essentially reparameterize) (52) by replacing the $\hat{x}_I(0)$ using (53). It is important to note that the latter is an identity and so we are not changing anything by doing so. However, the coefficients a_I remain to be determined and at our disposal. We use the freedom in their definition to eliminate the secular terms in (52) at the particular time $t = \tau$. We will here carry this out to $O(\epsilon)$, the $O(\epsilon^2)$ and higher calculations being fairly straightforward (though eventually complicated) extensions. To $O(\epsilon)$ one finds

$$a_0 = a_1 = -\epsilon\tau\widehat{dF}_0^{\,0} \tag{55}$$

Now, from (53), as $\hat{x}_I(0)$ is independent of τ so must be $\sum_J \hat{Z}_I^{\,J}(\tau)\hat{x}_J(\tau)$ which therefore must satisfy

$$\sum_J \hat{Z}_I^{\,J}(\tau+1)\hat{x}_J(\tau+1) = \sum_J \hat{Z}_I^{\,J}(\tau)\hat{x}_J(\tau) \tag{56}$$

Substituting in $\hat{Z}_I^{\,J}$ to $O(\epsilon)$ using (55) one finds

$$\hat{x}_I(\tau+1) = (1 + \epsilon\widehat{dF}_I^{\,I})\hat{x}_I(\tau) \tag{57}$$

which can be iterated to yield

$$\hat{x}_I(\tau) = (1 + \epsilon\widehat{dF}_I^{\,I})^\tau\hat{x}_I(0) \tag{58}$$

using as initial condition $\hat{x}_I(0)$. Substituting this expression into (52) and using our freedom to choose τ, we set $\tau = t$ to find

$$\hat{x}_0(t) = (1 + \epsilon F_0^{\,0})^t(\hat{x}_0(0) + \epsilon\frac{\widehat{dF}_0^{\,1}}{2p}\hat{x}_1(0)) - (1 - 2p)^t(1 + \epsilon F_0^{\,0})^t\epsilon\frac{\widehat{dF}_0^{\,1}}{2p}\hat{x}_1(0) \tag{59}$$

$$\hat{x}_1(t) = (1 - 2p)^t(1 + \epsilon F_1^{\,1})^t(\hat{x}_1(0) - \epsilon\frac{\widehat{dF}_0^{\,1}}{2p}\frac{(1-2p)}{2p}\hat{x}_0(0))$$

$$+ (1 + \epsilon F_0^{\,0})^t\epsilon\frac{(1-2p)\widehat{dF}_0^{\,1}}{2p}\hat{x}_0(0) \tag{60}$$

These expressions are equivalent to the exact solution (42), where the eigenvalues $\lambda_\pm$ and the amplitudes A_{I-} and A_{I+} have been expanded to $O(\epsilon)$. Thus, the renormalization group has resummed the diagrammatic perturbation theory and thereby gives a better approximation than the latter. These statements are true for any N, although we are only illustrating the one-bit problem. In the latter case, when considering the $\hat{P}_I$ or P_I, because to $O(\epsilon)$ the approximate eigenvalues are of the form $\lambda_i^{(0)}(1 + \epsilon F_0^{\,0})$ for $i = +$ or $-$, and because in the non-renormalization group resummed perturbation theory of section 6.2 the secular terms cancel, the RG and non-RG resummed answers are the same. This is not true beyond lowest order or for $N > 1$.

8   Conclusions

We have introduced and proposed perturbation theory as a candidate tool for analysing the dynamics of EAs. We showed that within the umbrella of perturbative methods there are many different implementations. In the context of mutation-selection EAs we briefly discussed one of the most familiar ones, where the eigenvalues and eigenvectors of the transition matrix are computed perturbatively. This methodology has various drawbacks. In particular, the computation of the perturbed eigenvectors beyond leading order is complicated. Additionally, it lacks intuitive transparency. To ameliorate some of these defects we considered a perturbative calculation of the population variables themselves (really we are computing the transition matrix of the Markov chain $G_I{}^J(t, t')$, which is then used to compute the $P_I(t)$ via $P_I(t) = \sum_J G_I{}^J(t, t')P_J(t')$), using diagrammatic methods familiar from field theory, showing how a pictorial representation of the processes that contribute to the production of a given string could be systematically constructed using a simple set of rules.

As a simple illustration of the results one might expect we showed how the approximate solutions were close to the exact solutions for a variety of landscapes and for strings of length 7, even when the perturbation was quite large, the approximation systematically improving as different orders in ϵ were considered. To make transparent exactly how the methodology works we also considered a simple one-bit example. We showed that a defect of the direct diagrammatic expansion is the existence of "secular" terms and then introduced the renormalization group, which was seen to eliminate these secular terms to give uniform approximations for all t.

We emphasise that this paper is merely an introduction to these techniques and, given the lack of space, a brief one at that. Although we used a toy one bit example to illustrate in as simple a context as possible how the different perturbative implementations work and how they approximate the exact solution, we also showed that using diagrammatic methods one could push on to more realistic problems. How simple it is to implement the renormalization group in that context remains to be seen, as when there exist degenerate unperturbed eigenvectors a non-diagonal matrix renormalization is necessary. As at heart we are calculating the transition matrix for the Markov chain it should be relatively straightforward to include in finite population effects. All in all, we believe there to be a huge space in which further work may be carried out to check to what extent perturbative methods can help narrow the expectation gap between theoreticians and practitioners.

Acknowledgements

CRS acknowledges support from: CONACyT project 30422-E, a DGAPA Sabbatical Fellowship, a Royal Irish Academy Visiting Professorship and hospitality and financial support from the Dublin Institute for Advanced Studies. CRS is grateful to Bill Langdon for comments on the manuscript and to Brian Dolan for useful discussions.

214 C.R. Stephens, A. Zamora, and A.H. Wright

References

1. C. R. Stephens and H. Waelbroeck. Schemata evolution and building blocks. *Evol. Comp.*, 7:109–124, 1999.
2. Riccardo Poli. Exact schema theory for genetic programming and variable-length genetic algorithms with one-point crossover. *Genetic Programming and Evolvable Machines*, 2(2):123–163, 2001.
3. C. R. Stephens. Some exact results from a coarse grained formulation of genetic dynamics. In *Proceedings of GECCO-2001)*, pages 631–638, San Francisco, California, USA, 7-11 July 2001. Morgan Kaufmann.
4. W. B. Langdon and R. Poli. *Foundations of Genetic Programming*. Springer Verlag, Berlin, New York, 2002.
5. C. R. Stephens and R. Poli. E C theory - in theory: Towards a unification of evolutionary computation theory. In A. Menon, editor, *Frontiers of Evolutionary Computation*, pages 129–156. Kluwer Academic Publishers, 2004.
6. Adam Prügel-Bennett and Jonathan L. Shapiro. An analysis of genetic algorithms using statistical mechanics. *Physical Review Letters*, 72:1305–1309, 1994.
7. Jonathan L. Shapiro. Statistical mechanics theory of genetic algorithm. In B. Naudts L. Kallel and A. Rogers, editors, *Theoretical Aspects of Evolutionary Computing*, pages 87–108, Berlin, Germany, 2001. Springer Verlag.
8. J. McCaskill M. Eigen and P. Schuster. The molecular quasi-species. *Adv. Chem. Phys.*, 75:149–263, 1989.
9. J. Hofbauer T. Nagylaki and P. Brunovsky. Convergence of multilocus systems under weak epistasis or weak selection. *J. Math. Biol.*, 38:103–133, 1999.
10. M. Rattray and J.L. Shapiro. Cumulant dynamics of a population under multiplicative selection, mutation and drift. *Theor. Pop. Biol.*, 60:17–32, 2001.
11. C. R. Reeves and J.E. Rowe. *Genetic Algorithms - Principles and Perspectives*. Kluwer Academic Publishers, 2003.
12. M.A. Akivis and V.V. Goldberg. *An Introduction to Linear Algebra and Tensors*. Dover Publications, Mineola, NY, 1977.
13. C. Chryssomalakos and C. R. Stephens. What basis for genetic dynamics? In Kalyanmoy Deb *et al*, editor, *Proceedings of GECCO 2004*, pages 1394–1402, Berlin, Germany, 2004. Springer Verlag.
14. Riccardo Poli. Exact schema theorem and effective fitness for GP with one-point crossover. In *Proceedings of the Genetic and Evolutionary Computation Conference*, pages 469–476, Las Vegas, July 2000. Morgan Kaufmann.
15. Christopher R. Stephens. The renormalization group and the dynamics of genetic systems. *Acta Phys. Slov.*, 52:515–524, 2003.
16. A. H. Wright J. E. Rowe, R. Poli and C. R. Stephens. A fixed point analysis of a gene pool GA with mutation. In *Proceedings of GECCO 2002*, San Francisco, USA, 2002. Morgan Kaufmann.
17. W. D. Lakin and D.A. Sanchez. *Topics in Ordinary Differential Equations*. Dover, Publications Inc., 1970.

Optimal Weighted Recombination

Dirk V. Arnold

Faculty of Computer Science
Dalhousie University
Halifax, NS, Canada B3H 1W5
dirk@cs.dal.ca

Abstract. Weighted recombination is a means for improving the local search performance of evolution strategies. It aims to make effective use of the information available, without significantly increasing computational costs per time step. In this paper, the potential speed-up resulting from using rank-based weighted recombination is investigated. Optimal weights are computed for the sphere model, and comparisons with the performance of strategies that do not make use of weighted recombination are presented. It is seen that unlike strategies that rely on unweighted recombination and truncation selection, weighted multirecombination evolution strategies are able to improve on the serial efficiency of the $(1 + 1)$-ES on the sphere. The implications of the use of weighted recombination for noisy optimization are studied, and parallels to the use of rescaled mutations are drawn. The cumulative step length adaptation mechanism is formulated for the case of an optimally weighted evolution strategy, and its performance is analyzed.

1 Introduction

In his seminal book Rechenberg [18] in 1973 presented the derivation of a law describing the progress rate of the $(1 + 1)$-ES on the high-dimensional sphere model. From that law, it can be seen numerically that for optimally adapted mutation strength, the normalized rate at which the optimum is approached equals 0.202. In the years that followed, evolution strategies evolved. The single-parent strategy was replaced by population-based strategies, and recombination was introduced. In 1996, Beyer [9] studied the performance on the sphere model of the $(\mu/\mu, \lambda)$-ES – a population-based strategy that uses multi-recombination. He made the surprising discovery that the serial efficiency of the $(\mu/\mu, \lambda)$-ES for optimally chosen population size parameters asymptotically approaches the same value of 0.202 that the $(1 + 1)$-ES had achieved more than two decades earlier. Moreover, while few theoretical results exist, there is evidence that none of the $(\mu/\rho \dagger \lambda)$-ES achieve a serial efficiency on the sphere model that exceeds that of the simple $(1 + 1)$-ES. Needless to say, this is not to imply that no progress had been made. Population-based strategies allow for parallelization, have greater adaptation capabilities, and are much superior when applied to noisy optimization problems. Nonetheless, the $(1 + 1)$-ES sets the benchmark for serial efficiency on the simple sphere model.

A.H. Wright et al. (Eds.): FOGA 2005, LNCS 3469, pp. 215–237, 2005.

Key to achieving a serial efficiency that exceeds that of the $(1 + 1)$-ES is to recognize that generally, all $(\mu/\rho \overset{+}{,} \lambda)$-ES discard information. Truncation selection leads to all of the selected offspring having the same influence on the progress of the strategy, irrespective of their relative ranks within the population. For example, for the $(\mu/\mu, \lambda)$-ES the influence of the best candidate solution equals that of the μth best. Similarly, all relative rank information from those offspring that are not selected to survive is discarded. Those candidate solutions are without influence on the step taken by the strategy, no matter whether they narrowly missed the cut or they missed it by a wide margin.

More complete use of the information gained by evaluating offspring candidate solutions can be made by weighting their influence in the recombination and selection process. Weights can be chosen such that they more carefully discriminate between good and bad candidate solutions than truncation selection does. The choice of weights can be based either on function values or on rank within the set of offspring generated. A strategy that uses function values to determine weights is the evolutionary gradient search strategy (EGS) proposed by Salomon [21]. EGS differs from evolution strategies not only in its reliance on function values rather than ordinal data, but also in its use of "negative information". The weight assigned to a candidate solution is proportional to the difference between that candidate solution's fitness and the fitness of the search point that it has been generated from. As a consequence, those offspring that improve on the previous time step's fitness receive positive weights, and offspring that are inferior to their parent receive negative weights and thus result in the strategy moving in the opposite direction. An investigation in [2] has shown that EGS is indeed capable of achieving serial efficiencies on the sphere model that exceed those of the $(1 + 1)$-ES. However, it has also been seen that the explicit rescaling of progress vectors that EGS performs hampers genetic repair, and that as a result EGS is generally inferior to the $(\mu/\mu, \lambda)$-ES in the presence of noise as well as if implemented on parallel computers.

Rank-based weighted recombination has been employed by Hansen and Ostermeier [15] in connection with their covariance matrix adaptation evolution strategy (CMA-ES), and it has also been used in the comparative review of evolutionary algorithms by Kern et al. [16]. In both references, it is suggested to assign positive weights of different magnitudes to the better 50 percent of the candidate solutions generated. A heuristic rule for choosing those weights is proposed. Without a reason being given, but probably in realization of the fact that the opposite of a bad direction is not always a good direction, the use of negative weights is discouraged. Zero weights are assigned to the inferior 50 percent of candidate solutions generated. In [15] it is noted that speed-up factors of less than two are observed compared to the $(\mu/\mu, \lambda)$-ES. A direct and systematic comparison between weighted and unweighted recombination is not performed.

The only attempt made so far to explore the consequences of the choice of weights analytically has been made by Rudolph [20]. For a weighted strategy that generates offspring by placing them on a sphere shell rather than by Gaussian mutations, Rudolph computes expressions for the progress rate on the sphere model. Those expressions involve expectations of joint beta order statistics and are difficult to determine in the general case. For that reason, Rudolph explores consequences of his results only for the case that the search space dimensionality equals three. Even for this special case the resulting expressions are too complicated to determine optimal weights. Rudolph does observe that the use of negative weights can have effects beneficial for the progress rate of the strategy, and that the serial efficiency of a strategy using unweighted recombination in connection with truncation selection can be exceeded by strategies that make use of weighted recombination.

It is the goal of this paper to obtain an improved understanding of the interplay of mutation, recombination, and selection in evolution strategies, and of the potential that weighted recombination has to speed up local search. In contrast to the aforementioned paper by Rudolph, focus here is on Gaussian mutations and the case that the search space dimensionality is high. This situation has the advantage of being comparatively well understood, and of allowing an analytical treatment. The results obtained are exact only in the limit of infinite search space dimensionality, but they do contribute to the understanding of the evolutionary processes in sufficiently high-dimensional spaces.

The remainder of this paper is organized as follows. The sphere model as an important environment for studying local search properties of direct optimization strategies as well as weighted multirecombination evolution strategies are introduced in Sect. 2. In Sect. 3, the quality gain of weighted multirecombination evolution strategies is computed and optimal weights are determined. Section 4 addresses the issue of how the performance of evolution strategies with optimally weighted multirecombination is affected by noise. It is seen that the issue of rescaled mutations raised by Rechenberg [19] and studied by Beyer [10, 11] arises naturally in connection with the choice of weights and the issue of genetic repair in multirecombination strategies. In Sect. 5, the cumulative step length adaptation mechanism is formulated for the case of the optimally weighted multirecombination evolution strategy, and its performance is analyzed. Finally, Sect. 6 concludes with a brief summary and directions for future research.

2   Preliminaries

In this section evolution strategies using weighted multirecombination for the minimization of functions $f : I\!\!R^N \to I\!\!R$ are formally introduced. Then the sphere model is briefly discussed as an important environment for learning about the behavior of local search algorithms.

2.1   Weighted Multirecombination Evolution Strategies

Weighted multirecombination evolution strategies repeatedly update a search point $\mathbf{x} \in I\!\!R^N$ using the following four steps:

1. Generate λ offspring candidate solutions $\mathbf{y}^{(i)} = \mathbf{x} + \sigma\mathbf{z}^{(i)}$, $i = 1, \ldots, \lambda$. The $\mathbf{z}^{(i)}$ are vectors consisting of N independent, standard normally distributed components and are referred to as mutation vectors. The nonnegative quantity σ is referred to as the mutation strength and determines the step length of the strategy.

2. Determine the objective function values $f(\mathbf{y}^{(i)})$ of the offspring candidate solutions and order the $\mathbf{y}^{(i)}$ according to those values. After ordering, index $k; \lambda$ refers to the kth best of the λ offspring (the kth smallest for minimization; the kth largest for maximization).
3. Compute the weighted average

$$\mathbf{z}^{(\text{avg})} = \sum_{k=1}^{\lambda} w_{k;\lambda} \mathbf{z}^{(k;\lambda)} \tag{1}$$

of the $\mathbf{z}^{(i)}$ vectors. The $w_{k;\lambda}$ are weights that depend on the rank of the corresponding candidate solution in the set of all offspring.
4. Replace the search point $\mathbf{x}$ by $\mathbf{x} + \sigma \mathbf{z}^{(\text{avg})}$.

The vector $\mathbf{z}^{(\text{avg})}$ defined in Eq. (1) is referred to as the progress vector. Clearly, $\sigma \mathbf{z}^{(\text{avg})}$ connects consecutive search points. Notice that for the particular choice of weights

$$w_{k;\lambda} = \begin{cases} 1/\mu & \text{if } 1 \leq k \leq \mu \\ 0 & \text{otherwise} \end{cases}, \tag{2}$$

the weighted multirecombination evolution strategy simply is the $(\mu/\mu, \lambda)$-ES. In that case, the search point $\mathbf{x}$ is the centroid of the population that consists of the μ best of the λ offspring candidate solutions generated. Also notice that the evolutionary gradient search strategy introduced in [21] does not entirely fit into the framework of rank-based weighted multirecombination as weights are chosen proportional to $f(\mathbf{x}) - f(\mathbf{y}^{(i)})$ rather than based on rank. Moreover, a normalization step is required between the averaging of mutation vectors and the update of the search point.

2.2   The Sphere Model

Since the early work of Rechenberg [18], the local performance of evolution strategies has commonly been studied on the quadratic sphere given by objective function

$$f(\mathbf{x}) = (\hat{\mathbf{x}} - \mathbf{x})^{\mathrm{T}}(\hat{\mathbf{x}} - \mathbf{x}), \qquad \mathbf{x} \in I\!\!R^N,$$

where the task is minimization and where $\hat{\mathbf{x}} \in I\!\!R^N$ is the optimizer. The sphere serves as a model for objective functions in the vicinity of well-behaved local optima. See [5] for a justification of the usefulness of such considerations and for possible generalizations. Possibly most important among the arguments presented is that strategies such as the CMA-ES described in [15] have been found to effectively transform a wide range of convex quadratic functions into the sphere, opening up the possibility that findings made for the sphere model have much wider-ranging significance.

In order to quantify the local performance of search strategies on the sphere, consider the effect of adding a vector $\sigma \mathbf{z}$ to the current search point $\mathbf{x}$. Multirecombination evolution strategies do so both when generating offspring candidate

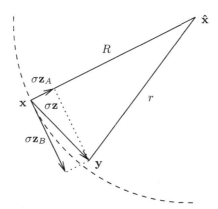

Fig. 1. Decomposition of a vector $\mathbf{z}$ into central component $\mathbf{z}_A$ and lateral component $\mathbf{z}_B$. Vector $\mathbf{z}_A$ is parallel to $\hat{\mathbf{x}} - \mathbf{x}$, vector $\mathbf{z}_B$ is in the hyperplane perpendicular to that. The starting and end points, $\mathbf{x}$ and $\mathbf{y} = \mathbf{x} + \sigma\mathbf{z}$, of vector $\sigma\mathbf{z}$ are at distances R and r from the optimizer $\hat{\mathbf{x}}$, respectively

solutions and when updating the search point at the end of an iteration. Denoting the respective distances of $\mathbf{x}$ and $\mathbf{y} = \mathbf{x} + \sigma\mathbf{z}$ from the optimizer by R and r, the difference $\delta(\mathbf{z}) = R^2 - r^2$ between objective function values $f(\mathbf{x}) = R^2$ and $f(\mathbf{y}) = r^2$ is referred to as the fitness advantage associated with vector $\mathbf{z}$ [1]. The fitness advantage associated with mutation vectors determines the ordering of the candidate solutions and thus the weights with which those mutation vectors enter recombination. The fitness advantage associated with the progress vector can be used for defining a performance measure for evolution strategies as seen below.

The commonly used approach to determining $\delta(\mathbf{z})$ on the sphere model relies on a decomposition of vector $\mathbf{z}$ that has been used in [12, 19] and that is illustrated in Fig. 1. A vector $\mathbf{z}$ originating at search space location $\mathbf{x}$ can be written as the sum of two vectors $\mathbf{z}_A$ and $\mathbf{z}_B$, where $\mathbf{z}_A$ is parallel to $\hat{\mathbf{x}} - \mathbf{x}$ and $\mathbf{z}_B$ is in the $(N-1)$-dimensional hyperplane perpendicular to that. The vectors $\mathbf{z}_A$ and $\mathbf{z}_B$ are referred to as the central and lateral components of vector $\mathbf{z}$, respectively. The signed length z_A of the central component of vector $\mathbf{z}$ is defined to equal $\|\mathbf{z}_A\|$ if $\mathbf{z}_A$ points towards the optimizer and to equal $-\|\mathbf{z}_A\|$ if it points away from it. Using elementary geometry, it can easily be seen that

$$r^2 = (R - \sigma z_A)^2 + \sigma^2\|\mathbf{z}_B\|^2,$$

and therefore, rearranging terms and noticing that $\|\mathbf{z}\|^2 = z_A^2 + \|\mathbf{z}_B\|^2$, that

$$\delta(\mathbf{z}) = R^2 - r^2$$
$$= 2R\sigma z_A - \sigma^2\|\mathbf{z}\|^2.$$

---

[1] While the notation adopted here is deliberately brief and does not reflect that explicitly, it is important to keep in mind that the fitness advantage $\delta(\mathbf{z})$ depends not only on vector $\mathbf{z}$ but also on the mutation strength σ

Introducing normalized quantities

$$\sigma^* = \sigma \frac{N}{R} \quad \text{and} \quad \delta^* = \delta \frac{N}{2R^2},$$

it follows

$$\delta^*(\mathbf{z}) = \sigma^* z_A - \frac{\sigma^{*2}}{2N} \|\mathbf{z}\|^2 \tag{3}$$

for the normalized fitness advantage associated with vector $\mathbf{z}$.

In order to compute the normalized fitness advantage associated with vector $\mathbf{z}$ using Eq. (3), both the squared length and the signed length of the central component of that vector need to be determined. For the case that $\mathbf{z}$ is a mutation vector, it is well known from [12] that z_A is standard normally distributed, and that $\|\mathbf{z}\|^2$ is χ_N^2-distributed. As for large N the χ_N^2-distribution tends to a normal distribution with mean N and with standard deviation $\sqrt{2N}$, it follows that the variance of $\|\mathbf{z}\|^2/N$ is of order $1/N$ and therefore that $\|\mathbf{z}\|^2/N$ can be approximated with unity provided that N is sufficiently large. In what follows, we will write $A \overset{N \to \infty}{=} B$ when B is obtained from A by making the simplification of replacing $\|\mathbf{z}\|^2/N$ by unity. Until step length adaptation is considered in Sect. 5, no further simplifications are required. The normalized fitness advantage associated with a mutation vector

$$\delta^*(\mathbf{z}) \overset{N \to \infty}{=} \sigma^* z_A - \frac{\sigma^{*2}}{2} \tag{4}$$

is asymptotically normally distributed with mean $-\sigma^{*2}/2$ and with variance σ^{*2}.

A commonly used performance measure for local search strategies is the quality gain which measures the rate at which the optimum is approached in the space of fitness values. It is defined as the expectation of the normalized fitness advantage associated with the progress vector and is thus

$$\Delta^* = \mathrm{E}\left[\delta^*\left(\mathbf{z}^{(\mathrm{avg})}\right)\right]$$

$$= \sigma^* \mathrm{E}\left[z_A^{(\mathrm{avg})}\right] - \frac{\sigma^{*2}}{2N} \mathrm{E}\left[\|\mathbf{z}^{(\mathrm{avg})}\|^2\right]. \tag{5}$$

With $\mathbf{z}^{(\mathrm{avg})}$ being a linear combination of mutation vectors, the considerations with regard to the scaling of the central and lateral components of mutation vectors made above ensure us that both summands on the right hand side of Eq. (5) remain finite as N approaches infinity.

Another common performance measure – the progress rate – measures the rate at which the optimizer is approached in search space and is known from [12] to agree asymptotically with the quality gain on the sphere model for high search space dimensionality provided that appropriate normalizations are used. Moreover, as a performance measure that takes computational costs into account, it is commonplace to define the serial efficiency η of evolution strategies as the maximal quality gain per evaluation of the objective function. As the number of objective function evaluations per time step is λ, the serial efficiency of an evolution strategy is

$$\eta = \frac{1}{\lambda} \max_{\sigma^*} \Delta^*. \tag{6}$$

Inherent in this definition are the assumptions that computational costs are dominated by the cost of evaluating the fitness function, and that evaluations need to be performed one after the other on a single processor.

3   Optimal Weighted Recombination

In this section an expression for the quality gain of the weighted multirecombination evolution strategy on the sphere model is derived that generalizes the corresponding result for the $(\mu/\mu, \lambda)$-ES obtained in [12, 19]. Then, optimal settings for the mutation strength and the recombination weights are computed, and consequences for the quality gain of the strategy are discussed.

3.1   Determining the Quality Gain

In order to determine the quality gain of the weighted multirecombination evolution strategy using Eq. (5), expected values of the squared length and of the signed length of the central component of the progress vector defined in Eq. (1) need to be computed. The progress vector's squared length is

$$
\begin{aligned}
\|\mathbf{z}^{(\text{avg})}\|^2 &= \sum_{i=1}^{N} \left(\sum_{k=1}^{\lambda} w_{k;\lambda} z_i^{(k;\lambda)} \right)^2 \\
&= \sum_{i=1}^{N} \sum_{k=1}^{\lambda} w_{k;\lambda}^2 z_i^{(k;\lambda)^2} + \sum_{i=1}^{N} \sum_{k=1}^{\lambda} \sum_{\substack{l=1 \\ l \neq k}}^{\lambda} w_{k;\lambda} w_{l;\lambda} z_i^{(k;\lambda)} z_i^{(l;\lambda)},
\end{aligned} \tag{7}
$$

where the $z_i^{(j)}$ are the components of the mutation vectors and as such standard normally distributed. The second term on the right hand side is a crosstalk term with mean zero. Thus, taking the expectation and exchanging the order of the summations in the first term it follows that

$$
\begin{aligned}
\frac{\mathrm{E}\left[\|\mathbf{z}^{(\text{avg})}\|^2\right]}{N} &= \sum_{k=1}^{\lambda} w_{k;\lambda}^2 \frac{\mathrm{E}\left[\|\mathbf{z}^{(k;\lambda)}\|^2\right]}{N} \\
&\stackrel{N \to \infty}{=} \sum_{k=1}^{\lambda} w_{k;\lambda}^2,
\end{aligned} \tag{8}
$$

where in the second step we have made use of the important fact noted in Sect. 2 that asymptotically, $\mathrm{E}[\|\mathbf{z}\|^2]/N \to 1$ for mutation vector $\mathbf{z}$.

As for the expected signed length of the central component of the progress vector, it follows from the definition of that vector in Eq. (1) that

$$E\left[z_A^{(\text{avg})}\right] = \sum_{k=1}^{\lambda} w_{k;\lambda} E\left[z_A^{(k;\lambda)}\right],$$

where of course $z_A^{(k;\lambda)}$ is the signed length of the central component of the mutation vector that corresponds to the kth best offspring candidate solution. In order to compute the expectations, it is important to recall from Sect. 2 that the $z_A^{(i)}$ are standard normally distributed. From Eq. (4) it follows that the signed lengths of the central components of the mutation vectors determine the fitness of the corresponding offspring candidate solutions in that the offspring with the kth largest value of z_A is the kth fittest. Thus, in the limit of infinite search space dimensionality, $z_A^{(k;\lambda)}$ is the $(\lambda + 1 - k)$th order statistic of a sample of λ independent realizations of a standard normally distributed random variate. According to [7], the probability density function of $z_A^{(k;\lambda)}$ is

$$p_{k;\lambda}(x) = \frac{1}{\sqrt{2\pi}} \frac{\lambda!}{(\lambda - k)!(k - 1)!} e^{-\frac{1}{2}x^2} [\Phi(x)]^{\lambda-k} [1 - \Phi(x)]^{k-1}, \qquad (9)$$

where $\Phi(x)$ denotes the cumulative distribution function of the standardized normal distribution. It thus follows that the expected value of the signed length of the central component of the progress vector is

$$E\left[z_A^{(\text{avg})}\right] \overset{N\to\infty}{=} \sum_{k=1}^{\lambda} w_{k;\lambda} E_{k;\lambda}, \qquad (10)$$

where

$$E_{k;\lambda} = E\left[z_A^{(k;\lambda)}\right] = \int_{-\infty}^{\infty} x p_{k;\lambda}(x)\,\mathrm{d}x$$

denotes the expectation of the $(\lambda + 1 - k)$th order statistic and can easily be obtained by numerical integration.

Using Eqs. (8) and (10) in Eq. (5), it follows that the quality gain of the weighted multirecombination evolution strategy is

$$\Delta^* \overset{N\to\infty}{=} \sigma^* \sum_{k=1}^{\lambda} w_{k;\lambda} E_{k;\lambda} - \frac{\sigma^{*2}}{2} \sum_{k=1}^{\lambda} w_{k;\lambda}^2. \qquad (11)$$

Note that for the choice of weights in Eq. (2), Eq. (11) agrees with the quality gain law for the $(\mu/\mu, \lambda)$-ES derived in [12, 19]. Figure 2 compares predictions made using Eq. (11) with measurements obtained in runs in finite-dimensional search spaces of a $(3/3, 10)$-ES and a weighted evolution strategy for which the choice of weights is motivated in Sect. 3.2. It can be seen that the agreement

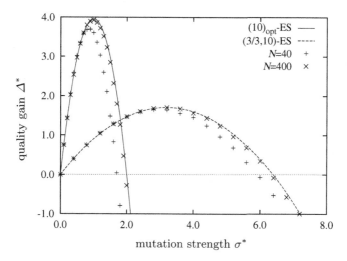

Fig. 2. Quality gain Δ^* of a $(10)_{\text{opt}}$-ES and of a $(3/3, 10)$-ES plotted against the mutation strength σ^*. The solid and dashed curves have been obtained from Eq. (11) for the choices of weights in Eqs. (2) and (15), respectively. The crosses mark measurements obtained in runs of the strategies in search spaces with $N = 40$ (+) and $N = 400$ ($\times$)

between predictions and measurements is good provided that the search space dimensionality is sufficiently high[2].

3.2   Optimal Parameter Settings

Of course, it is desirable to choose the strategy's parameters such that the quality gain is maximized. Demanding that the derivative of Eq. (11) with respect to σ^* equals zero yields optimal normalized mutation strength

$$\sigma^* \stackrel{N \to \infty}{=} \frac{\sum_{k=1}^{\lambda} w_{k;\lambda} E_{k;\lambda}}{\sum_{k=1}^{\lambda} w_{k;\lambda}^2}. \tag{12}$$

Reinserting this result into Eq. (11), the quality gain of the strategy for optimally adapted mutation strength is

$$\Delta^* \stackrel{N \to \infty}{=} \frac{1}{2} \frac{\left(\sum_{k=1}^{\lambda} w_{k;\lambda} E_{k;\lambda}\right)^2}{\sum_{k=1}^{\lambda} w_{k;\lambda}^2}. \tag{13}$$

---

[2] In Fig. 2 as well in the figures below, the quality gain has been measured using definition

$$\Delta^* = -\frac{N}{2} \log \frac{r^2}{R^2}$$

rather than that in Eq. (5). Both definitions agree in the limit $N \to \infty$, and the difference is small for large enough N. While mathematically not as convenient, the definition used for the measurements is more useful for describing the progress of evolution strategies on the sphere model over many time steps

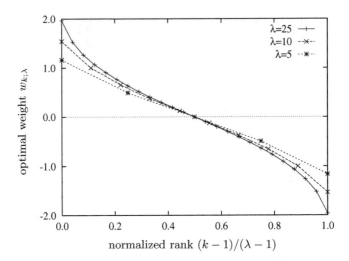

Fig. 3. Optimal weights $w_{k;\lambda} = E_{k;\lambda}$ plotted against the rank k of a candidate solution in the set of offspring for several values of λ. The ranks have been scaled linearly to fall in the range from zero to one. Note that only the points, not the connecting lines, are of practical significance

Optimal weights $w_{k;\lambda}$ can now be determined by computing the derivatives of Eq. (13) with respect to $w_{k;\lambda}$ for $k = 1, \ldots, \lambda$. Demanding that all derivatives be zero yields the system of equations

$$E_{k;\lambda} \sum_{l=1}^{\lambda} w_{l;\lambda}^2 = w_{k;\lambda} \sum_{l=1}^{\lambda} w_{l;\lambda} E_{l;\lambda}, \qquad k = 1, \ldots, \lambda. \qquad (14)$$

Clearly, the system can be solved by setting

$$w_{k;\lambda} = E_{k;\lambda} \qquad \text{for } k = 1, \ldots, \lambda, \qquad (15)$$

and it is easily seen that the corresponding extremum really is a maximum. Therefore, optimal weights of the multirecombination evolution strategy on the infinite-dimensional sphere model are given by the first moments of the order statistics of the standardized normal distribution. We will refer to the strategy with optimally chosen weights as $(\lambda)_{\text{opt}}$-ES.

The dependence of the optimal weights on the rank within the set of candidate solutions is illustrated in Fig. 3. It can be seen that in order to achieve maximal progress on the sphere model, half of the offspring should enter recombination with positive weights, the other half should receive negative weights. Optimal weights are symmetric in that for every positive weight, there is a negative weight of equal value. This is in contrast to the behavior of EGS that assigns negative weights to the majority of the offspring generated as in convex environments, most will be inferior to their parent. Also note that the curves in Fig. 3 differ strongly from the step curves defined by Eq. (2) that describe the choice of weights characterizing the $(\mu/\mu, \lambda)$-ES. Finally, it is worth mentioning the good correspondence between the left half of the curves in Fig. 3 and the (presumably empirically based) recommendations with regard to the choice of weights made in [15].

Inserting the optimal weights given in Eq. (14) into Eq. (13), it follows that the maximal quality gain of the $(\lambda)_{\text{opt}}$-ES is

$$\Delta^* \overset{N \to \infty}{=} \frac{1}{2} \sum_{k=1}^{\lambda} E_{k;\lambda}^2. \tag{16}$$

Defining

$$W_\lambda = \sum_{k=1}^{\lambda} E_{k;\lambda}^2$$

and using results on properties of order statistics from [7], it can be seen that W_λ/λ asymptotically approaches unity as λ increases. Thus, the serial efficiency of the $(\lambda)_{\text{opt}}$-ES defined in Eq. (6) asymptotically approaches a value of 0.5, nearly two and a half times that of both the $(\mu/\mu, \lambda)$-ES and the $(1 + 1)$-ES. Figure 2 illustrates that that performance advantage can indeed be observed in runs of evolution strategies. The curve for the $(10)_{\text{opt}}$-ES peaks at a value about 2.3 times as large as that of the $(3/3, 10)$-ES, and most of that performance advantage is present in the measurements for $N = 40$ and $N = 400$ as well. Finally, the dependence of the serial efficiency on the number of offspring generated per time step is illustrated in Fig. 4. It can be seen that the $(\lambda)_{\text{opt}}$-ES solidly outperforms not only the $(\mu/\mu, \lambda)$-ES, but it also has a higher serial efficiency than EGS for all but the smallest values of λ.

4   Noise

As many real-world optimization problems are plagued by noise, the assumption that the fitness of a candidate solution can be determined exactly often is an idealization. In order to study the effects of noisy fitness evaluations on the performance of optimization strategies, it is frequently assumed that noise can be modeled by means of an additive Gaussian term. That is, it is assumed that the evaluation of a candidate solution $\mathbf{y}$ yields a value that is normally distributed with mean $f(\mathbf{y})$ and with variance σ_ϵ^2, where σ_ϵ is referred to as the noise strength and may vary with the location in search space. See [1] for comprehensive results with regard to the effects of noise on various $(\mu/\rho \overset{+}{,} \lambda)$-ES.

An evolution strategy that has been found to be particularly robust with regard to the effects of noise is the $(\mu/\mu, \lambda)$-ES. In [3] it has been seen that this robustness is to be attributed to the genetic repair effect. The term "genetic repair" has been introduced by Beyer [8, 12] and refers to statistical error correction properties inherent in the multirecombination procedure. Typically, genetic repair affords the ability to operate with mutation strengths that increase (for the sphere model roughly linearly) with the number of candidate

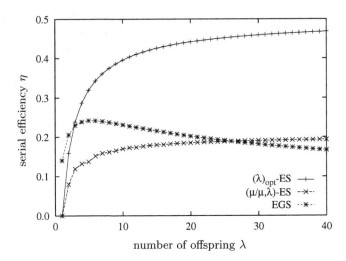

Fig. 4. Serial efficiency η of strategies on the sphere model in the limit $N \to \infty$ plotted against the number of offspring λ generated per time step. The curves represent results for the $(\lambda)_{\mathrm{opt}}$-ES described by Eq. (16), the $(\mu/\mu, \lambda)$-ES analyzed in [12], and EGS studied in [2]

solutions generated per time step. The accompanying increase in quality gain is also roughly linear in λ, opening up the possibility of linear speed-up in a parallel implementation. In the presence of noise, the increased mutation strengths have been found to yield the additional benefit of reducing the noise-to-signal ratio $\vartheta = \sigma_\epsilon^*/\sigma^*$ that the strategy operates under. As seen in a comparison with other direct search strategies in [5], that benefit can be very substantial.

In the light of the results from the previous section, it seems interesting to ask whether the $(\lambda)_{\mathrm{opt}}$-ES is capable of outperforming the $(\mu/\mu, \lambda)$-ES in the presence of noise as it does in its absence. At first sight, it appears that the $(\lambda)_{\mathrm{opt}}$-ES is not able to benefit from genetic repair the way the $(\mu/\mu, \lambda)$-ES does. From Eq. (12) with the choice of weights $w_{k;\lambda} = E_{k;\lambda}$, it follows that the optimal normalized mutation strength of the $(\lambda)_{\mathrm{opt}}$-ES in the absence of noise is unity and thus does not increase with increasing λ. Indeed, optimal performance of the $(10)_{\mathrm{opt}}$-ES in Fig. 2 is achieved at much smaller mutation strengths than for the $(3/3, 10)$-ES. However, the system of equations (14) is solved not only by the choice of weights in Eq. (15), but also by the assignment

$$w_{k;\lambda} = \frac{E_{k;\lambda}}{\kappa}, \qquad k = 1, \ldots, \lambda, \tag{17}$$

for any $\kappa > 0$ [3]. With this modified choice of weights, it follows from Eq. (13) that the optimal quality gain in the absence of noise according to Eq. (16) is

---

[3] Negative κ also solve the system of equations, but correspond to extrema of the quality gain that are minima rather than maxima

unchanged. However, considering Eq. (12), it is clear that the mutation strength at which this quality gain is attained is κ and can thus be large if κ is chosen to be large.

The effect of the scaling of the weights in Eq. (17) is reminiscent of the use of rescaled mutations in the $(1, \lambda)$-ES proposed in [19] and analyzed in [10, 11]. The idea behind using rescaled mutations is to generate offspring using a high mutation strength, but to update the search point using a much smaller step length. A large mutation strength has the advantage of affording a strong signal component for selection that can outweigh any noise that is present, and to thus yield a good search direction. However, it is also likely to lead to a set of offspring all of which are inferior to the parent they are generated from. It is thus only the direction, not the length of the step that is used by the strategy. An evolution strategy using rescaled mutations updates the search point by using a progress vector that is reduced by some factor compared to the mutation vectors.

It has been seen in [3] that the genetic repair effect resulting from multirecombination has the effect of providing an implicit rescaling. For mutation vectors, $\|\mathbf{z}\|^2/N$ asymptotically tends to unity. For the $(\mu/\mu, \lambda)$-ES, $\|\mathbf{z}^{(\mathrm{avg})}\|^2/N$ asymptotically approaches $1/\mu$. Similarly, for $(\lambda)_{\mathrm{opt}}$-ES with the choice of weights in Eq. (17), $\|\mathbf{z}^{(\mathrm{avg})}\|^2/N$ according to Eq. (8) asymptotically approaches W_λ/κ^2. The choice of κ for the $(\lambda)_{\mathrm{opt}}$-ES is thus similar to the choice of μ for the $(\mu/\mu, \lambda)$-ES in that it affords control over the amount of implicit rescaling inherent in the multirecombination process. Generally, larger values of κ can be expected to afford greater robustness in the presence of noise as they allow operating with a larger mutation strength, thus strengthening the signal and thereby reducing the noise-to-signal ratio.

In order to derive a quality gain law for the $(\lambda)_{\mathrm{opt}}$-ES in the presence of noise from Eq. (5), expected values of the overall squared length and of the signed length of the central component of the progress vector need to be computed in a fashion analogous to Sect. 3. Equation (8) for the squared length of the progress vector still holds as its derivation is unaffected by the presence of noise. The computation of the expected signed length of the progress vector's central component is less straightforward. For the purpose of selection, the candidate solutions are ordered according to their noisy fitness values. However, it is the *true* fitness values that determine the signed lengths of the central components of the respective mutation vectors. Technically, those signed lengths are concomitants of the order statistics. See [13] for an introduction to the topic and see [1, 3, 12] for the application to the problem of selection under Gaussian fitness noise. In the latter references it is shown that the probability density function of the concomitant $z_A^{(k;\lambda)}$ of the $(\lambda + 1 - k)$th order statistic is

$$p_{k;\lambda}(x) = \frac{1}{2\pi\vartheta} \frac{\lambda!}{(\lambda - k)!(k - 1)!} e^{-\frac{1}{2}x^2} \int_{-\infty}^{\infty} \exp\left(-\frac{1}{2}\left(\frac{y - x}{\vartheta}\right)^2\right)$$
$$\left[\Phi\left(\frac{y}{\sqrt{1 + \vartheta^2}}\right)\right]^{\lambda - k} \left[1 - \Phi\left(\frac{y}{\sqrt{1 + \vartheta^2}}\right)\right]^{k-1} dy,$$

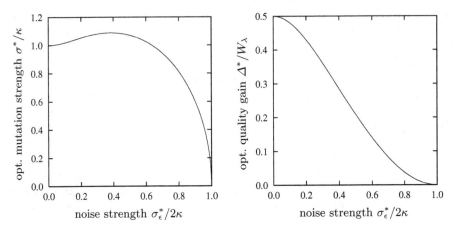

Fig. 5. Optimal mutation strength and corresponding quality gain of the $(\lambda)_{\text{opt}}$-ES plotted against the noise strength σ_ϵ^*. Due to the scaling of the axes, the curves are independent of the choice of λ and κ

where $\vartheta = \sigma_\epsilon^*/\sigma^*$ denotes the noise-to-signal ratio that the strategy operates under, and where $\sigma_\epsilon^* = \sigma_\epsilon N/2R^2$ is the normalized noise strength. Note that no assumptions with regard to the dependence of the noise strength on the location in search space need to be made as long as only individual time steps are considered. Using this density to replace Eq. (9), simple calculations analogous to those in [3] show that

$$\mathrm{E}\left[z_A^{(k;\lambda)}\right] \stackrel{N\to\infty}{=} \frac{E_{k;\lambda}}{\sqrt{1+\vartheta^2}},$$

and therefore that in generalization of Eq. (10),

$$\mathrm{E}\left[z_A^{(\text{avg})}\right] \stackrel{N\to\infty}{=} \frac{1}{\sqrt{1+(\sigma_\epsilon^*/\sigma^*)^2}} \sum_{k=1}^{\lambda} w_{k;\lambda} E_{k;\lambda}. \tag{18}$$

Thus, using Eqs. (8) and (18) in Eq. (5) and choosing the weights according to Eq. (17) it follows that the quality gain of the $(\lambda)_{\text{opt}}$-ES on the sphere model in the presence of Gaussian noise is

$$\Delta^* \stackrel{N\to\infty}{=} \frac{W_\lambda}{\kappa} \left[\frac{\sigma^{*2}}{\sqrt{\sigma^{*2}+\sigma_\epsilon^{*2}}} - \frac{\sigma^{*2}}{2\kappa} \right]. \tag{19}$$

The dependence of the optimal mutation strength and of the resulting quality gain on the noise strength is illustrated in Fig. 5. The graphs look the same as the corresponding graphs for the $(\mu/\mu, \lambda)$-ES in [3] except for the different scaling of the axes. It can be inferred from the figures that while the $(\mu/\mu, \lambda)$-ES is capable of achieving positive quality gain up to a noise strength of $\sigma_\epsilon^* = 2\mu c_{\mu/\mu,\lambda}$, where

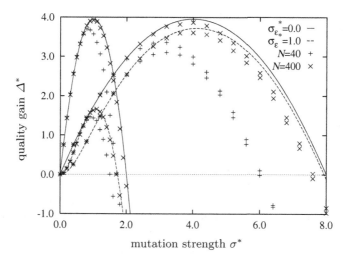

Fig. 6. Quality gain Δ^* of the $(10)_{\mathrm{opt}}$-ES plotted against the mutation strength σ^* for noise strengths $\sigma_\epsilon^* = 0.0$ (solid curves) and $\sigma_\epsilon^* = 1.0$ (dashed curves). The curves have been obtained from Eq. (19) and represent results for the limit case $N \to \infty$. The narrower curves on the left assume $\kappa = 1.0$ whereas the wider ones on the right reflect results for $\kappa = 4.0$. The crosses mark measurements obtained in runs of the strategies in search spaces with $N = 40$ ($+$) and $N = 400$ ($\times$)

$c_{\mu/\mu,\lambda}$ is the $(\mu/\mu, \lambda)$-ES progress coefficient defined in [12], the $(\lambda)_{\mathrm{opt}}$-ES does not need to stagnate up to a noise strength of $\sigma_\epsilon^* = 2\kappa$.

It is important to note however that practically, finite search space dimensionalities set limits on the useful parameter values in both cases. As the degree of accuracy of the quality gain law of the $(\mu/\mu, \lambda)$-ES in [3] decreases with increasing μ and λ, so does that of the $(\lambda)_{\mathrm{opt}}$-ES in Eq. (19) when λ and κ are increased. Figure 6 illustrates the dependence of the quality gain on the mutation strength for the $(10)_{\mathrm{opt}}$-ES with two settings of the parameter κ and noise strengths $\sigma_\epsilon^* = 0.0$ and $\sigma_\epsilon^* = 1.0$. While the accuracy of the predictions clearly decreases with increasing κ, it can also be seen that the larger choice of κ is indeed strongly preferable for the case of nonzero noise strength even for $N = 40$.

5   Cumulative Step Length Adaptation

In the considerations so far, the mutation strength has always been treated as an external parameter. In practice, of course, it needs to be adapted continually by the strategy, making the evolutionary algorithm together with the fitness environment it operates in a dynamic system. One mechanism for the adaptation of the mutation strength is the cumulative step length adaptation procedure introduced by Ostermeier et al. [17]. In this section, that procedure is formulated for the $(\lambda)_{\text{opt}}$-ES with the choice of weights from Eq. (17) and then studied for the sphere model.

The goal of cumulative step length adaptation is to minimize correlations between successive steps. For that purpose, an exponentially fading record of the most recently taken steps is kept by accumulating progress vectors. Specifically, N-dimensional vector $\mathbf{s}$ is defined by $\mathbf{s}^{(0)} = \mathbf{0}$ and

$$\mathbf{s}^{(t+1)} = (1 - c)\mathbf{s}^{(t)} + \kappa\sqrt{\frac{c(2 - c)}{W_\lambda}}\mathbf{z}^{(\text{avg})}, \tag{20}$$

where t indicates time. The cumulation parameter c is set to $1/\sqrt{N}$ according to a recommendation made in [14]. The above definition differs from that in [14] (but parallels that in [15] for the case of weighted recombination) in that a different coefficient is used to weight $\mathbf{z}^{(\text{avg})}$. This is necessary in order to account for the differences between the $(\mu/\mu, \lambda)$-ES and the $(\lambda)_{\text{opt}}$-ES. Recall from Sect. 3 that for the $(\lambda)_{\text{opt}}$-ES, $\|\mathbf{z}^{(\text{avg})}\|^2/N$ asymptotically tends to W_λ/κ^2. It is easy to verify that the choice of coefficient in Eq. (20) ensures that the distribution of the components of the accumulated progress vector $\mathbf{s}$ tends to standardized normality if the ordering of candidate solutions according to fitness values is random (as is the case in flat fitness landscapes, as well as in the presence of excessive amounts of noise).

As in [6] (and in a minor variation from [14, 15]), the mutation strength is then adapted according to

$$\sigma^{(t+1)} = \sigma^{(t)} \exp\left(\frac{\|\mathbf{s}^{(t+1)}\|^2 - N}{2DN}\right), \tag{21}$$

where the damping parameter D is set to $1/c$ as suggested in [14]. As a result of Eq. (21), the mutation strength is increased if the squared length of $\mathbf{s}$ exceeds N, which is a sign of positive correlations in the sequence of most recently taken steps. Conversely, the mutation strength is decreased if the squared length of $\mathbf{s}$ is less than N, which indicates negative correlations. It is important to realize that Eq. (21) is a prescription for modifying mutation strengths rather than normalized mutation strengths, and that no knowledge of the current location in search space is required in order to apply it.

As an evolution strategy with cumulative step length adaptation together with the environment it operates in forms a stochastic dynamic system, an analysis of its performance is substantially more complicated than the analyses presented in earlier sections that consider individual time steps only. While the only simplification made so far is to replace terms of the form χ_N^2/N with unity, stronger assumptions need to be made in what follows. In particular, it will be assumed that fluctuations of the state variables, such as the normalized mutation strength or the squared length of the accumulated progress vector, can be ignored, and that the dynamic equations can be written in terms of average values (thus effectively eliminating stochastic aspects). Rather than attempting to identify probability distributions of the state variables, we will set out to determine a fixed point in the deterministic mapping of average values. Moreover, at several points, terms that become increasingly irrelevant as the search space dimensionality increases are dropped. Identifying such terms is not always trivial and indeed sometimes relies on the (unproven) assumption that fluctuations of the state variables can be ignored. Any results that are obtained will therefore need to be confirmed in computer experiments.

A further aspect that we had been able to ignore up to now is the dependence of the noise strength on the location in search space. As long as only individual time steps are considered, all candidate solutions are sufficiently close to each other in order to tacitly assume that they are subjected to the same amount of noise. If the performance of the strategy is to be characterized over extended periods of time, this is no longer possible. The considerations below assume constant normalized noise strength $\sigma_\epsilon^* = \sigma_\epsilon N / 2R^2$ and thus that the noise strength decreases as the optimizer is approached. While not always reasonable, that assumption captures the important case of a relative error of measurement that occurs for example when using measurement devices that are accurate to a certain fixed percentage of the quantity they measure. Other forms of the dependency of noise on the location in search space do not lead to behavior that can be characterized by a fixed point and thus require different approaches for their analysis.

As a result of assuming that the normalized noise strength is constant, the environment is scale-invariant in that the distance R from the optimizer does not appear in the equations that describe the evolution of the system. The following analysis closely parallels that presented in [1, 6] for the case of the $(\mu/\mu, \lambda)$-ES and proceeds in three steps:

1. The accumulated progress vector **s** is decomposed into its central and lateral components, and Eq. (20) is used to derive recursive equations for the overall squared length $\|\mathbf{s}\|^2$ and for the signed length s_A of the central component of that vector.
2. Expectations are taken in order to arrive at average values and terms that become irrelevant in the limit $N \to \infty$ are dropped.
3. It is made use of the scale-invariance properties of the quantities considered by demanding that their average values do not change from one time step to the next.

The result of that procedure are two equations that can be used to determine (approximate) average values of $\|\mathbf{s}\|^2$ and s_A. The derivation occupies a considerable amount of space without adding any important insights. As it is closely analogous to the derivation for the $(\mu/\mu, \lambda)$-ES in [1, 6], we refrain from presenting detailed calculations here. The resulting equations read

$$\|\mathbf{s}\|^2 = (1-c)^2\|\mathbf{s}\|^2 + 2(1-c)\kappa\sqrt{\frac{c(2-c)}{W_\lambda}} s_A z_A^{(\text{avg})} + c(2-c)\frac{\kappa^2}{W_\lambda}\|\mathbf{z}^{(\text{avg})}\|^2 \quad (22)$$

and

$$s_A = (1-c)s_A + \kappa\sqrt{\frac{c(2-c)}{W_\lambda}}\left(z_A^{(\text{avg})} - \sigma^*\frac{\|\mathbf{z}^{(\text{avg})}\|^2}{N}\right). \quad (23)$$

It is understood that in these equations, equality is merely approximate, and that all quantities stand for their respective average values. They differ from their respective equivalents for the $(\mu/\mu, \lambda)$-ES only in the coefficients. Using the relationships $z_A^{(\mathrm{avg})} = W_\lambda/(\sqrt{1 + \vartheta^2}\kappa)$ and $\|\mathbf{z}^{(\mathrm{avg})}\|^2/N = W_\lambda/\kappa^2$ that follow from Eqs. (18) and (8) with Eq. (17), it follows that solving Eq. (23) for the expected signed length of the central component of the accumulated progress vector yields

$$s_A = \sqrt{\frac{W_\lambda(2 - c)}{c}} \left(\frac{1}{\sqrt{1 + \vartheta^2}} - \frac{\sigma^*}{\kappa} \right).$$

Inserting this result in Eq. (22) and rearranging terms yields

$$\|\mathbf{s}\|^2 = N + \frac{2(1 - c)}{c} \frac{W_\lambda}{\sqrt{1 + \vartheta^2}} \left(\frac{1}{\sqrt{1 + \vartheta^2}} - \frac{\sigma^*}{\kappa} \right) \qquad (24)$$

for the expected squared length of the accumulated progress vector.

With the characterization of the accumulated progress vector thus obtained, Eq. (21) can now be used to determine the average mutation strength that the strategy seeks to attain. The target mutation strength of the strategy is the mutation strength that cumulative step length adaptation does not affect a change for. For the $(\lambda)_{\mathrm{opt}}$-ES with cumulative step length adaptation, the target mutation strength is the mutation strength for which the argument to the exponential function in Eq. (21) equals zero. Using Eq. (24) and the fact that $\vartheta = \sigma_\epsilon^*/\sigma^*$, it follows that that mutation strength is

$$\sigma^* = \kappa\sqrt{1 - \left(\frac{\sigma_\epsilon^*}{\kappa} \right)^2}. \qquad (25)$$

The dependence of the target mutation strength on the noise strength is illustrated and compared with the optimal mutation strength derived in Sect. 4 in the left hand graph of Fig. 7. While the shape of the curves is the same as in the corresponding graph for the $(\mu/\mu, \lambda)$-ES in [6], it is important to note that the scaling of the axes is different. In both cases, the target mutation strength agrees with the optimal mutation strength only in the case of no noise being present. This is the case that cumulative step length adaptation was designed and its coefficients were chosen for. For nonzero noise strengths, target mutation strengths are consistently below optimal mutation strengths. However, it is also clear that due to the scaling of the horizontal axis, by increasing κ it is possible to move closer to the left hand edge of the graph, thereby operating in a regime where there is a good agreement between target mutation strength and optimal mutation strength. The same effect can be achieved for the $(\mu/\mu, \lambda)$-ES by increasing both μ and λ in equal proportions.

Finally, it is important to emphasize that as the $(\mu/\mu, \lambda)$-ES, the $(\lambda)_{\mathrm{opt}}$-ES never actually attains its target mutation strength. As adaptation is gradual rather than instantaneous, and as the distance to the optimizer continually decreases, the strategy will always be "behind" its target. Expanding the exponential function in Eq. (21) into a Taylor series, taking the decrease in distance

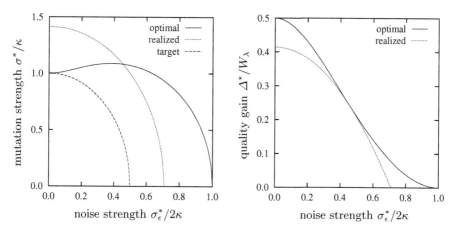

Fig. 7. Mutation strength σ^* and resulting quality gain Δ^* plotted against the noise strength σ_ϵ^*. In both graphs, the solid curves represent the optimal values from Fig. 5 and the dotted curves correspond to the values realized by the strategy and described by Eqs. (26) and (27), respectively. The dashed curve in the left hand graph is the target mutation strength given in Eq. (25)

to the optimizer into account, dropping all but the first terms, and demanding stationarity in the sense that the normalized mutation strength does not change yields equation

$$\sigma^* = \sigma^* \left(1 + \frac{\Delta^*}{N} + \frac{\|\mathbf{s}\|^2 - N}{2DN} \right).$$

Inserting Eqs. (19) and (24) and solving for σ^*, it follows that the average mutation strength actually realized by the strategy is

$$\sigma^* = \kappa \sqrt{2 - \left(\frac{\sigma_\epsilon^*}{\kappa} \right)^2} \tag{26}$$

if $\sigma_\epsilon^* \leq \sqrt{2}\kappa$, and it is zero if $\sigma_\epsilon^* > \sqrt{2}\kappa$. Inserting this result in Eq. (19) it follows that the resulting average quality gain of the $(\lambda)_{\text{opt}}$-ES with cumulative step length adaptation is

$$\Delta^* = \frac{\sqrt{2} - 1}{2} W_\lambda \left(2 - \left(\frac{\sigma_\epsilon^*}{\kappa} \right)^2 \right) \tag{27}$$

for $\sigma_\epsilon^* \leq \sqrt{2}\kappa$, and it is zero for $\sigma_\epsilon^* > \sqrt{2}\kappa$. Figure 8 illustrates for the case of a $(10)_{\text{opt}}$-ES that the accuracy of the predictions afforded by Eq. (26) is quite good even for small values of N while good agreement of the measurements of the quality gain with Eq. (27) requires very high search space dimensionalities. Similar results had been observed for the $(\mu/\mu, \lambda)$-ES in [1], and better agreement had been achieved by taking some N-dependent terms into account.

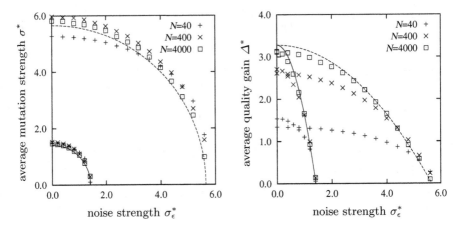

Fig. 8. Average mutation strength σ^* and average quality gain Δ^* of the $(10)_{\mathrm{opt}}$-ES with cumulative step length adaptation plotted against the noise strength σ_ϵ^*. The solid lines have been obtained from Eqs. (26) and (27) for $\kappa = 1.0$, the dashed lines for $\kappa = 4.0$. The crosses mark measurements obtained in runs of the strategies in search spaces with $N = 40$ (+), $N = 400$ (×), and $N = 4000$ ($\square$)

Both the mutation strength and the corresponding quality gain according to Eqs. (26) and (27) are illustrated in Fig. 7. While the mutation strength realized by the strategy generally differs from the optimal mutation strength, the right hand graph shows that the loss in quality gain is quite acceptable provided that the strategy operates not too close to the right hand edge of the graphs. For the $(\mu/\mu, \lambda)$-ES, the recipe for achieving this is to increase μ and λ; for the $(\lambda)_{\mathrm{opt}}$-ES, it is to increase κ.

6   Summary and Conclusions

In this paper, the behavior of weighted multirecombination evolution strategies has been studied on the infinite-dimensional sphere model. Optimal rank-based weights have been computed, and it has been found that optimal performance is achieved if those weights are set to equal the expected values of the order statistics of the standardized normal distribution. The performance of the resulting strategy – referred to as $(\lambda)_{\mathrm{opt}}$-ES – has been analyzed, and it was seen that unlike the $(\mu/\mu, \lambda)$-ES, the $(\lambda)_{\mathrm{opt}}$-ES is capable of exceeding the serial efficiency of the $(1 + 1)$-ES by a factor of roughly two and a half. It has then been found that the $(\lambda)_{\mathrm{opt}}$-ES in its original form does not benefit from genetic repair in the sense that a larger number of offspring generated per time step allows it to operate with larger mutation strengths. However, the strategy can be modified by scaling all weights using a common factor κ. While that factor is without influence on the performance of the strategy if there is no noise present, it has been found to be able to contribute positively to the strategy's robustness in the presence of Gaussian fitness noise. The scaling of weights has been likened to the

Table 1. Comparison of properties of $(\mu/\mu, \lambda)$-ES and $(\lambda)_{\text{opt}}$-ES on the infinite-dimensional sphere model

	$(\mu/\mu, \lambda)$-ES	$(\lambda)_{\text{opt}}$-ES
quality gain Δ^*	$\dfrac{\sigma^{*2} c_{\mu/\mu,\lambda}}{\sqrt{\sigma^{*2} + \sigma_\epsilon^{*2}}} - \dfrac{\sigma^{*2}}{2\mu}$	$\dfrac{W_\lambda}{\kappa}\left(\dfrac{\sigma^{*2}}{\sqrt{\sigma^{*2} + \sigma_\epsilon^{*2}}} - \dfrac{\sigma^{*2}}{2\kappa}\right)$
optimal σ^* (no noise)	$\mu c_{\mu/\mu,\lambda}$	κ
optimal Δ^* (no noise)	$\mu c_{\mu/\mu,\lambda}^2/2 \;(\overset{\lambda\to\infty}{\longrightarrow} 0.202\lambda)$	$W_\lambda/2 \;(\overset{\lambda\to\infty}{\longrightarrow} 0.5\lambda)$
maximal σ_ϵ^*	$2\mu c_{\mu/\mu,\lambda}$	2κ

idea of using rescaled mutations to which it is similar in effect, but from which it differs in that no explicit rescaling is required. Rather, the possibility of making large trial steps and at the same time small search steps is an implicit result of weighted multirecombination in combination with an appropriate choice of weights. Finally, it has been seen that by virtue of a simple modification, the cumulative step length adaptation mechanism works for the $(\lambda)_{\text{opt}}$-ES as well as it does for the $(\mu/\mu, \lambda)$-ES, and that good mutation strength settings can be arrived at by choosing κ sufficiently large. Table 1 summarizes some of the most important findings with regard to the performance of the $(\lambda)_{\text{opt}}$-ES on the sphere model and contrasts them with the corresponding results for the $(\mu/\mu, \lambda)$-ES.

Finally, it is important to emphasize that all results in this paper have been derived under the assumption of infinite search space dimensionality. The findings help provide a good intuitive understanding of the influence of the parameters λ and κ, of the issues involved in the choice of weights, and of the consequences of that choice for genetic repair and the performance of multirecombination evolution strategies. However, it has also been seen in computer experiments that the accuracy of the predictions can decrease with increasing κ, and that the recommendation to work with a large κ in the presence of noise has limits in finite-dimensional search spaces. Similar findings have been made for the choice of μ and λ in the $(\mu/\mu, \lambda)$-ES, and an improved approximation for the noisy case has been derived in [4]. That approximation replaces χ_N^2/N not with unity but instead with a normally distributed term with mean 1 and with variance $2/N$. A similar investigation for the $(\lambda)_{\text{opt}}$-ES would help determine optimal settings for λ and κ in finite-dimensional search spaces, and it could be used for verifying to what degree the performance advantages predicted can be realized.

Acknowledgments

The author is supported by the Natural Sciences and Engineering Research Council of Canada (NSERC). This paper was born out of discussions at the 2004 Dagstuhl Seminar on the Theory of Evolutionary Algorithms.

References

1. D. V. Arnold. *Noisy Optimization with Evolution Strategies*. Genetic Algorithms and Evolutionary Computation Series. Kluwer Academic Publishers, Boston, 2002.
2. D. V. Arnold. An analysis of evolutionary gradient search. In *Proc. of the 2004 IEEE Congress on Evolutionary Computation*, pages 47–54. IEEE Press, Piscataway, NJ, 2004.
3. D. V. Arnold and H.-G. Beyer. Local performance of the $(\mu/\mu_I, \lambda)$-ES in a noisy environment. In W. N. Martin and W. M. Spears, editors, *Foundations of Genetic Algorithms 6*, pages 127–141. Morgan Kaufmann Publishers, San Francisco, 2001.
4. D. V. Arnold and H.-G. Beyer. Performance analysis of evolution strategies with multi-recombination in high-dimensional $I\!R^N$-search spaces disturbed by noise. *Theoretical Computer Science*, 289(1):629–647, 2002.
5. D. V. Arnold and H.-G. Beyer. A comparison of evolution strategies with other direct search methods in the presence of noise. *Computational Optimization and Applications*, 24(1):135–159, 2003.
6. D. V. Arnold and H.-G. Beyer. Performance analysis of evolutionary optimization with cumulative step length adaptation. *IEEE Transactions on Automatic Control*, 49(4):617–622, 2004.
7. N. Balakrishnan and C. R. Rao. Order statistics: An introduction. In N. Balakrishnan and C. R. Rao, editors, *Handbook of Statistics*, volume 16, pages 3–24. Elsevier, Amsterdam, 1998.
8. H.-G. Beyer. Toward a theory of evolution strategies: On the benefit of sex – the $(\mu/\mu, \lambda)$-theory. *Evolutionary Computation*, 3(1):81–111, 1995.
9. H.-G. Beyer. On the asymptotic behavior of multirecombinant evolution strategies. In H.-M. Voigt, W. Ebeling, I. Rechenberg, and H.-P. Schwefel, editors, *Parallel Problem Solving from Nature – PPSN IV*, pages 122–133. Springer Verlag, Heidelberg, 1996.
10. H.-G. Beyer. Mutate large, but inherit small! On the analysis of rescaled mutations in $(\tilde{1}, \tilde{\lambda})$-ES with noisy fitness data. In A. E. Eiben, T. Bäck, M. Schoenauer, and H.-P. Schwefel, editors, *Parallel Problem Solving from Nature – PPSN V*, pages 109–118. Springer Verlag, Heidelberg, 1998.
11. H.-G. Beyer. Evolutionary algorithms in noisy environments: Theoretical issues and guidelines for practice. *Computer Methods in Mechanics and Applied Engineering*, 186:239–267, 2000.
12. H.-G. Beyer. *The Theory of Evolution Strategies*. Natural Computing Series. Springer Verlag, Heidelberg, 2001.
13. H. A. David and H. N. Nagaraja. Concomitants of order statistics. In N. Balakrishnan and C. R. Rao, editors, *Handbook of Statistics*, volume 16, pages 487–513. Elsevier, Amsterdam, 1998.
14. N. Hansen. *Verallgemeinerte individuelle Schrittweitenregelung in der Evolutionsstrategie*. Mensch & Buch Verlag, Berlin, 1998.
15. N. Hansen and A. Ostermeier. Completely derandomized self-adaptation in evolution strategies. *Evolutionary Computation*, 9(2):159–195, 2001.
16. S. Kern, S. D. Müller, N. Hansen, D. Büche, J. Ocenasek, and P. Koumoutsakos. Learning probability distributions in continuous evolutionary algorithms – A comparative review. *Natural Computing*, 3(1):77–112, 2004.
17. A. Ostermeier, A. Gawelczyk, and N. Hansen. Step-size adaptation based on non-local use of selection information. In Y. Davidor, H.-P. Schwefel, and R. Männer, editors, *Parallel Problem Solving from Nature – PPSN III*, pages 189–198. Springer Verlag, Heidelberg, 1994.

18. I. Rechenberg. *Evolutionsstrategie – Optimierung technischer Systeme nach Prinzipien der biologischen Evolution.* Friedrich Frommann Verlag, Stuttgart, 1973.
19. I. Rechenberg. *Evolutionsstrategie '94.* Friedrich Frommann Verlag, Stuttgart, 1994.
20. G. Rudolph. *Convergence Properties of Evolutionary Algorithms.* Verlag Dr. Kovač, Hamburg, 1997.
21. R. Salomon. Evolutionary search and gradient search: Similarities and differences. *IEEE Transactions on Evolutionary Computation,* 2(2):45–55, 1998.

On the Prediction of the Solution Quality
in Noisy Optimization*,**

Hans-Georg Beyer[1] and Silja Meyer-Nieberg[2]

[1] Department of Computer Science,
Vorarlberg University of Applied Sciences,
Achstr. 1, A-6850 Dornbirn, Austria
hans-georg.beyer@fh-vorarlberg.ac.at
[2] Department of Computer Science XI,
University of Dortmund,
D-44221 Dortmund, Germany
meyer@Ls11.cs.uni-dortmund.de

Abstract. Noise is a common problem encountered in real-world optimization. Although it is folklore that evolution strategies perform well
in the presence of noise, even their performance is degraded. One effect
on which we will focus in this paper is the reaching of a steady state that
deviates from the actual optimal solution.
The quality gain is a local progress measure, describing the expected
one-generation change of the fitness of the population. It can be used
to derive evolution criteria and steady state conditions which can be
utilized as a starting point to determine the final fitness error, i.e. the
expected difference between the actual optimal fitness value and that of
the steady state. We will demonstrate the approach by determining the
final solution quality for two fitness functions.

1 Introduction

Noise is often encountered in real-world optimization situations. The sources
can be manifold. Noise can stem from the use of stochastic simulation tools,
measurement errors, or can be due to the fact that it is not possible to sample
data from the entire space. Sometimes it is also introduced as a means to obtain
more robust solutions [10, 12, 20].

We will consider the effects of noise on evolution strategies (ES). These
nature-inspired search strategies aim at optimizing the objective or fitness function F by applying the evolutionary principles of recombination, mutation, and
selection. Although ES and other evolutionary algorithms are generally assumed

---

* This work was supported by the Deutsche Forschungsgemeinschaft (DFG) through
the Collaborative Research Center SFB 531 at the University of Dortmund and
by the Research Center for Process- and Product-Engineering at the Vorarlberg
University of Applied Sciences
** We would like to thank especially the anonymous reviewer # 2 for pointing out some
inconsistencies in the first version of this paper

A.H. Wright et al. (Eds.): FOGA 2005, LNCS 3469, pp. 238–259, 2005.

to be robust against the effects of noise [2, 10, 11, 15, 16], even their performance can be degraded. The convergence velocity is reduced and the evolution strategy is unable to get arbitrarily close to the actual optimum ending up in a steady state at some distance, instead. Therefore, a final fitness error, defined as the expected distance of the actual optimal value and the steady state fitness, is encountered.

There are two local progress measures describing the course of the evolution process. The first (not considered in this paper) is the progress rate describing the expected one-generation change in the space of the object parameters. The second, the quality gain, operates on the fitness space instead. Both progress measures can be used to derive evolution criteria that have to be met if progress towards the optimum is to be guaranteed. On the other hand, these criteria also lead to a description of the steady state since no further changes on average are expected there.

We will present an approach to determine the final fitness error based on stationarity conditions gained from the quality gain. For quadratic functions and the progress rate, the final fitness error was already obtained in [5]. An important point in the derivation was the introduction of the *equipartition assumption.* Once the ES has reached the steady state, the weighted components appear to contribute on average the same to the final fitness error. Using this assumption it was possible to attain a simple formula for the final solution quality.

This paper is organized as follows. In the next paragraph, we will introduce the quality gain for $(1, \lambda)$-ES maximizing a fitness function under noisy fitness evaluations. Using the quality gain, we will derive a necessary evolution criterion that ensures progress towards the optimum – finally obtaining a stationarity condition for the steady state. Afterwards, we will show how this condition can be used to determine the final fitness error.

To illustrate the general approach, we will finally consider two fitness functions as examples – biquadratic fitness functions of the form $F_1(\mathbf{y}) = - \sum_{i=1}^{N} c_i y_i^4$ and a variant of the L_1-norm with $F_2(\mathbf{y}) = - \sum_{i=1}^{N} |y_i|$. The biquadratic function class might be considered as a local continuous plateau model whereas F_2 is of interest because its isometric plot equals a rotated N-dimensional hypercube with a closed success region and with exceedingly low success probabilities close to the corners.

2   Evolution Criteria and Steady State

In the following, we consider $(1, \lambda)$-ES trying to maximize the fitness function F under noisy fitness evaluations. Since the parent population consists of one member, only the mutation and selection processes have to be taken into account. Based on the parent, λ offspring are created by adding normally distributed mutation vectors to the parental object vector. In general, all components of the mutation vector have the same standard deviation σ also called the mutation strength. The offspring with the seemingly best fitness value is then selected as the parent of the next generation.

Considering an object vector $\mathbf{y}$, the quality change $Q(\mathbf{x})$ induced by a mutation vector $\mathbf{x}$ is $Q(\mathbf{x}) = F(\mathbf{y} + \mathbf{x}) - F(\mathbf{y})$. But since the fitness evaluations are disturbed by noise, only the perceived fitness values $\tilde{F}(\mathbf{x}) = F(\mathbf{x}) + \epsilon$ can be observed. The noise term ϵ is assumed to be normally distributed with $E[\epsilon] = 0$ and σ_ϵ as standard deviation, which is also called the noise strength.

Under these conditions, it can be shown [6] that the quality gain, i.e. the expected one-generation change of the fitness, becomes

$$\overline{\Delta Q}_{1,\lambda} \simeq \frac{S_Q^2}{\sqrt{S_Q^2 + \sigma_\epsilon^2}} c_{1,\lambda} + M_Q, \tag{1}$$

provided that the dimension N of the search space is sufficiently large and that the higher order cumulants of the mutation-induced fitness distribution do asymptotically vanish. The parameter $c_{1,\lambda}$ is the so-called progress coefficient [4, p.72] and is defined as the expected value of the λth order statistic of a standard normally distributed variable, i.e. $c_{1,\lambda} := \frac{\lambda}{\sqrt{2\pi}} \int_{-\infty}^{\infty} u e^{-\frac{u^2}{2}} \Phi^{\lambda-1}(u)\,du$. The variables S_Q^2 and M_Q are the variance and the expected value, respectively, of the local quality change.

A positive quality gain guarantees progress towards the optimum since the quality increases from one generation to the next. Therefore, the condition $\overline{\Delta Q}_{1,\lambda} \geq 0$ constitutes a sufficient evolution criterion.

Let us assume that $M_Q \leq 0$ which is a typical characteristic of many test functions. The sufficient evolution criterion is therefore satisfied for $S_Q^2/\sqrt{S_Q^2 + \sigma_\epsilon^2} \geq |M_Q|/c_{1,\lambda}$. The inequality can be solved for S_Q^2 leading to

$$S_Q^2 \geq \frac{M_Q^2}{2c_{1,\lambda}^2} \left(1 + \sqrt{1 + \left(\frac{2c_{1,\lambda}\sigma_\epsilon}{M_Q} \right)^2} \right). \tag{2}$$

Since $\left(1 + \sqrt{1 + (2c_{1,\lambda}\sigma_\epsilon/M_Q)^2} \right) > 2c_{1,\lambda}\sigma_\epsilon/|M_Q|$, a necessary evolution criterion can be obtained as

$$S_Q^2 \geq \frac{|M_Q|}{c_{1,\lambda}} \sigma_\epsilon. \tag{3}$$

Only if (3) is fulfilled, progress towards the optimal value is possible. On the other hand, the case "$=$" in (3) leads to the stationarity condition $\overline{\Delta Q}_{1,\lambda} = 0$, provided that $S_Q^2 \ll \sigma_\epsilon^2$. In this case, S_Q^2 can be neglected in the square root in (1) and the condition $\overline{\Delta Q}_{1,\lambda} = 0$ simplifies to $(S_Q^2/\sigma_\epsilon)c_{1,\lambda} + M_Q = 0$. Thus, we obtain

$$S_Q^2 \simeq \frac{|M_Q|}{c_{1,\lambda}} \sigma_\epsilon \tag{4}$$

as steady state condition. Equation (4) can serve as a starting point to derive the final fitness error $E[\Delta F]$ if the noise strength is sufficiently large or – since the variance S_Q^2 generally strongly depends on the mutation strength – σ is sufficiently small.

3   How to Calculate the Final Fitness Error

The final fitness error is defined as the residual fitness error $\mathrm{E}[\Delta F] = \hat{F} - \mathrm{E}[F]$, where $\hat{F} = \max_{\mathbf{y}} F(\mathbf{y})$. In general, an expression for $\mathrm{E}[F]$ has to be found. Considering separable functions of the form $F(\mathbf{y}) = \sum_{i=1}^{N} f_i(y_i)$, the task simplifies to determining expressions for $\mathrm{E}[f_i(y_i)]$. As a starting point, the stationarity condition (4), $S_Q^2 = |M_Q|\sigma_\epsilon/c_{1,\lambda}$, can be used. Since the local quality is of the form $Q(\mathbf{x}, \mathbf{y}) = \sum_{i=1}^{N} q_i(x_i, y_i)$, the expected value and the variance are functions of type $M_Q = \sum_{i=1}^{N} \mathrm{E}[q_i(x_i, y_i)]$ and $S_Q^2 = \sum_{i=1}^{N} \mathrm{E}[q_i^2(x_i, y_i)] - \mathrm{E}^2[q_i(x_i, y_i)]$. The expectation is taken over the normally distributed variable x_i. Therefore, the expected values are functions of y_i and the mutation strength σ, i.e. $M_Q = \sum_{i=1}^{N} m_i(y_i, \sigma)$ and $S_Q^2 = \sum_{i=1}^{N} s_i(y_i, \sigma)$.

The main task is now to relate $\sum_{i=1}^{N} \mathrm{E}[f_i(y_i)]$ with $\sum_{i=1}^{N} \mathrm{E}[m_i(y_i, \sigma)]$ and $\sum_{i=1}^{N} \mathrm{E}[s_i(y_i, \sigma)]$ by means of the stationarity condition (4) and to eliminate the mutation strength in the equation. The tractability of this approach depends on the fitness function considered.

So far, we were able to determine the final fitness error for quadratic, biquadratic functions, the variant of the L_1-norm, and to test the ranges of the applicability for the bit-counting function OneMax.

As mentioned before, the final fitness error was already derived for quadratic fitness functions in [5] using a stationarity condition obtained by considering the progress rate. Therefore, we will only give a short outline of the determination of the final solution quality for this function type.

The fitness function is given by $F(\mathbf{y}) = \mathbf{b}^T\mathbf{y} - \mathbf{y}^T\mathbf{Q}\mathbf{y}$ where $\mathbf{y}$ and $\mathbf{b}$ are N-dimensional real-valued vectors and $\mathbf{Q}$ is a positive definite (symmetric) matrix. After performing a principal axis transformation [5], the final fitness error $\mathrm{E}[\Delta F]$ can be expressed as $\mathrm{E}[\Delta F] = \sum_{i=1}^{N} q_i \mathrm{E}[(y_i - \hat{y}_i)^2]$, where $\hat{\mathbf{y}} = \arg\max F(\mathbf{y})$, q_i is the ith eigenvalue of $\mathbf{Q}$, and $y_i = \mathbf{e}_i^T\mathbf{y}$ with $\mathbf{e}_i$ the ith eigenvector.

The local quality $Q(\mathbf{x}) = F(\mathbf{y}+\mathbf{x}) - F(\mathbf{y})$ is given by $Q(\mathbf{x}) = (2\mathbf{Q}(\hat{\mathbf{y}} - \mathbf{y}))^T\mathbf{x} - \mathbf{x}^T\mathbf{Q}\mathbf{x}$. The expected value and variance of Q were obtained in [4, p.122f] as $M_Q = -\sigma^2\mathrm{Tr}[\mathbf{Q}]$ and $S_Q^2 = 4\sigma^2\|\mathbf{Q}(\hat{\mathbf{y}} - \mathbf{y})\|^2 + 2\sigma^4\mathrm{Tr}[\mathbf{Q}^2]$. Inserting these expressions into (4), we get –assuming smallness of the mutation strength σ at the steady state

$$\|\mathbf{Q}(\hat{\mathbf{y}} - \mathbf{y})\|^2 \simeq \frac{\mathrm{Tr}[\mathbf{Q}]}{4c_{1,\lambda}}\sigma_\epsilon. \tag{5}$$

This agrees with the result found in [5]. Following the same assumptions used there, we can give an approximation for the expected final fitness error by relating (5) with $\mathrm{E}[\Delta F] = \sum_{i=1}^{N} q_i \mathrm{E}[(y_i - \hat{y}_i)^2]$ as

$$\mathrm{E}[\Delta F] = \frac{\sigma_\epsilon N}{4c_{1,\lambda}}. \tag{6}$$

Equation (6) is surprisingly simple and does not depend on the specific form of F. The validity of (6) was already tested in [5], yielding good results. One of the most important assumptions in the derivation of (6) is the so-called equipartition assumption stating that the weighted contribution of all components to the final fitness error can be supposed to be the same on average. In nearly all derivations of the final solution quality, this assumption is a crucial point simplifying the calculations. In the following, we will illustrate the general approach by two further examples.

3.1   The Biquadratic Case

In this section, we consider functions of the type $F_1(\mathbf{y}) = -\sum_{i=1}^{N} c_i y_i^4$, yielding the local quality $Q(\mathbf{x}) = \sum_{i=1}^{N}(-4c_i)y_i^3 x_i - 6c_i y_i^2 x_i^2 - 4c_i y_i x_i^3 - c_i x_i^4$. The final fitness error is given by $\mathrm{E}[\Delta F] = \sum_{i=1}^{N} c_i \mathrm{E}[y_i^4]$. Thus, we need to find expressions for the fourth moment of y_i using the steady state condition (4). To calculate (4) in the biquadratic case, we need the expected value and the variance of the local quality which are given as [7]

$$M_Q = -3\sigma^2 \sum_{i=1}^{N} c_i(2y_i^2 + \sigma^2) \tag{7}$$

$$S_Q^2 = \sum_{i=1}^{N} 16c_i^2 y_i^6 \sigma^2 + 168c_i^2 y_i^4 \sigma^4 + 384c_i^2 y^2 \sigma^6 + 96c_i^2 \sigma^8. \tag{8}$$

Considering only small mutation strengths, the variance simplifies to $S_Q^2 \simeq \sigma^2 \sum_{i=1}^{N} 16c_i^2 y_i^6$. Thus, the steady state condition (4), $S_Q^2 \simeq |M_Q|\sigma_\epsilon/c_{1,\lambda}$, reads

$$\sum_{i=1}^{N} 16c_i^2 y_i^6 \simeq \frac{3\sigma_\epsilon}{c_{1,\lambda}} \sum_{i=1}^{N} c_i(2y_i^2 + \sigma^2). \tag{9}$$

We will consider the limit case of $\sigma \to 0$. Therefore, the addends containing the mutation strength will be neglected

$$\sum_{i=1}^{N} 16c_i^2 y_i^6 \simeq \frac{6\sigma_\epsilon}{c_{1,\lambda}} \sum_{i=1}^{N} c_i y_i^2. \tag{10}$$

This resulting condition contains the y_i-variates which obey certain (unknown) steady state distributions. Assuming statistical independence of the y_i at the steady state, there still remains the determination of the $p_i(y_i)$ steady state density functions. The approach presented here does not allow for a derivation of the p_i from first principles. Instead, we will use an *ansatz*: As a first approximation, we assume that the y_i are normally distributed with zero mean and standard deviation σ_i

$$y_i \sim \mathcal{N}(0, \sigma_i^2). \tag{11}$$

In order to obtain σ_i, we make reference to our equipartion principle. That is, we again assume that the components Δf_i in the fitness sum

$$\Delta F = \sum_{i=1}^{N} \Delta f_i = \sum_{i=1}^{N} c_i y_i^4 \tag{12}$$

obey the same distribution and, therefore,

$$\forall i = 1, \ldots, N : \mathrm{E}[\Delta f_i] = c_i \overline{y_i^4} = K. \tag{13}$$

Thus, we get with (11), $\mathrm{E}[\Delta f_i] \approx 3 c_i \sigma_i^4 \approx K$ and

$$\sigma_i^2 \approx \sqrt{\frac{K}{3c_i}}, \tag{14}$$

where K is a constant to be determined using the steady state condition (10). Taking the expected value in (10), we get

$$\sum_{i=1}^{N} 16 \cdot 15 c_i^2 \sigma_i^6 \approx \frac{6\sigma_\epsilon}{c_{1,\lambda}} \sum_{i=1}^{N} c_i \sigma_i^2. \tag{15}$$

and plugging (14) into (15) results in

$$\sum_{i=1}^{N} 16 \cdot 5\sqrt{c_i} K \approx \frac{6\sigma_\epsilon}{c_{1,\lambda}} \sum_{i=1}^{N} \sqrt{c_i}. \tag{16}$$

This condition must hold for all choices of c_i and therefore it must hold for each single addend in (16). Thus, we get

$$K \approx \frac{3\sigma_\epsilon}{40 c_{1,\lambda}}. \tag{17}$$

Now, consider the expected value of (12) which is the final fitness error we are interested in $\mathrm{E}[\Delta F] := \sum_{1=1}^{N} c_i \overline{y_i^4}$. Using (13), we see that $\mathrm{E}[\Delta F] = NK$ and inserting (11) we finally end up with

$$\mathrm{E}[\Delta F] \approx \frac{3N\sigma_\epsilon}{40 c_{1,\lambda}}. \tag{18}$$

Equation (18) was obtained by applying an equipartition assumption stating that the contribution of the weighted fitness components to the final fitness error can be supposed to be the same once the steady state is reached. Furthermore, the limit case $\sigma \to 0$ was considered. As in the quadratic case, (18) does not depend on any terms of $F_1(\mathbf{y}) = -\sum_{i=1}^{N} c_i y_i^4$, that is, the final fitness error is independent of the specific c_i-values.

In the following, we will examine the predictive quality of (18). Although it was developed for $(1, \lambda)$-ES, we will extend it to $(\mu/\mu_I, \lambda)$-ES by replacing $c_{1,\lambda}$

with $\mu c_{\mu/\mu,\lambda}$. The progress coefficient $c_{\mu/\mu,\lambda}$ is defined as the expectation of the average over the μ largest samples of a population of λ random samples which are chosen according to the standard normal distribution [4, p.216]. The expectation $E[\Delta F]$ then corresponds to the final fitness error of the centroid of the parental population. As test functions, $F_{1.1}(\mathbf{y}) = -\sum_{i=1}^{N} y_i^4$, $F_{1.2}(\mathbf{y}) = -\sum_{i=1}^{N} i y_i^4$, and $F_{1.3}(\mathbf{y}) = -\sum_{i=1}^{N} i^2 y_i^4$ were chosen.

The plots in Fig. 1 show the dependency of the final fitness error $E[\Delta F]$ on the number of parents μ for some $(\mu/\mu_I, 60)$-ES runs. Depicted are the values predicted by (18) and the actually measured ones. The experimental values were obtained by averaging over 500,000 ($N = 30$) and 900,000 ($N = 100$) generations in the steady state region. The values for $\mu = 58$ and $\mu = 59$ are not shown

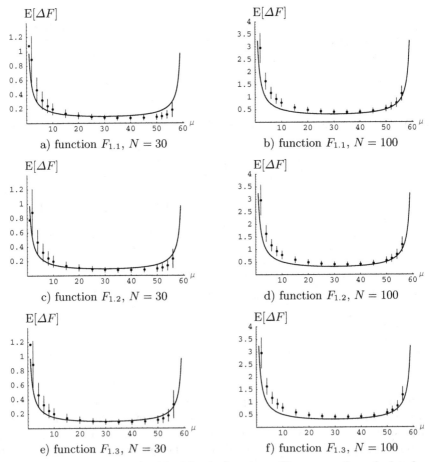

Fig. 1. Final fitness error for biquadratic functions. Shown are the calculated and experimental values for $E[\Delta F]$ of $(\mu/\mu_I, 60)$-ES with $\mu =$1, 2, 4, 6, 8, 10, 15, 20, 25, 30, 35, 40, 45, 50, 52, 54, 56. The vertical bars indicate the $\pm$ standard deviations. The noise strength is $\sigma_\epsilon = 1$

because of convergence problems for these ES. The standard mutative σ self-adaptation (σSA, [3, 9, 18, 19]) was used as σ-control rule in all experiments.

The overall agreement between equation and experiment is rather good although some deviations can be observed. The experimentally found final fitness error of the $(1, 60)$ evolution strategy is significantly higher than the theoretically obtained lower bound. A similar case was already observed in [5] in the case of quadratic test functions. $(1, \lambda)$ evolution strategies using σSA tend to premature stagnation which is the consequence of a very fast reduction of the mutation strength.

This can also be observed in Fig. 2 showing exemplary runs of an $(1, 60)$-ES and of a $(30/30_I, 60)$-ES optimizing test function $F_{1.2}$. In the case of the second ES, σ is reduced to a value around 0.03, where it starts to fluctuate. As a consequence, ΔF also approaches its steady state a short time later. In contrast to this, the mutation strength is reduced faster and further in the case of the $(1, 60)$-ES reaching values that are much closer to zero. Therefore, ΔF stagnates very soon. For shorter periods, σ can attain higher values though being unable to stabilize at this level. The reasons for this behavior are not fully understood.

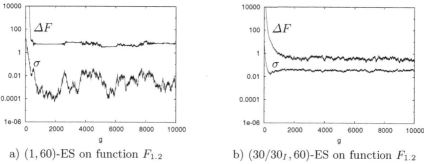

a) $(1, 60)$-ES on function $F_{1.2}$ b) $(30/30_I, 60)$-ES on function $F_{1.2}$

Fig. 2. ΔF and σ-dynamics of an $(1, 60)$ and a $(30/30_I, 60)$ ES using σSA. The noise strength is $\sigma_\epsilon = 1$

The following Fig. 3 shows the average value of the weighted fitness components $\langle \Delta f_i \rangle$ for the functions $F_{1.1}$-$F_{1.3}$. The experimental values were obtained by averaging over a total of $15,000,000$ generations with 30 restarts. The plots for $F_{1.2}$ and $F_{1.3}$ show a trend in the data with decreasing values from the lower weighted components to the higher weighted ones which does not actually seem to support the equipartion assumption. Therefore, the equipartion assumption for the biquadratic functions $F(\mathbf{y}) = \sum_{i=1}^{N} c_i y_i^4$ will have to be investigated more closely in the future. Possible reasons might be a very slow convergence of the single components or numerical instabilities due to the fourth power of the y_i that appears in the fitness functions. We also investigated the weighted fitness components obtained by $(1, 60)$-ES runs, where a minimal value for the mutation strength ($\sigma_{min} = 0.01$) was introduced. The values were obtained by sampling over $40,000,000$ generations with 80 restarts. As one can see in Fig. 4, there is a good agreement with our equipartition assumption in the case of the

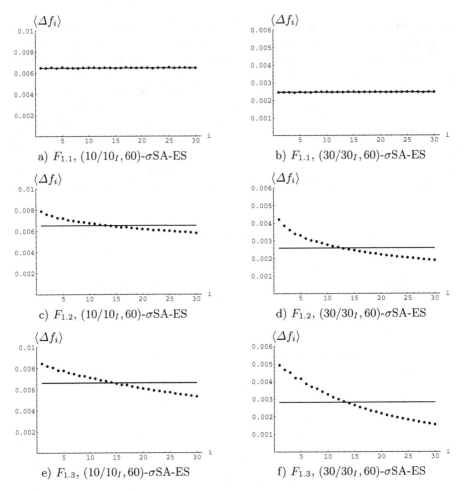

Fig. 3. Average value of the weighted fitness contributions $\langle \Delta f_i \rangle$ of the functions $F_{1.1}$-$F_{1.3}$ for $N = 30$. The values were obtained by sampling over $15,000,000$ generations with 30 restarts. Shown as lines are the estimates $\frac{1}{N} \sum_{i=1}^{N} \langle \Delta f_i \rangle$

$(1, \lambda)$-ES. This is in contrast to the $\mu > 1$ cases and raises the question why the intermediate recombination destroys the equipartition of the fitness components.

It is interesting to see, though, that the average value over all components resembles the estimate of $F_{1.1}$. Therefore, the error introduced by assuming that the contributions of components are the same seems to be small.

Figure 5 shows histograms for $(30/30_I, 60)$-σSA-ES on the functions $F_{1.1}$-$F_{1.3}$. Shown are the values of the first, 15th, and the 30th fitness component. The values were obtained by sampling over a total of $20,000,000$ generations with 40 restarts and grouped in 1000 intervals from 0-0.005. In general, all fitness components lead to skew distributions with a distinctive tail. Again differences can be observed in the cases of $F_{1.2}$ and $F_{1.3}$, where the higher weighted components tend more towards smaller values.

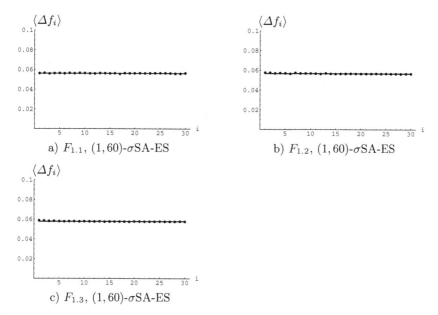

Fig. 4. Average value of the weighted fitness contributions $\langle \Delta f_i \rangle$ of the functions $F_{1.1}$-$F_{1.3}$ for $N = 30$. The values were obtained by sampling over $40,000,000$ generations with 80 restarts. Shown as lines are the estimates $\frac{1}{N} \sum_{i=1}^{N} \langle \Delta f_i \rangle$

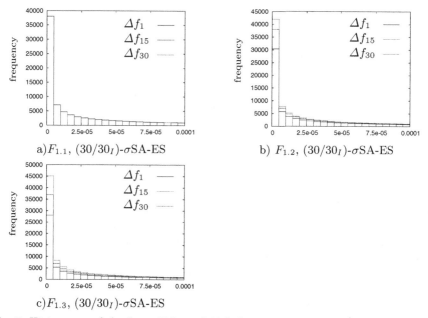

Fig. 5. Histograms of the first, 15th, and 30th fitness component. The noise strength is $\sigma_\epsilon = 1$. The histograms comprise 1000 classes between 0 and 0.005. Shown is the section up to 0.0001

3.2 L_1-Norm Case

We will now consider the function $F_2(\mathbf{y}) = -\sum_{i=1}^N |y_i|$. In order to obtain an expression for the final fitness error which is given as $E[\Delta F] = \sum_{i=1}^N E[|y_i|]$, we will follow a similar approach as before. First, we have to derive the specific steady state criterion $S_Q^2 = |M_Q|\sigma_\epsilon/c_{1,\lambda}$. The required expected value M_Q and the variance S_Q^2 can be given after some straightforward calculations [7] as

$$M_Q = \sum_{i=1}^N |y_i| - y_i(2\Phi_{0,\sigma^2}(y_i) - 1) - 2\sigma^2\phi_{0,\sigma^2}(y_i) \quad \text{and} \tag{19}$$

$$S_Q^2 = N\sigma^2 + 4\sum_{i=1}^N y_i^2 \Phi_{0,\sigma^2}(y_i)[1 - \Phi_{0,\sigma^2}(y_i)]$$

$$-4\sigma^2 \sum_{i=1}^N \phi_{0,\sigma^2}(y_i)y_i[2\Phi_{0,\sigma^2}(y_i) - 1] - 4\sigma^4 \sum_{i=1}^N \phi_{0,\sigma^2}^2(y_i), \tag{20}$$

where $\phi_{0,\sigma^2}(x_i) := \frac{1}{\sqrt{2\pi}\sigma}e^{-\frac{x_i^2}{2\sigma^2}}$ and Φ_{0,σ^2} is the corresponding distribution function. The steady state criterion requires the absolute value $|M_Q|$ which is obtained as $|M_Q| = \sum_{i=1}^N y_i(2\Phi_{0,\sigma^2}(y_i) - 1) + 2\sigma^2\phi_{0,\sigma^2}(y_i) - |y_i|$ (see Appendix 5.1). To simplify the calculations, we introduce an upper bound for S_Q^2 noting that $S_Q^2 \le N\sigma^2 - 4\sigma^4 \sum_{i=1}^N \phi_{0,\sigma^2}^2(y_i)$ (see Appendix 5.1). Considering the steady state criterion $S_Q^2 = |M_Q|\sigma_\epsilon/c_{1,\lambda}$, it follows that

$$N\sigma^2 - 4\sigma^4 \sum_{i=1}^N \phi_{0,\sigma^2}^2(y_i) \ge \frac{\sigma_\epsilon}{c_{1,\lambda}} \sum_{i=1}^N y_i(2\Phi_{0,\sigma^2}(y_i) - 1)$$

$$+2\sigma^2\phi_{0,\sigma^2}(y_i) - |y_i|. \tag{21}$$

In the following let $z_i = y_i/\sigma$. Equation (21) thus simplifies to

$$N\sigma - \frac{2\sigma}{\pi}\sum_{i=1}^N e^{-z_i^2} \ge \frac{\sigma_\epsilon}{c_{1,\lambda}}\Big(\sum_{i=1}^N z_i(2\Phi(z_i) - 1) + \frac{2}{\sqrt{2\pi}}e^{-\frac{z_i^2}{2}} - |z_i|\Big), \tag{22}$$

where $\Phi(z_i) = \Phi_{0,1}(z_i)$. Applying the equipartition assumption to $F_2(\mathbf{y}) = -\sum_{i=1}^N |y_i|$, we postulate $E[|y_i|] = E[|y_k|]$ for $k \ne i$ or $E[|z_i|] = E[|z_k|]$ for $z_i = y_i/\sigma$, respectively. In addition, we assume $E[h(z_i)] = E[h(z_k)]$, where $i \ne k$ and h is one of the functions in (22). Furthermore, since our approach again does not allow for a determination of the densities of the z_i from first principles, we assume the z_i to be normally distributed with expected value zero and variance σ_z^2. In this context, the final fitness error $E[\Delta F] = \sum_{i=1}^N E[|y_i|] = \sigma \sum_{i=1}^N E[|z_i|]$ is given as $E[\Delta F] = N\sigma E[|z|] \approx N\sigma\sigma_z\sqrt{2/\pi}$ and the expected values of the terms of (22) become

$$E[|z|] \approx \sqrt{\frac{2}{\pi}}\sigma_z,$$

$$E[z(2\Phi(z)-1)] \approx \sqrt{\frac{2}{\pi}}\frac{\sigma_z^2}{\sqrt{\sigma_z^2+1}},$$

$$\frac{2}{\sqrt{2\pi}}E[e^{-\frac{z^2}{2}}] \approx \sqrt{\frac{2}{\pi}}\frac{1}{\sqrt{\sigma_z^2+1}}, \quad \text{and}$$

$$\frac{2\sigma}{\pi}E[e^{-z^2}] \approx \frac{2}{\pi\sqrt{2\sigma_z^2+1}}\sigma_z. \tag{23}$$

Thus, assuming that the inequality still holds, we obtain for the expected value of (22)

$$N\sigma\left(1-\frac{2}{\pi\sqrt{2\sigma_z^2+1}}\right) \gtrsim \frac{\sigma_\epsilon}{c_{1,\lambda}}N\left(\sqrt{\frac{2}{\pi}}\sqrt{\sigma_z^2+1}-\sigma_z\sqrt{\frac{2}{\pi}}\right), \tag{24}$$

from which a relationship between the mutation strength in the steady state and the variance of z_i, i.e.

$$\sigma \gtrsim \frac{\sigma_\epsilon}{c_{1,\lambda}}\sqrt{\frac{2}{\pi}}\left(\frac{\sqrt{\sigma_z^2+1}-\sigma_z}{(1-\frac{2}{\pi\sqrt{2\sigma_z^2+1}})}\right), \tag{25}$$

can be derived. Inequality (25) can be used to determine a lower bound for the final fitness error since

$$E[\Delta F] \approx N\sqrt{\frac{2}{\pi}}\sigma_z\sigma \gtrsim N\frac{2\sigma_\epsilon}{c_{1,\lambda}\pi}\left(\sigma_z\frac{\sqrt{\sigma_z^2+1}-\sigma_z}{(1-\frac{2}{\pi\sqrt{2\sigma_z^2+1}})}\right). \tag{26}$$

The standard deviation σ_z is unknown. In order to proceed, we consider (25) noting that its right hand side is a monotonically decreasing function of σ_z. Assuming smallness of the mutation strength in the steady state, σ_z has to be sufficiently high. Especially, we assume $\sigma_z \geq 1$.

Hence, we are interested in finding a lower bound for $h(x) = (\sqrt{x^4+x^2}-x^2)/(1-\frac{2}{\pi\sqrt{2x^2+1}})$ for $x \geq 1$. As one can easily see, $h(x) \to 0.5$ for $x \to \infty$ since the denominator $1-\frac{2}{\pi\sqrt{2x^2+1}}$ approaches 1 and the numerator is given by $\sqrt{x^4+x^2}-x^2 = (\sqrt{x^4+x^2}-x^2)(\sqrt{x^4+x^2}+x^2)/(\sqrt{x^4+x^2}+x^2) = 1/(1+\sqrt{1+1/x^2})$ which goes to 0.5. As it can be shown in Appendix 5.1, we have $h(x) \geq 0.5$ for all $x \geq 1$. Therefore, we can give a lower bound for the final fitness error as

$$E[\Delta F] \gtrsim \frac{\sigma_\epsilon}{\pi c_{1,\lambda}}N. \tag{27}$$

To derive (27), we used again a variant of the equipartition assumption, first introduced in [5]. Furthermore, we had to assume that the variables y_i or standardized variables $z_i = y_i/\sigma$, respectively, obey a normal distribution. Under this assumption we could derive (25), which gives a function of the noise strength and the standard deviation of z as a lower bound for the mutation strength. Equation (25) can further be used to obtain an estimate for the final inequality (27). Assuming that $\sigma_z \geq 1$, $h(\sigma_z)$ is always higher than its limit value which can then be used to derive the lower bound.

As it has been done for biquadratic functions, the final fitness error can be easily extended to $(\mu/\mu_I, \lambda)$-ES by substituting $c_{1,\lambda}$ with $\mu c_{\mu/\mu,\lambda}$. The applicability of the resulting lower bound, i.e. $E[\Delta F] \geq N\sigma_\epsilon/(\pi\mu c_{\mu/\mu_I,\lambda})$, was investigated by conducting experiments with $(\mu/\mu_I, 60)$-ES for a noise strength of 1 using a 100-, 200-, and a 300-dimensional search space. Unless stated otherwise, the experimental values were obtained by sampling over $900,000$ generations in the steady state. As σ control rules, σSA and CSA (cumulative step-length adaptation [1, 13, 14]) were used. In the latter case, a σ_{min} value was introduced in order to prevent premature convergence.

As one can see in Fig. 6, the lower bound is not violated and seems to be a reasonable approximation of the real final fitness error. In the case of the ES using σSA, the approximation error decreases from around 40-50% for $N = 100$ to 30-40% for $N = 200$. The approximation error for higher-dimensional search spaces is generally smaller. Again $(1, \lambda)$-ES show premature convergence and their results are therefore not always depicted.

In general, the CSA-algorithm seems to lead towards smaller values especially for higher parental numbers where the deviation of the ES using σSA increases.

Since we assumed the $z_i = y_i/\sigma$ to be normally distributed with the same variance σ_z^2, we also investigated the validity of this assumption. Figure 7 shows relative frequencies for some $(\mu/\mu_I, \lambda)$-ES using σSA for $N = 100$ and $N = 200$. The values were obtained by sampling over a total of $800,000$ generations in the steady state with eight restarts grouping the values in 509 ($N = 100$) and 209 ($N = 200$) intervals ranging from -15.5 to 15.5. Also shown are the relative frequencies gained by sampling from normal distributions with the experimentally found σ_z-values.

The normal distribution seems to be only a rough approximation of the actual distribution function since the curves are generally steeper and narrower. Therefore, we introduce the double exponential distribution with the density function $p(z_i) = 1/(2\alpha)e^{-|z_i|/\alpha}$ as a more appropriate choice. To obtain an estimate of the final fitness error, we start again with Equation (22), i.e.

$$N\sigma - \frac{2\sigma}{\pi}\sum_{i=1}^{N}e^{-z_i^2} \geq \left(\sum_{i=1}^{N}z_i(2\Phi(z_i) - 1) + \frac{2}{\sqrt{2\pi}}e^{-\frac{z_i^2}{2}} - |z_i|\right)\frac{\sigma_\epsilon}{c_{1,\lambda}}.$$

In order to continue, we need to determine the expected values of the terms in (22) which are obtained after some calculations as

$$E[|z_i|] \approx \alpha,$$

$$E[z_i(2\Phi(z_i) - 1)] \approx 2(\alpha - \frac{1}{\alpha})\Phi(-\frac{1}{\alpha})e^{\frac{1}{2\alpha^2}} + \sqrt{\frac{2}{\pi}},$$

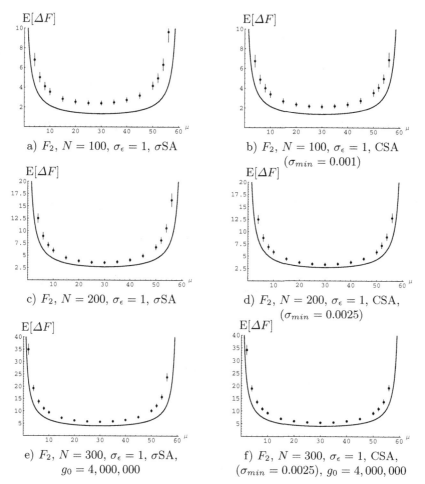

Fig. 6. $E[\Delta F]$ for some $(\mu/\mu_I, 60)$-ES optimizing function F_2. The equation for the final fitness error was gained using the normal distribution in the derivation. Shown are the results obtained by (27) (solid curves) and the experimentally obtained values. As control rules for the mutation strength adaptation, σSA and CSA were used. The vertical bars indicate the $\pm$ standard deviations

$$\frac{2}{\sqrt{2\pi}}E[e^{-\frac{z_i^2}{2}}] \approx \frac{2}{\alpha}\Phi(-\frac{1}{\alpha})e^{\frac{1}{2\alpha^2}}, \quad \text{and}$$

$$\frac{2\sigma}{\pi}E[e^{-z_i^2}] \approx \frac{2}{\sqrt{\pi}\alpha}\sigma\Phi(-\frac{1}{\sqrt{2}\alpha})e^{\frac{1}{4\alpha^2}}. \tag{28}$$

Therefore, the expected value of (22) becomes

$$\sigma\left(1 - \frac{2}{\sqrt{\pi}\alpha}\Phi(-\frac{1}{\sqrt{2}\alpha})e^{\frac{1}{4\alpha^2}}\right) \gtrless \frac{\sigma_\epsilon}{c_{1,\lambda}}\left(2(\alpha - \frac{1}{\alpha})\Phi(-\frac{1}{\alpha})e^{\frac{1}{2\alpha^2}} + \sqrt{\frac{2}{\pi}}\right)$$

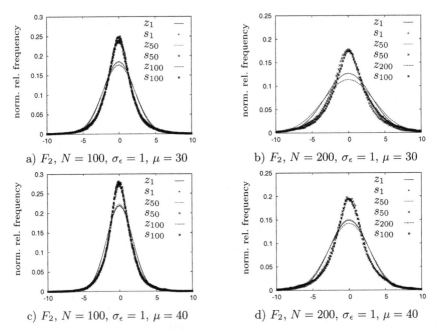

a) F_2, $N = 100$, $\sigma_\epsilon = 1$, $\mu = 30$

b) F_2, $N = 200$, $\sigma_\epsilon = 1$, $\mu = 30$

c) F_2, $N = 100$, $\sigma_\epsilon = 1$, $\mu = 40$

d) F_2, $N = 200$, $\sigma_\epsilon = 1$, $\mu = 40$

Fig. 7. Relative frequencies for function F_2 for some $(\mu/\mu_I, 60)$-ES using σSA for $N = 100$ and $N = 200$. Also included are the corresponding relative frequencies s_i drawn from normal distributions using the experimentally found standard deviations

$$+ \frac{\sigma_\epsilon}{c_{1,\lambda}} \left(\frac{2}{\alpha} \Phi(-\frac{1}{\alpha}) e^{\frac{1}{2\alpha^2}} - \alpha \right)$$

$$\Rightarrow \sigma \gtrsim \frac{\sigma_\epsilon}{c_{1,\lambda}} \left(\frac{2\alpha\Phi(-\frac{1}{\alpha}) e^{\frac{1}{2\alpha^2}} + \sqrt{\frac{2}{\pi}} - \alpha}{1 - \frac{2}{\sqrt{\pi}\alpha} \Phi(-\frac{1}{\sqrt{2}\alpha}) e^{\frac{1}{4\alpha^2}}} \right). \quad (29)$$

The final fitness error $E[\Delta F] = \sum_{i=1}^{N} E[|y_i|]$ is given as $E[\Delta F] \approx N\sigma\alpha$. Therefore, we obtain the inequality

$$E[\Delta F] \gtrsim N \frac{\sigma_\epsilon}{c_{1,\lambda}} \left(\frac{\alpha^2(2\Phi(-\frac{1}{\alpha}) e^{\frac{1}{2\alpha^2}} - 1) + \sqrt{\frac{2}{\pi}}\alpha}{1 - \frac{2}{\sqrt{\pi}\alpha} \Phi(-\frac{1}{\sqrt{2}\alpha}) e^{\frac{1}{4\alpha^2}}} \right). \quad (30)$$

This lower bound for the final fitness error depends on the parameter α which is unknown. Therefore, we consider the function

$$h(x) = \frac{x^2(2\Phi(-\frac{1}{x}) e^{\frac{1}{2x^2}} - 1) + \sqrt{\frac{2}{\pi}}x}{1 - \frac{2}{\sqrt{\pi}x} \Phi(-\frac{1}{\sqrt{2}x}) e^{\frac{1}{4x^2}}}. \quad (31)$$

The plot of $h(x)$ is shown in Fig. 8. We have $h(x) \to 0.5$ for $x \to \infty$ which can be easily shown by applying l'Hospital's Rule [7]. When using the normal

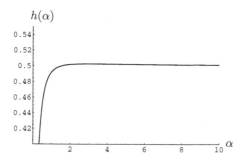

Fig. 8. The function $h(\alpha)$ (31) which appears in the derivation of the alternative final fitness error (32) for function F_2 using the double exponential distribution. It approaches 0.5 for $\alpha \to \infty$

distribution, we assumed $\sigma_z \geq 1$. Since the variance of a double exponentially distributed variable z is given as $\mathrm{Var}[z] = 2\alpha^2$, we use $\alpha \geq 1/\sqrt{2}$ as an equivalent assumption. Since $h(x) \geq h(1/\sqrt{2}) \geq 0.47715$ for all $x \geq 1/\sqrt{2}$ (see Appendix 5.1), we obtain the lower bound

$$E[\Delta F] \gtrsim 0.47715 N \frac{\sigma_\epsilon}{c_{1,\lambda}}. \tag{32}$$

To obtain (32), we made analog assumptions as in the derivation of (27). That is, we introduced the same equipartition assumption for the z_i and tried to determine an estimation for the final fitness error by finding a lower bound for (31). The only difference is the assumption of a different distribution function.

The predictive quality of (32) will be assessed in the following. Figure 9 shows the results obtained by $E[\Delta F] \gtrsim 0.47715 N \sigma_\epsilon / (\mu c_{\mu/\mu,\lambda})$ (32) using the double exponential distribution instead of the normal distribution in the derivation. Generally, the approximation quality seems to be better compared with (27), although the lower bound is violated for $N = 200$ and $N = 300$. This might be a hint that the double exponential distribution is not an appropriate choice for higher-dimensional search spaces although the predictions of (32) do not show a further decrease when switching from $N = 200$ to $N = 300$.

4   Conclusions and Outlook

In this paper, we considered the final solution quality of evolution strategies that are disturbed by fitness noise of constant variance. As a starting point for the derivation, the quality gain, a local performance measure, was chosen. As in the case of the progress rate [5], it can be used to derive evolution criteria and to characterize the steady state. Assuming a sufficiently large noise strength or a sufficiently small mutation strength, respectively, the steady state condition (4), i.e. $S_Q^2 = |M_Q|\sigma_\epsilon/c_{1,\lambda}$, can be obtained. Equation (4) can be utilized in turn for the determination of the final solution quality, i.e. final fitness error.

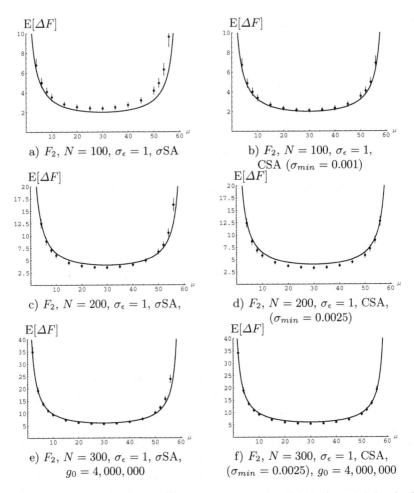

Fig. 9. $E[\Delta F]$ for some $(\mu/\mu_I, 60)$-ES optimizing function F_2. The equation for the final fitness error was gained using the double exponential distribution in the derivation. Shown are the results obtained by (32) (solid curves) and the experimentally obtained values. As control rules for the mutation strength adaptation, σSA and CSA were used. The vertical bars indicate the $\pm$ standard deviations

In this paper two examples – the biquadratic functions and the L_1-norm – were considered. For these two function types and the class of quadratic functions the final fitness error can be given as $E[\Delta F] = \sigma_\epsilon N/(Cc_{1,\lambda})$. The constant C depends only on the fitness function type but not on the specific function itself. Although the formula is surprisingly simple, the prediction quality is reasonably good. Since the predictions of the final fitness error are so similar, optimal values of the final solution quality are attained for a μ/λ-ratio of around $1/2$. This is not the optimal choice as to the convergence velocity for which $\mu/\lambda \approx 0.27$ is usually recommended [5]. But since the curve of the final fitness error is quite flat in the interval $\mu/\lambda = 1/3$ to $\mu/\lambda = 2/3$, one of these values can also be used without expecting a serious increase of the final fitness error.

The equations for the final fitness error were derived for $(1, \lambda)$ evolution strategies. We considered the $(\mu/\mu_I, \lambda)$ variants by a simple modification of the respective equations. Although the prediction quality is generally quite good as we have seen, a formal derivation of the final fitness error for $(\mu/\mu_I, \lambda)$-ES still needs to be developed.

We applied the standard noise model assuming that the noise can be modeled simply by adding a normally distributed random variable to the fitness function. While this model is used in most publications, there are several cases where it might not be applicable and other forms of modeling will have to be used. As an example, we refer to the optimization of aerodynamic structures [17], where the noise has to be modeled inside the fitness function. The investigations will have to be extended to develop equations for these cases, which will require the application of novel techniques like the ones which were developed in [8].

A central and crucial point in the determination of the final fitness error was the equipartition assumption first introduced in [5]. Variants of it have been applied for nearly all test functions considered so far. An open problem remains in the case of the biquadratic functions, where different contributions of the weighted fitness components have been observed. This might be due to setting $\mu > 1$, since $(1, \lambda)$-ES do not show the same trend on the data as $(\mu/\mu_I, \lambda)$-ES. The exact reasons for this behavior will have to be investigated.

5   Appendix

5.1   Final Fitness Error of Function $F_2(y) = -\sum_{i=1}^{N} |y_i|$

Inequality 1: $S_Q^2 \leq N\sigma^2 - 4\sigma^4 \sum_{i=1}^{N} \phi_{0,\sigma^2}^2(y_i)$

In order to prove the first inequality, we consider a single addend of S_Q^2, i.e.
$\sigma^2 + 4y^2 \Phi_{0,\sigma^2}(y)[1 - \Phi_{0,\sigma^2}(y)] - 4\sigma^2 \phi_{0,\sigma^2}(y)y[2\Phi_{0,\sigma^2}(y) - 1] - 4\sigma^4 \phi_{0,\sigma^2}^2(y)$, and show that

$$y^2 \Phi_{0,\sigma^2}(y)[1 - \Phi_{0,\sigma^2}(y)] \leq \sigma^2 \phi_{0,\sigma^2}(y)y[2\Phi_{0,\sigma^2}(y) - 1]. \tag{33}$$

Since the terms on both sides of the inequality are symmetric, it suffices to show the validity of (33) for $y \geq 0$. Since (33) is fulfilled for $y = 0$, we consider $y > 0$ in the following. Using the transformation $z = \sigma y$, (33) simplifies to

$$z\Phi(z)(1 - \Phi(z)) = z\Phi(z)\Phi(-z) \leq \phi(z)(2\Phi(z) - 1), \tag{34}$$

where $\Phi(z) = \Phi_{0,1}(z)$ and $\phi(z) = \phi_{0,1}(z)$. Considering $f(z) := z\Phi(z)\Phi(-z) - \phi(z)(2\Phi(z) - 1)$, we see that $f(0) = 0$ and that $f(z) \to 0$ for $z \to \infty$.

We need to show that $f(z) \leq 0$ for all $z > 0$. If f has only a minimum on $(0, \infty)$, the inequality is proven.

The first derivative of f is given as $f'(z) = \Phi(z)\Phi(-z) - 2\phi^2(z) := h(z) - g(z)$ with $f'(0) < 0$ and $\lim_{z \to \infty} f'(z) = 0$.

The second derivative is given by $f''(z) = h''(z) - g''(z) = -\phi(z)(2\Phi(z) - 1) + 4z\phi^2(z)$.

We will first consider the zero points of $f''(z)$, i.e. the solutions of

$$h''(z) = g''(z) \Longleftrightarrow 2\Phi(z) - 1 = 4z\phi(z).$$

Naturally, we have $f''(0) = 0$. Setting $k(z) = 2\Phi(z) - 1$ and $l(z) = 4z\phi(z)$, we note that k with $k'(z) = 2\phi(z)$ is monotonically increasing and approaches one. In contrast to this, l with $l'(z) = 4\phi(z)(1 - z^2)$ grows until $z = 1$ and then decreases monotonically approaching zero. We also note that $l'(z) \geq k'(z)$ for $z \leq 1/\sqrt{2}$ whereas $l'(z) \leq k'(z)$ for $z \geq 1/\sqrt{2}$. Therefore, there is exactly one $z_0 > 1/\sqrt{2}$, with $l(z_0) = k(z_0)$ or $f''(z_0) = 0$. Since $f''(z) \geq 0$ for $z \leq z_0$ and $f''(z) \leq 0$ for $z_0 \leq z$, $f'(z)$ has its only maximum in z_0.

We know that $f'(z_0) > 0$, because f' decreases monotonically towards zero for $z \geq z_0$. Since $f'(0) < 0$, there is exactly one $z_l \in (0, z_0)$ with $f'(z_l) = 0$. Therefore, f has only one extremum (minimum) in $(0, \infty)$ which is necessarily smaller than zero.

Equation 2: $|M_Q| = \sum_{i=1}^{N} y_i(2\Phi_{0,\sigma^2}(y_i) - 1) + 2\sigma^2\phi_{0,\sigma^2}(y_i) - |y_i|$

To show the validity of this equation, it suffices to prove $y(2\Phi_{0,\sigma^2}(y) - 1) + 2\sigma^2\phi_{0,\sigma^2}(y) \geq |y|$. Since all terms are symmetric, we will only consider $y \geq 0$.

Using the transformation $z = \sigma y$ again, the inequality is given as $2z\Phi(z) + 2\phi(z) \geq 2z$. Considering the function $f(z) := z\Phi(z) + \phi(z) - z = z(\Phi(z) - 1) + \phi(z)$, we see that $f(0) > 0$ and $f(z) \to 0$ for $z \to \infty$. The derivative is given by $f'(z) = \Phi(z) - 1 \leq 0$. The function f is therefore monotonically decreasing approaching zero in the limit case which shows $f(z) \geq 0$ for all $z \geq 0$.

Inequality 3: $h(x) = \dfrac{\sqrt{x^4+x^2}-x^2}{1-\frac{2}{\pi\sqrt{2x^2+1}}} \geq 0.5$ for $x \geq 1$ which can be shown as follows.

$$\sqrt{x^4 + x^2} - x^2 \geq \frac{1}{2}\left(1 - \frac{2}{\pi\sqrt{2x^2 + 1}}\right)$$

$$\Longleftrightarrow \sqrt{2x^2 + 1}(\sqrt{x^4 + x^2} - x^2) \geq \frac{1}{2}\left(\sqrt{2x^2 + 1} - \frac{2}{\pi}\right)$$

$$\Longleftrightarrow 2\sqrt{2x^6 + 3x^4 + x^2} - (2x^2 + 1)\sqrt{2x^2 + 1} \geq -\frac{2}{\pi}$$

$$\Longleftrightarrow \sqrt{(2x^2 + 1)^3 - (2x^2 + 1)} - \sqrt{(2x^2 + 1)^3} \geq -\frac{2}{\pi} \quad (35)$$

Setting $t = 2x^2 + 1$, we consider $g(t) = \sqrt{t^3 - t} - \sqrt{t^3}$, $t \geq 3$. We will show that g is a monotonically increasing function. Since $g(3) > -2/\pi$, this will prove $h(x) \geq 0.5$ for all $x \geq 1$. The derivative is given as $g'(t) = \frac{3t^2-1}{2\sqrt{t^3-t}} - \frac{3t^2}{2\sqrt{t^3}}$. Setting $g'(t) = 0$ leads to

$$\frac{3t^2 - 1}{\sqrt{t^3 - t}} = \frac{3t^2}{\sqrt{t^3}}$$

$$\Rightarrow (3t^2 - 1)\sqrt{t^3} = 3t^2\sqrt{t^3 - t}$$

$$\Longleftrightarrow (9t^4 - 6t^2 + 1)t^3 = 9t^4(t^3 - t)$$
$$\Longleftrightarrow 9t^4 - 6t^2 + 1 = 9t^4 - 9t^2 \text{ since } t \geq 3$$
$$\Rightarrow \ 3t^2 + 1 = 0 \tag{36}$$

Therefore, $g'(t) \neq 0$ for all $t \geq 3$. Since $g'(3) \geq 0$, the function g is monotonically increasing which proves $h(x) \geq 0.5$ for all $x \geq 1$.

Alternative distribution for F_5: The function $h(x) = (x^2(2\Phi(-\frac{1}{x})e^{\frac{1}{2x^2}} - 1) + \sqrt{\frac{2}{\pi}}x)/(1 - \frac{2}{\sqrt{\pi}x}\Phi(-\frac{1}{\sqrt{2}x})e^{\frac{1}{4x^2}})$ (31) will be considered in the following.

It is to be shown that $h(x) \geq 0.47715$ for all $x \geq 1/\sqrt{2}$. Setting $f(x) = x^2(2\Phi(-1/x)e^{\frac{1}{2x^2}} - 1) + \sqrt{2/\pi}x$ and $g(x) = 1 - 2/(\sqrt{\pi}x)\Phi(-1/(\sqrt{2}x))e^{\frac{1}{4x^2}}$, we know that $h(x) \geq f(x)$.

Setting $c = 0.47715$, let us consider $l(x) = f(x) - cg(x)$ for $x \in (1/\sqrt{2}, \infty)$.

$$l(x) = x^2(2\Phi(-\frac{1}{x})e^{\frac{1}{2x^2}} - 1) + \sqrt{\frac{2}{\pi}}x - c + c\frac{2}{\sqrt{\pi}x}\Phi(-\frac{1}{\sqrt{2}x})e^{\frac{1}{4x^2}} \geq 0$$
$$\Longleftrightarrow (2\Phi(-\frac{1}{x})e^{\frac{1}{2x^2}} - 1) + \frac{\sqrt{2}}{\sqrt{\pi}x} - \frac{c}{x^2} + c\frac{2}{\sqrt{\pi}x^3}\Phi(-\frac{1}{\sqrt{2}x})e^{\frac{1}{4x^2}} \geq 0 \tag{37}$$

Setting $s = 1/x$, we now consider $h(s) = 2\Phi(-s)e^{\frac{s^2}{2}} - 1 + \frac{s\sqrt{2}}{\sqrt{\pi}} - s^2c + c\frac{2s^3}{\sqrt{\pi}}\Phi(-\frac{s}{\sqrt{2}})$ $e^{\frac{s^2}{4}}$ on $I = (0, \sqrt{2})$. We decompose I into smaller intervals, i.e. $I = \bigcup_k I_k$ and show that $h(s) \geq 0$ for all $s \in I_k = [b_k, b_{k+1}]$. Since $h(s)$ is given by $h(s) = h(b_k) + \int_{b_k}^s h'(t)\,\mathrm{d}t$, we will derive a lower bound for $h'(t)$ which can be determined as $h'(t) = t(2\Phi(-t)e^{\frac{t^2}{2}} + \frac{c}{\sqrt{\pi}}\Phi(-\frac{t}{\sqrt{2}})e^{\frac{t^2}{4}}(6t + t^3) - 2c - \frac{c}{\pi}t^2) = tu(t)$.

Since $\Phi(-t)e^{\frac{t^2}{2}}$ is a decreasing function, we have $\Phi(-t)e^{\frac{t^2}{2}} \geq \Phi(-b_{k+1})e^{\frac{b_{k+1}^2}{2}} = w_{k+1}$ and $\Phi(-\frac{t}{\sqrt{2}})e^{\frac{t^2}{4}} \geq \Phi(-\frac{b_{k+1}}{\sqrt{2}})e^{\frac{b_{k+1}^2}{4}} = v_{k+1}$. Thus, we obtain $u(t) \geq 2w_{k+1} + \frac{c}{\sqrt{\pi}}v_{k+1}(6t + t^3) - 2c - \frac{c}{\pi}t^2$ as a lower bound for u on I_k. In order to find a minimal value for that lower bound $\tilde{u}$ in I_k, we consider the possible zero points of $\tilde{u}'(t) = \frac{c}{\sqrt{\pi}}v_{k+1}(6 + 3t^2) - \frac{2c}{\pi}t$ which are given by $t_{1,2} = \frac{1}{3\sqrt{\pi}v_{k+1}} \pm \sqrt{\frac{1}{9\pi v_{k+1}^2} - 2}$. The root is only a real number if $1/(9\pi v_{k+1}^2) \geq 2$ or $1/(18\pi) \geq v_{k+1}^2$. The smallest possible value for v_{k+1} is $\Phi(-1)\sqrt{e}$ which is greater than $1/\sqrt{18\pi}$. Therefore, $\tilde{u}'$ does not have any zero points on I_k and $\tilde{u}$ is a monotone function. Depending on the sign of $\tilde{u}'$, $\tilde{u}$ assumes its minimum either at b_k or at b_{k+1}. Setting $c_k = \min_{t \in I_k} \tilde{u}(t)$, a lower bound for $h(s)$ is thus given by

$$h(s) \geq h(b_k) + \int_{b_k}^s tc_k\,\mathrm{d}t = h(b_k) + \frac{c_k}{2}(s^2 - b_k^2). \tag{38}$$

Depending on the sign of c_k, we finally obtain the lower bound

$$h(s) \geq h_k = \begin{cases} h(b_k) & \text{if } c_k > 0 \\ h(b_k) + \frac{c_k}{2}(b_{k+1}^2 - b_k^2) & \text{if } c_k < 0 \end{cases}. \tag{39}$$

Table 1. The lower bounds h_k for the interval decomposition considered. Also shown are the values for the derivative of $\tilde{u}$ and the c_k-values

b_k	b_{k+1}	$\tilde{u}'(b_k + (b_{k+1} - b_k)/2)$	c_k	h_k
0.	0.05	0.777974	0.00702329	0.
0.05	0.1	0.743357	0.0084933	0.0000573665
0.1	0.15	0.7114	0.00939039	0.00022919
0.15	0.2	0.681914	0.00976756	0.000512478
0.2	0.25	0.654726	0.00967363	0.00090116
0.25	0.3	0.629675	0.00915364	0.0013865
0.3	0.35	0.606612	0.00824899	0.00195748
0.35	0.4	0.585401	0.00699779	0.00260111
0.4	0.45	0.565914	0.00543505	0.00330276
0.45	0.5	0.548037	0.00359291	0.00404641
0.5	0.55	0.531658	0.00150086	0.00481491
0.55	0.6	0.51668	-0.000814083	0.00556674
0.6	0.65	0.503009	-0.00332715	0.00624933
0.65	0.7	0.490558	-0.00601568	0.00688195
0.7	0.75	0.479249	-0.00885894	0.00744421
0.75	0.8	0.469008	-0.011838	0.00791558
0.8	0.85	0.459764	-0.0149355	0.00827546
0.85	0.9	0.451456	-0.0181358	0.00850329
0.9	0.95	0.444023	-0.0214243	0.00857861
0.95	1	0.437411	-0.0247881	0.00848115
1	1.05	0.431568	-0.0282151	0.00819085
1.05	1.1	0.426447	-0.0316945	0.00768791
1.1	1.05	0.422003	-0.0352165	0.00695286
1.15	1.2	0.418195	-0.038772	0.00596657
1.2	1.25	0.414985	-0.0423531	0.00471029
1.25	1.3	0.412336	-0.0459523	0.00316564
1.3	1.35	0.410214	-0.049563	0.00131466
1.35	1.37	0.413664	-0.038805	0.0017404
1.37	1.39	0.413097	-0.0402718	0.00087403
1.39	1.4	0.41428	-0.0368903	0.000607786
1.4	1.41	0.414053	-0.0376248	0.000142087
1.41	1.414	0.414844	-0.03543814214607244	$5.27635490428548 * 10^{-6}$
1.414	$\sqrt{2}$	0.415395	-0.0338832	$5.246888030107535 * 10^{-6}$

The intervals chosen and the values of h_k are given in Table 1. All values were obtained by using Mathematica. As one can see, $h_k \geq 0$ for all k which shows $h(s) \geq 0$ on I.

References

1. D. V. Arnold. *Noisy Optimization with Evolution Strategies*. Kluwer Academic Publishers, Dordrecht, 2002.

2. D. V. Arnold and H.-G. Beyer. On the Benefits of Populations for Noisy Optimization. *Evolutionary Computation*, 11(2):111–127, 2003.
3. T. Bäck, U. Hammel, and H.-P Schwefel. Evolutionary computation: comments on the history and current state. *IEEE Transactions on Evolutionary Computation*, 1(1):3–17, 1997.
4. H.-G. Beyer. *The Theory of Evolution Strategies*. Natural Computing Series. Springer, Heidelberg, 2001.
5. H.-G. Beyer and D. V. Arnold. The Steady State Behavior of $(\mu/\mu_I, \lambda)$-ES on Ellipsoidal Fitness Models Disturbed by Noise. In E. Cantú-Paz et al., editor, *GECCO-2003: Proceedings of the Genetic and Evolutionary Computation Conference*, pages 525–536, Berlin, Germany, 2003. Springer.
6. H.-G. Beyer and S. Meyer-Nieberg. On the quality gain of $(1, \lambda)$-ES under fitness noise. In X. Yao and H.-P. Schwefel et al., editors, *Parallel Problem Solving from Nature – PPSN VIII, Proc. Eighth Int'l Conf., Birmingham*, pages 1–10, Berlin, 2004. Springer.
7. H.-G. Beyer and S. Meyer-Nieberg. Predicting the Solution Quality in Noisy Optimization. Series CI 160/04, SFB 531, University of Dortmund, 2004.
8. H.-G. Beyer, M. Olhofer, and B. Sendhoff. On the Behavior of $(\mu/\mu_I, \lambda)$-ES Optimizing Functions Disturbed by Generalized Noise. In K. De Jong, R. Poli, and J. Rowe, editors, *Foundations of Genetic Algorithms, 7*, pages 307–328, San Francisco, CA, 2003. Morgan Kaufmann.
9. H.-G. Beyer and H.-P. Schwefel. Evolution Strategies: A Comprehensive Introduction. *Natural Computing*, 1(1):3–52, 2002.
10. J. Branke. *Evolutionary Optimization in Dynamic Environments*. Kluwer Academic Publishers, Dordrecht, 2001.
11. J.M. Fitzpatrick and J.J. Grefenstette. Genetic Algorithms in Noisy Environments. In P. Langley, editor, *Machine Learning: Special Issue on Genetic Algorithms*, volume 3, pages 101–120. Kluwer Academic Publishers, Dordrecht, 1988.
12. U. Hammel and T. Bäck. Evolution Strategies on Noisy Functions. How to Improve Convergence Properties. In Y. Davidor, R. Männer, and H.-P. Schwefel, editors, *Parallel Problem Solving from Nature, 3*, pages 159–168, Heidelberg, 1994. Springer-Verlag.
13. N. Hansen. *Verallgemeinerte individuelle Schrittweitenregelung in der Evolutionsstrategie*. Doctoral thesis, Technical University of Berlin, Berlin, 1998.
14. N. Hansen and A. Ostermeier. Completely Derandomized Self-Adaptation in Evolution Strategies. *Evolutionary Computation*, 9(2):159–195, 2001.
15. B. L. Miller. *Noise, Sampling, and Efficient Genetic Algorithms*. PhD thesis, University of Illinois at Urbana-Champaign, Urbana, IL 61801, 1997. IlliGAL Report No. 97001.
16. V. Nissen and J. Propach. On the Robustness of Population-Based Versus Point-Based Optimization in the Presence of Noise. *IEEE Transactions on Evolutionary Computation*, 2(3):107–119, 1998.
17. M. Olhofer, T. Arima, T. Sonoda, M. Fischer, and B. Sendhoff. Aerodynamic Shape Optimisation Using Evolution Strategies. In I. Parmee and P. Hajela, editors, *Optimisation in Industry*, pages 83–94. Springer Verlag, 2002.
18. I. Rechenberg. *Evolutionsstrategie '94*. Frommann-Holzboog Verlag, Stuttgart, 1994.
19. H.-P. Schwefel. *Numerical Optimization of Computer Models*. Wiley, Chichester, 1981.
20. S. Tsutsui and A. Ghosh. Genetic Algorithms with a Robust Solution Searching Scheme. *IEEE Transactions on Evolutionary Computation*, 1(3):201–208, 1997.

Rigorous Runtime Analysis of the (1+1) ES: 1/5-Rule and Ellipsoidal Fitness Landscapes

Jens Jägersküpper*

Dept. of Computer Science 2, Univ. Dortmund, 44221 Dortmund, Germany
`jj@Ls2.cs.uni-dortmund.de`

Abstract. We consider the (1+1) Evolution Strategy, a simple evolutionary algorithm for continuous optimization problems, using so-called Gaussian mutations and the 1/5-rule for the adaptation of the mutation strength. Here, the function $f: \mathbb{R}^n \to \mathbb{R}$ to be minimized is given by a quadratic form $f(x) = x^\top Q x$, where $Q \in \mathbb{R}^{n \times n}$ is a positive definite diagonal matrix and x denotes the current search point. This is a natural extension of the well-known SPHERE-function ($Q = I$). Thus, very simple unconstrained quadratic programs are investigated, and the question is addressed how Q effects the runtime. For this purpose, quadratic forms

$$f(x) \;=\; \xi \cdot \left(x_1{}^2 + \cdots + x_{n/2}{}^2 \right) + x_{n/2+1}{}^2 + \cdots + x_n{}^2$$

with $\xi = \omega(1)$, i.e. $1/\xi \to 0$ as $n \to \infty$, and $\xi = \mathrm{poly}(n)$ are investigated exemplarily. It is proved that the optimization very quickly stabilizes and that, subsequently, the runtime (defined as the number of f-evaluations) to halve the approximation error is $\Theta(\xi \cdot n)$. Though $\xi \cdot n = \mathrm{poly}(n)$, this result actually shows that the evolving search point indeed creeps along the "gentlest descent" of the ellipsoidal fitness landscape.

1 Introduction

Finding – or at least approximating – an optimum of a given function $f: S \to \mathbb{R}$ is one of the fundamental problems – in theory as well as in practice. Methods for solving continuous optimization problems, e. g. $S = \mathbb{R}^n$, are usually classified into first-order, second-order, and zeroth-order methods depending on whether they utilize the gradient (the first derivative) of the objective function, the gradient and the Hessian (the second derivative), or neither of the two.

> Note that here "continuous" relates to the search space rather than to f, and that, unlike in math programming, throughout this paper "n" denotes the number of dimensions of the search space and *not* the number of optimization steps; "d" generally denotes a distance in the n-dimensional search space.

A zeroth-order method is also called *derivative-free* or *direct search method*. Newton's method is the example of a second-order method; first-order methods

---

* Supported by the German Research Foundation (DFG) as part of the research center "Computational Intelligence" (SFB 531)

A.H. Wright et al. (Eds.): FOGA 2005, LNCS 3469, pp. 260–281, 2005.

can be (sub)classified into Quasi-Newton, steepest descent, and conjugate gradient methods. Classical zeroth-order methods try to approximate the gradient in order to plug this estimate into a first-order method. Finally, amongst the "modern" zeroth-order methods, evolutionary algorithms (EAs) come into play. EAs for continuous optimization, however, are usually subsumed under the term *evolution(ary) strategies (ESs)*. Obviously, in general we cannot expect zeroth-order methods to out-perform first-order methods or even second-order methods.

However, when information about the gradient is not available, for instance if f relates to a property of some workpiece and is given by simulations or even by real-world experiments, first-order (and also second-order) methods just cannot by applied. As the approximation of the gradient usually involves $\Omega(n)$ f-evaluations, a single optimization step of a classical zeroth order-method is computationally intensive, especially if f is given implicitly by simulations. In practical optimization, especially in mechanical engineering, this is often the case, and particularly in this field EAs become more and more widely used. However, the enthusiasm in practical EAs has led to an unclear variety of very sophisticated and problem-specific EAs. Unfortunately, from a theoretical point of view, the development of such EAs is solely driven by practical success and the aspect of a theoretical analysis is left aside. In other words, – concerning EAs – theory has not kept up with practice, and thus, we should not try to analyze the algorithmic runtime of the most sophisticated EA en vogue, but concentrate on very basic, or call them "simple", EAs in order to build a sound and solid basis for EA-theory.

For discrete search spaces, essentially $\{0,1\}^n$, such a theory has been developed successfully since the mid-1990s (cf. Wegener (2001) and Droste et al. (2002)). Recently, first results for non-artificial but well-known problems have been obtained (namely for the maximum matching problem by Giel and Wegener (2003), for sorting and the shortest-path problem by Scharnow et al. (2002), and for the minimum-spanning tree by Neumann and Wegener (2004)).

The situation for continuous evolutionary optimization is different. Here, the vast majority of the results are based on empiricism, i.e., experiments are performed and their outcomes are interpreted, which leads to a theory in the sense of physics rather than computer science. Also convergence properties of EAs have been studied to a considerable extent (e.g. Rudolph (1997), Greenwood and Zhu (2001), Bienvenue and Francois (2003)). A lot of results have been obtained by analyzing a simplifying model of the stochastic process induced by the EA, for instance by letting the number of dimensions approach infinity. Unfortunately, such results rely on experimental validation as a justification for the simplifications/inaccuracies introduced by the modeling. In particular Beyer has obtained numerous results that focus on local performance measures (*progress rate, fitness gain*; cf. Beyer (2001b)), i.e., the effect of a single mutation (or, more generally, of a single transition from one generation to the next) is investigated. Best-case assumptions concerning the mutation adaptation in this single step then provide estimates of the maximum gain a single step may yield. However, when one aims at analyzing the (1+1) ES as an algorithm, rather than a model of the stochastic process induced, a different, more algorithmic approach is needed. In 2003 a first theoretical analysis of the expected runtime, given by the number of function evaluations, of the $(1+1)$ ES using the 1/5-rule was presented (Jägersküpper, 2003). The function/fitness landscape considered therein is the well-know SPHERE-function, given by $\text{SPHERE}(x) := \sum_{i=1}^{n} x_i^2 = x^\top I x$, and the multi-step behavior that the $(1+1)$ ES bears when using the 1/5-rule for the adaptation of the mutation strength is rigorously analyzed. As mentioned in the abstract, the present paper will extend this result to a broader class of functions. One may guess that an ellipsoidal landscape is similar to the ridge-function scenario (especially to the parabolic ridge). Beyer (2001a) focuses on local measures for this fitness landscape. However, since ridge functions are unbounded, i.e. there is no optimum, and there is no need for adaptation, from an algorithmic point of view – when one is interested in adaptation mechanisms and how they work – ellipsoidal fitness landscapes are more challenging.

Finally note that, regarding the approximation error, for unconstrained optimization it is generally not clear how the runtime can be measured (solely) with respect to the absolute error of the approximation. In contrast to discrete and finite problems, the initial error is generally not bounded, and hence, the question how many steps it takes to get into the ε-ball around an optimum does not make sense without specifying the starting conditions. Hence, we must consider the runtime with respect to the relative improvement of the approximation. Given that the (relative) progress that a step yields becomes steady-state, considering the number of steps/f-evaluations to halve the approximation error is a natural choice. For the SPHERE-function, Jägersküpper (2003) gives a proof that the 1/5-rule makes the $(1+1)$ ES perform $\Theta(n)$ steps to halve the distance from the optimum and, in addition, that this is asymptotically the best possible w.r.t. isotropically distributed mutation vectors, i.e., for any adaptation of isotropic mutations, the expected number of f-evaluations is $\Omega(n)$ (moreover, for any constant $\varepsilon > 0$, $O(n^{1-\varepsilon})$ f-evaluations suffice only with an exponentially small probability).

The Algorithm

We will concentrate on the $(1+1)$ evolution strategy *($(1+1)$ ES)*, which dates back to the mid 1960s (cf. Rechenberg (1973) and Schwefel (1995)). This simple EA uses solely mutation due to a single-individual population, where here "individual" is just a synonym for "search point". Let $c \in \mathbb{R}^n$ denote the current individual. Given a starting point, i.e. an initialization of c, the $(1+1)$ ES performs the following evolution loop:

1. Choose a random mutation vector $m \in \mathbb{R}^n$, where the distribution of m may depend on the course of the optimization process.
2. Generate the mutant $c' \in \mathbb{R}^n$ by $c' := c + m$.
3. IF $f(c') \leq f(c)$ THEN c' becomes the current individual ($c := c'$) ELSE c' is discarded (c unchanged).
4. IF the stopping criterion is met THEN output c ELSE goto 1.

Since a worse mutant (with respect to the function to be minimized) is always discarded, the (1+1) ES is a randomized hill climber, and the selection rule is called *elitist selection*. Fortunately, for the type of results we are after we need not define a reasonable stopping criterion. How the mutation vectors are generated must be specified, though. Originally, the mutation vector $\boldsymbol{m} \in \mathbb{R}^n$ is generated by firstly generating a *Gaussian mutation* vector $\widetilde{\boldsymbol{m}} \in \mathbb{R}^n$ each component of which is independently standard normal distributed; subsequently, this vector is scaled by the multiplication with a scalar $s \in \mathbb{R}_{>0}$, i.e. $\boldsymbol{m} = s \cdot \widetilde{\boldsymbol{m}}$. Gaussian mutations are the most common type of mutations (for the search space $\mathbb{R}^n$) and, therefore, will be considered here. The crucial property of a Gaussian mutation is that $\widetilde{\boldsymbol{m}}$, and with it $\boldsymbol{m}$, is isotropically distributed, i. e., $\boldsymbol{m}/|\boldsymbol{m}|$ is uniformly distributed upon the unit hypersphere and the length of the mutation, namely the random variable $|\boldsymbol{m}|$, is independent of the direction $\boldsymbol{m}/|\boldsymbol{m}|$.

> The state of the art in mutation adaptation seems to be the *covariance matrix adaptation (CMA)* (Hansen and Ostermeier, 1996) where $s \cdot \mathbf{B} \cdot \widetilde{\boldsymbol{m}}$ makes up the mutation vector with a matrix $\mathbf{B} \in \mathbb{R}^{n \times n}$ which is also adapted. Unlike $\mathbf{B} = t \cdot \mathbf{I}$ for some scalar t, the mutation vector is not isotropically distributed.

The question that naturally arises is how the scaling factor s is to be chosen. Obviously, the smaller the approximation error, i. e., the closer $\boldsymbol{c}$ is to an optimum, the shorter $\boldsymbol{m}$ needs to be for a further improvement of the approximation to be possible. Unfortunately, the algorithm does not know about the current approximation error, but can utilize only the knowledge obtained by f-evaluations. Based on experiments and rough calculations for two function scenarios (namely SPHERE and a corridor function), Rechenberg proposed the *1/5-(success-)rule*. The idea behind this adaptation mechanism is that in a step of the (1+1) ES the mutant should be accepted with probability 1/5. Hereinafter, a mutation that results in $f(\boldsymbol{c}') \leq f(\boldsymbol{c})$ is called *successful*, and hence, when talking about a mutation, *success probability* denotes the probability that the mutant $\boldsymbol{c}' = \boldsymbol{c} + \boldsymbol{m}$ is at least as good as $\boldsymbol{c}$. Obviously, when elitist selection is used, the success probability of a step equals the probability that the mutation is accepted in this step. If every step was successful with probability 1/5, we would observe that on the average one fifth of the mutations are successful. Thus, the 1/5-rule works as follows: the optimization process is observed for n steps without changing s; if more than one fifth of the steps in this observation phase have been successful, s is doubled, otherwise, s is halved.

> Various implementations of the 1/5-rule can be found in the literature, yet in fact, one result of (Jägersküpper, 2003) is that the order of the runtime is indeed not affected as long as the observation lasts $\Theta(n)$ steps and the scaling factor s is multiplied by a constant greater than 1 resp. by a positive constant smaller than 1.

The Function Scenario

In this section we will have a closer look at the fitness landscape under consideration and preview isotropic mutations in this scenario. Note that, as minimization is considered, "function value" ("f-value") will be used rather than "fitness". Since the optimum function value is 0, the current approximation error is defined as $f(c)$, the f-value of the current individual. As mentioned in the abstract, we are going to consider the fitness landscapes induced by certain positive definite quadratic forms.

At first glance, one might guess that mixed terms like $3x_1x_2$ may crucially affect the fitness landscape induced by a positive definite quadratic form $x^\top Qx$. However, this is not the case. First note that w.l.o.g. we may assume Q to be symmetric (by balancing Q_{ij} with Q_{ji} for $i \neq j$). Furthermore, any symmetric matrix can by diagonalized since it has n eigen vectors. Namely, eigendecomposition yields $Q = RDR^{-1}$ for a diagonal matrix D and an orthogonal matrix R.

Note that an orthogonal matrix R corresponds to a orthonormal transformation, which is merely a (possibly improper) rotation; then R^{-1} is the corresponding "anti-rotation".

Thus, the quadratic form equals $x^\top RDR^{-1}x$, and since $x^\top R = (R^\top x)^\top$, we have $(R^\top x)^\top D(R^{-1}x)$. As $R^\top = R^{-1}$ for an orthogonal matrix, the quadratic form equals $(R^{-1}x)^\top D(R^{-1}x)$. Thus, investigating $x^\top Qx$ using the standard basis for $\mathbb{R}^n$ (given by I) is the same as investigating $x^\top Dx$ using the orthonormal basis given by R. Finally note that the inner product is independent of the orthonormal basis that we use (because $(Rx)^\top(Rx) = x^\top R^\top Rx = x^\top R^{-1}Rx = x^\top Ix = x^\top x$). Consequently, we may assume that Q is a diagonal matrix each entry of which is positive. In other words, when talking about positive definite quadratic forms we are in fact talking about functions of the form $f_n(x) = \sum_{i=1}^{n} \xi_i \cdot x_i^2$ with $\xi_i > 0$, and we may even assume $\xi_n \geq \cdots \geq \xi_1$.

As mentioned in the abstract, we exemplarily restrict ourselves to the following class of (sequences of) quadratic forms, where $n \in 2\mathbb{N}$ and $1/\xi \to 0$ as $n \to \infty$:

$$f_n(x) \quad := \quad \xi \cdot \left(x_1^2 + \cdots + x_{n/2}^2\right) + x_{n/2+1}^2 + \cdots + x_n^2$$

Hence, $f_n(x) = \xi \cdot \text{SPHERE}_{n/2}(y) + \text{SPHERE}_{n/2}(z)$ where $y := (x_1, \ldots, x_{n/2})$ and $z := (x_{n/2+1}, \ldots, x_n)$. Thus, the aim is to minimize the sum of two separate sphere functions, in $S_1 = \mathbb{R}^{n/2}$ resp. $S_2 = \mathbb{R}^{n/2}$, having weight ξ resp. 1, i.e., $f(x) = \xi \cdot |y|^2 + |z|^2$, where $|\cdot|$ denotes the length of a vector in Euclidean space (Euclidean norm). Recall that the mutation vector m equals $s \cdot \widetilde{m}$. As each component of $\widetilde{m}$ is independently standard normal distributed, $m_1 := (m_1, \ldots, m_{n/2})$ and $m_2 := (m_{n/2+1}, \ldots, m_n)$ are two independent $(n/2)$-dimensional Gaussian mutations which are respectively scaled by the same factor s. Obviously, m_1 only affects y, whereas m_2 only affects z, and thus, the f-value of the mutant equals $\xi \cdot |y + m_1|^2 + |z + m_2|^2$.

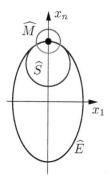

Let $d_1 := |\boldsymbol{y}|$ and $d_2 := |\boldsymbol{z}|$ denote the distance from the origin/optimum in S_1 resp. S_2. Since Gaussian mutations as well as SPHERE are invariant with respect to rotations of the coordinate system, we may rotate S_1 and S_2 such that we are located at $(d_1, 0, \ldots, 0) \in S_1$ resp. $(0, \ldots, 0, d_2) \in S_2$. In other words, we may assume w. l. o. g. that the current search point is located at $(d_1, 0, \ldots, 0, d_2) \in \mathbb{R}^n$, i. e., that it lies in the x_1-x_n-plane. In fact, we have just described a projection $\widehat{\ }: \mathbb{R}^n \to \mathbb{R}^2$. Note that due to the properties of f and Gaussian mutations this projection only conceals irrelevant information, i. e., all information relevant to the analysis is preserved. Thus, we can concentrate on the 2D-projection as depicted in the figure. For some arguments, however, it is crucial to keep in mind that this projection is based on the fact that the current search point, and also its mutant, can be assumed to lie in the x_1-x_n-plane w. l. o. g. (obviously, for the mutant to lie in this plane, S_1 and S_2 must almost surely (a. s.) be re-rotated).

In the next section some of the results presented in (Jägersküpper, 2003), which will be used here, will be shortly restated. In Section 3 the crucial properties of a single mutation in the considered fitness landscape are discussed, and in the subsequent section we will have a closer look at the adaptation, i. e., the multi-step behavior of the (1+1) ES will be analyzed for the considered function class/fitness landscape. We end with some concluding remarks in Section 5.

2   Preliminaries

In this section some notions and notations are introduced. Furthermore, the results obtained for the SPHERE-scenario in (Jägersküpper, 2003) that we will use are recapitulated; for more details cf. (Jägersküpper, 2002).

Definition 1. *A probability $p(n)$ is **exponentially small** in n if for a constant $\varepsilon > 0$, $p(n) = \exp(-\Omega(n^\varepsilon))$. An event $A(n)$ happens **with overwhelming probability (w. o. p.)** with respect to n if $\mathsf{P}\{\neg A(n)\}$ is exponentially small in n.*
*A statement $Z(n)$ holds **for n large enough** if $(\exists n_0 \in \mathbb{N})(\forall n \geq n_0)\, Z(n)$.*

Recall the following asymptotics: $g(n) = O(h(n))$ iff there exists a positive constant κ such that $g(n) \leq \kappa \cdot h(n)$ for n large enough; $g(n) = \Omega(h(n))$ iff $h(n) = O(g(n))$; $g(n) = \Theta(h(n))$ iff $g(n)$ is both $O(h(n))$ and $\Omega(h(n))$; for $h(n), g(n) > 0$, we have $g(n) = o(h(n))$ iff $g(n)/h(n) \to 0$ as $n \to \infty$ and $g(n) = \omega(h(n))$ iff $h(n) = o(g(n))$. As we are interested in how the runtime depends on n, the dimensionality of the search space, all asymptotics are w. r. t. to this parameter (unless stated differently).

Let $\boldsymbol{c} \in \mathbb{R}^n - \{\boldsymbol{0}\}$ denote a search point and $\boldsymbol{m}$ a scaled Gauss mutation. Note that $\text{SPHERE}(\boldsymbol{c}) = |\boldsymbol{c}|^2$ (recall that $|\boldsymbol{c}|$ is the L^2-norm (Euclidian length) of $\boldsymbol{c}$). The analysis of the (1+1) ES for SPHERE has shown that for n large enough

$$\mathsf{P}\{|\boldsymbol{c} + \boldsymbol{m}| \leq |\boldsymbol{c}| \mid |\boldsymbol{m}| = \ell\} \geq \varepsilon \text{ for a constant } \varepsilon \in (0, \tfrac{1}{2}) \iff \ell = O(|\boldsymbol{c}|/\sqrt{n}),$$

i. e., the mutant of c is closer to a predefined point, here the origin, with probability $\Omega(1)$ iff the length of the isotropic mutation vector is at most an $O(1/\sqrt{n})$-fraction of the distance between c and this point. On the other hand,

$$\mathsf{P}\{|c + m| \le |c| \mid |m| = \ell\} \le \varepsilon \text{ for a constant } \varepsilon \in (0, \tfrac{1}{2}) \iff \ell = \Omega(|c|/\sqrt{n}),$$

in other words, the mutant obtained by an isotropic mutation of c is closer to a predefined point, here again the origin, with a constant probability strictly smaller than $1/2$ iff the length of the mutation vector is at least an $\Omega(1/\sqrt{n})$-fraction of the distance between c and this point. (The actual constant ε respectively correlates with the constant in the O-notation resp. in the Ω-notation.)

The expected length of m equals $s \cdot \mathsf{E}[|\widetilde{m}|] = s \cdot \sqrt{n} \cdot (1 - \Theta(1/n))$ since $|\widetilde{m}|$ is χ-distributed (with n degrees of freedom). Moreover, with $\bar{\ell} := \mathsf{E}[|m|]$ we have $\mathsf{P}\{||m| - \bar{\ell}| \ge \delta \cdot \bar{\ell}\} \le \delta^{-2}/(2n - 1)$ for $\delta > 0$, in other words, there is only small deviation in the length of a mutation; e. g., with probability $1 - O(1/n)$ the mutation vector's actual length differs from its expected length by no more than $\pm 1\%$.

Concerning the mutation adaptation by the 1/5-rule for SPHERE, we know that there exists a constant $p_h \in (0, 1/5)$ such that if the success probability of the mutation in the first step of an observation phase is smaller than p_h, then w. o. p. less than 1/5 of the steps in this phase are successful so that the scaling factor is halved. Analogously, a constant $p_d \in (1/5, 1/2)$ exists such that if the first step of a phase is successful with probability at least p_d, then w. o. p. more than 1/5 of the steps in this phase are successful so that s is doubled. This can be used to show that the 1/5-rule in fact ensures that each step is successful with a probability in $[a, b] \subset (0, 1/2)$ for two constants a, b.

Let $\Delta := |c| - |c'|$ denote the spatial gain towards the origin, the optimum of SPHERE, in a step. For SPHERE, a mutation is accepted (by elitist selection) iff $\Delta \ge 0$. Consequently, negative gains are zeroed out so that the expected spatial gain of a step is $\mathsf{E}[\Delta \cdot \mathbb{1}_{\{\Delta \ge 0\}}]$. We know that $\mathsf{E}[\Delta \cdot \mathbb{1}_{\{\Delta \ge 0\}}]$ is $O(\bar{\ell}/\sqrt{n})$ and – however the scaling factor is chosen/adapted – also $O(|c|/n)$. Furthermore, $\mathsf{E}[\Delta \cdot \mathbb{1}_{\{\Delta \ge 0\}} \mid s = \Theta(|c|/n)]$ is $\Omega(\bar{\ell}/\sqrt{n})$ and $\Omega(|c|/n)$, i. e., the distance from the optimum is expected to decrease by an $\Theta(1/n)$-fraction if s is chosen/adapted appropriately. Furthermore, in this situation for any constant $\kappa > 0$ the distance decreases (at least) by an κ/n-fraction with probability $\Omega(1)$.

Recall that $\bar{\ell} = s \cdot \sqrt{n} \cdot (1 - \Theta(1/n))$. Thus, when scaled Gaussian mutations are used, "$s = \Theta(|c|/n)$" is equivalent to "$\bar{\ell} = \Theta(|c|/\sqrt{n})$" which is again equivalent to "$\exists$ constant $\varepsilon > 0$ such that for n large enough $\mathsf{P}\{\Delta \ge 0\} \in [\varepsilon, 1/2 - \varepsilon]$" since $\mathsf{P}\{|m| = \Theta(\bar{\ell})\} = 1 - O(1/n)$. The equivalance of these three events/conditions will be of great help in the argumentation.

3   Gain in a Single Step

In this section we now take a closer look at the properties of a Gaussian mutation in the ellipsoidal fitness landscape under consideration. Since $\xi = \omega(1)$, $\xi > 1$ for n large enough, and therefore, we assume $\xi > 1$ in the following. Furthermore,

"f" will also be used as an abbreviation of the f-value of the current individual and "f'" stands for the mutant's f-value.

Recall that $f = \xi d_1^2 + d_2^2$ (for the current search point) and $f' = \xi d_1'^2 + d_2'^2$ (for its mutant), where $d_1' := |y + m_1|$ and $d_2' := |z + m_2|$. The crucial point to the analysis is the answer to the question how d_1, d_2 and the scaling factor s – and with it $|m|$ – relate when the success probability of a step, i.e. the probability that the mutant is accepted, is about $1/5$. In other words, how does the length of the mutation vector depend on d_1 and d_2, and how do d_1 and d_2 relate. Since $\nabla \widehat{f}(d_1, d_2) = (\xi 2 d_1, 2 d_2)^\top$, for a search point satisfying $d_1/d_2 = 1/\xi$ an infinitesimal change of d_1 has the same effect on f as an infinitesimal change of d_2. Though the length of a mutation is not infinitesimal, this may be taken as an indicator that the ratio d_1/d_2 will stabilize when using isotropic mutations, and indeed, it turns out that the process stabilizes w.r.t. $d_1/d_2 = \Theta(1/\xi)$. In this section, we will see that near the gentlest descent in our ellipsoidal fitness landscape, namely for $d_1/d_2 = O(1/\xi)$, a mutation succeeds with a constant probability greater than 0 but smaller than $1/2$ iff the scaling factor s is $\Theta((\sqrt{f}/n)/\xi)$. Furthermore, asymptotically tight bounds on the expected f-gain of a single step in such a situation will be obtained. Therefore, we will show that a mutation of a search point c for which $d_1/d_2 = O(1/\xi)$ with a mutation using a scaling factor $s = \Theta((\sqrt{f}/n)/\xi)$ in the ellipsoidal fitness landscape is "similar" to the mutation of a search point x in the SPHERE scenario with $\text{SPHERE}(x) = \Theta(f/\xi^2)$ (using the same scaling factor).

We start our analysis at a point c with $\widehat{c} = (0, \phi)$, i.e. $d_1 = 0$ and $d_2 = \phi$, so that $f = \phi^2$. Consequently, $\widehat{c}$ is located at a point with gentlest descent w.r.t. all points with f-value ϕ^2, and hence, the curvature of the 2D-curve given by the projection $\widehat{E}$ of the n-ellipsoid $E := \{x \mid f(x) = f(c)\} \subset \mathbb{R}^n$, is maximum at $\widehat{c}$. By a simple application of differential geometry, we get that the curvature of this 2D-curve at $\widehat{c}$ equals ξ/ϕ. Consequently, the radius of the osculating circle ($\widehat{S}$ in the figure) equals ϕ/ξ. As this circle $\widehat{S}$ actually lies in the x_1-x_n-plane, it is the equator of an n-sphere S with radius ϕ/ξ (the center of which lies on the x_n-axis, just like the current search point c). Note that this sphere lies completely inside E such that $S \cap E = \{c\}$. Thus, the probability that a mutation hits inside S is a lower bound on the probability that $f' \leq f$, i.e.,

$$
\begin{aligned}
\mathsf{P}\{f' \leq f\} &= \mathsf{P}\{c + m \text{ lies inside } E\} \\
&\geq \mathsf{P}\{c + m \text{ lies inside } S\} \\
&= \mathsf{P}\{|x + m| \leq |x| \text{ for some } x \text{ with } |x| = \text{radius of } \widehat{S} = \phi/\xi\} \\
&= \mathsf{P}\{\text{SPHERE}(x + m) \leq \text{SPHERE}(x) \mid \text{SPHERE}(x) = (\phi/\xi)^2\}.
\end{aligned}
$$

In fact, our argumentation yields that the above (in)equalities hold for any fixed length ℓ of the mutation vector m, i.e., if the probabilities are conditioned on the event $\{|m| = \ell\}$, respectively. Since ℓ is arbitrary here and the radius of S is independent of ℓ, they remain valid when this condition is dropped.

For an upper bound on the probability that a mutation hits inside E, consider a mutation (vector) having length $\ell < 2\phi$ (since for $\ell \geq 2\phi$, E lies inside M).

Let $M = \{x \mid |c - x| = \ell\} \subset \mathbb{R}^n$ denote the mutation sphere consisting of all potential mutants. Then $\widehat{M}$ is a circle (cf. the figure above) with radius ℓ centered at $\widehat{c}$. (Note that, though $c' = c + m$, given $|m| = \ell$, is uniformly distributed upon M, $\widehat{c'}$ is *not* uniformly distributed upon $\widehat{M}$). Now consider the curvature at a point in $\widehat{E} \cap \widehat{M} = \{z_1, z_2\}$ (there are exactly two points of intersection since $0 < \ell < 2\phi$). Simple differential geometry shows that the curvature at z_i is $\kappa_\ell = \Theta(\xi/\phi)$ if $\ell = O(\phi/\xi)$. As the curvature at any point of $\widehat{E}$ that lies inside $\widehat{M}$ is greater than κ_ℓ (since $\xi > 1$), $\widehat{c}$ as well as z_i lie inside the osculating circle at z_{3-i} which has radius $r_\ell := 1/\kappa_\ell = \Theta(\phi/\xi)$ if $\ell = O(\phi/\xi)$. Thus, there is also a circle with radius r_ℓ passing through $\widehat{c}$ such that z_1 and z_2 lie inside this circle. Therefore, the circle passing through z_1, z_2, and $\widehat{c}$ has a radius smaller than r_ℓ, and again, this circle actually lies in the x_1-x_n-plane of the search space and is the image of the n-sphere having this circle as an equator. Hence,

$$P\{f' \le f \mid |m| = \ell\}$$
$$\le \quad P\{\mathrm{SPHERE}(x + m) \le \mathrm{SPHERE}(x) \mid \mathrm{SPHERE}(x) = (\alpha\,\phi/\xi)^2, |m| = \ell\}$$

where $\alpha = \Theta(1)$ if $\ell = O(\phi/\xi)$. (Besides, $\alpha \searrow 1$, i.e. $r_\ell \searrow \phi/\xi$, as $\ell \searrow 0$.)

Recall that we assumed $\widehat{c} = (0, \phi) \in \mathbb{R}^2$, i.e. $d_1 = 0$ and $d_2 = \phi$, in the above argumentation. The estimates we have made for the bounds on the probability of a mutation hitting inside the n-ellipsoid E, however, remain valid as long as $d_1/d_2 = O(1/\xi)$: Since ξ/ϕ is the maximum curvature of $\widehat{E}$, there is always a circle $\widehat{S}$ with radius ϕ/ξ lying inside $\widehat{E}$ such that $\widehat{S} \cap \widehat{E} = \{\widehat{c}\}$, and since $\widehat{S}$ is in fact an equator of an n-sphere S, S lies completely inside E such that $S \cap E = \{c\}$. For the upper bound, we must merely consider the z_i at which the curvature is smaller, and indeed, it turns out that as long as $d_1/d_2 = O(1/\xi)$ and $\ell = O(\phi/\xi)$, κ_ℓ remains $\Theta(\xi/\phi)$.

Hence, when $f(c) = \phi^2$ such that c satisfies $d_1/d_2 = O(1/\xi)$, we are in a situation resembling (w.r.t. the success probability of a mutation) the minimization of SPHERE at a point having distance $\Theta(\phi/\xi)$ from the optimum/origin. Concerning the 1/5-rule, we then know (cf. Section 2) that

$$\exists \text{ constant } \varepsilon > 0 \text{ such that for } n \text{ large enough } P\{f' \le f\} \in [\varepsilon, 1/2 - \varepsilon]$$
$$\iff \quad s = \Theta((\phi/\xi)/n) \iff \bar{\ell} = \Theta((\phi/\xi)/\sqrt{n})$$

where ε correlates with the two multiplicative constants within the Θ-notation.

Thus, we are now going to investigate the gain of a step when $f = \phi^2$ and $s = \Theta((\phi/\xi)/n)$. As we have seen above, there exists an n-sphere S with radius $r = \phi/\xi$ lying completely in E such that $S \cap E = \{c\}$. Again owing to the results for SPHERE, we know that a mutation having length $\ell = \Theta(r/\sqrt{n})$ hits with probability $\Omega(1)$ a hyperspherical cap $C \subset M$ containing all points of M that are at least $\Omega(r/n)$ closer to the center of S than c. Consequently, with probability $\Omega(1)$ the mutant lies inside E such that its distance from E is $\Theta(r/n)$, i.e. $\Theta((\phi/\xi)/n)$. If we pessimistically assume that this spatial gain were realized along the gentlest descent of f, i.e. $d_1 = 0$ and $d'_1 = 0$ so that $d'_2 = d_2 - \Theta((\phi/\xi)/n)$, we obtain that with probability $\Omega(1)$

$$
\begin{aligned}
f' &\le (\phi - \Theta((\phi/\xi)/n))^2 \\
&= \phi^2 - 2\alpha\phi^2/(\xi n) + \alpha^2\phi^2/(\xi n)^2 \text{ for some } \alpha = \Theta(1) \\
&= \phi^2 - \underbrace{\alpha(2 - \alpha/(\xi n))}\,\phi^2/(\xi n) \\
&= \phi^2 - \quad \Theta(1) \quad \phi^2/(\xi n) \\
&= f - \Theta(f/(\xi n)).
\end{aligned}
$$

Let $c'' := \arg\min\{f(c), f(c')\}$ denote the search point that gets selected by elitist selection. Since mutants with a larger f-value are rejected, i.e. $f'' \le f$, this implies for the expected f-gain of a step

$$
\mathsf{E}\Big[f'' \mid s = \Theta((\sqrt{f}/n)/\xi)\Big] = f - \Omega(f/(\xi n)).
$$

Due to the pessimistic assumptions, this lower bound on the f-gain just derived is valid only for $s = \Theta((\sqrt{f}/n)/\xi))$, yet it holds independently of the ratio d_1/d_2. A spatial gain of $\Theta(f/(\xi n))$ could result in a much larger f-gain, though. If $d_1/d_2 = O(1/\xi)$, however, the f-gain is also $O(f/(\xi n))$ as we will see. Therefore, let $d_1 = \alpha \cdot \phi/\xi$ with $\alpha = O(1)$ and still $f = \xi \cdot d_1^2 + d_2^2 = \phi^2$. Owing to the argumentation for the upper bound on the success probability of a step, we know that there is an n-sphere S with radius $r = \Theta(\phi/\xi)$ such that $c \in S$ and $I := M \cap E \in S$, where I is the boundary of the hyperspherical cap $C \subset M$ lying inside E. Owing to the results for SPHERE, we know that $\mathsf{E}\big[\mathrm{dist}(c', I) \cdot \mathbb{1}_{\{c' \in C\}}\big] = O(r/n)$ for any choice of the scaling factor, i.e., even if the length of the mutation vector were magically chosen such that the expected distance of the selected search point c'' from the center of S is minimized. In other words, we know that if a mutation hits inside E, its expected distance from E is $O(r/n) = O((\phi/\xi)/n)$ anyway. Thus, if we optimistically assume that the spatial gain were realized completely in S_1, i.e. completely on the ξ-weighted SPHERE$_{n/2}$, (so that $d_2' = d_2$, implying $d_2'' = d_2$), we obtain

$$
\begin{aligned}
\mathsf{E}\big[\xi\, d_1''^2 + d_2''^2 \mid d_1/d_2 = O(1/\xi)\big] &\ge \xi\left(d_1 - O((\phi/\xi)/n)\right)^2 + d_2^2 \\
&= \xi\left(\alpha\phi/\xi - O((\phi/\xi)/n)\right)^2 + d_2^2 \\
&\ge \xi\left((\alpha\phi/\xi)^2 - 2\alpha(\phi/\xi) \cdot O((\phi/\xi)/n)\right) + d_2^2 \\
&= \xi\, d_1^2 - O(\phi^2/(\xi n)) + d_2^2
\end{aligned}
$$

and hence,

$$
\mathsf{E}[f'' \mid d_1/d_2 = O(1/\xi)] = \phi^2 - O(\phi^2/(\xi n)) = f - O(f/(\xi n)).
$$

This upper bound on the expected f-gain of a step holds only for $d_1/d_2 = O(1/\xi)$, yet independently of (the distribution of) $|m|$, which is converse to the lower bound. However, altogether we have proved the following:

Lemma 1. *Consider a step of the (1+1) ES. If $d_1/d_2 = O(1/\xi)$ in this step, then there exists a constant $\varepsilon > 0$ such that for n large enough $\mathsf{P}\{f' \le f\} \in [\varepsilon, 1/2 - \varepsilon]$ iff $s = \Theta((\sqrt{f}/n)/\xi)$.*
If $d_1/d_2 = O(1/\xi)$ and $s = \Theta((\sqrt{f}/n)/\xi)$ in this step, then $\mathsf{E}[f - f''] = \Theta((f/n)/\xi)$, and furthermore, $f - f'' = \Omega((f/n)/\xi)$ with probability $\Omega(1)$.

4   Multi-step Behavior

The results just obtained imply that if $d_1/d_2 = O(1/\xi)$ during a phase of n steps (an observation phase of the 1/5-rule) and $s = \Theta((\sqrt{f}/n)/\xi)$, i.e. $\mathsf{P}\{f' \le f\} \in [\varepsilon, 1/2 - \varepsilon]$ for a constant $\varepsilon > 0$, at the beginning of this phase, then we expect $\Theta(n)$ steps each of which reduces the f-value by $\Theta(f/(\xi n))$. By Chernoff bounds, there are $\Omega(n)$ such steps w.o.p., and thus, the f-value, and with it the approximation error, is reduced w.o.p. by an $\Theta(1/\xi)$-fraction in this phase. Consequently, after $\Theta(\xi)$ consecutive phases, w.o.p. the approximation error is halved – if during all these phases $d_1/d_2 = O(1/\xi)$. Since, up to now, the argumentation completely bases on the results for SPHERE, even the argumentation on the 1/5-rule can be adopted, which directly yields the following result (cf. Theorem 2 in (Jägersküpper, 2003) or Theorem 3 in (Jägersküpper, 2002)):

Theorem 1. *If $d_1/d_2 = O(1/\xi)$ in the complete optimization process and the initialization satisfies $s = \Theta((\sqrt{f(c)}/n)/\xi)$, then w.o.p. the number of steps/ f-evaluations to reduce the initial f-value/approximation error to a 2^{-t}-fraction, $t = poly(n)$, is $\Theta(t \cdot \xi \cdot n)$.*

Obviously, the assumption "$d_1/d_2 = O(1/\xi)$ in the complete optimization process" lacks any justification and is, therefore, objectionable. It must be replaced by a much weaker assumption on the starting conditions only. Thus, the crucial point in the analysis is the question why should the ratio d_1/d_2 remain $O(1/\xi)$ (once this is the case). This crucial question will be tackled in the remainder of this paper.

Let $\Delta_1 := d_1 - d_1'$ and $\Delta_2 := d_2 - d_2'$ denote the spatial gain of the mutant towards the origin in S_1 resp. S_2. Then d_1'/d_2' for the mutant is smaller than d_1/d_2 for its parent iff $\Delta_1/d_1 > \Delta_2/d_2$. Unfortunately, Δ_1 and Δ_2 correlate because m_1 and m_2 use the same scaling factor s, and furthermore, we must take selection into account since only certain combinations of Δ_1 and Δ_2 will be accepted. To see which combinations become accepted note that

$$f' = \xi(d_1 - \Delta_1)^2 + (d_2 - \Delta_2)^2 = \xi d_1^2 - \xi 2d_1\Delta_1 + \xi\Delta_1^2 + d_2^2 - 2d_2\Delta_2 + \Delta_2^2,$$

and hence,

$$f' \le f \iff f' - f \le 0 \iff -\xi 2d_1\Delta_1 + \xi\Delta_1^2 - 2d_2\Delta_2 + \Delta_2^2 \le 0.$$

Let α be defined by $\alpha/\xi = d_1/d_2$. Then the latter inequality is equivalent to

$$-2\alpha d_2\Delta_1 + \xi\Delta_1^2 - 2d_2\Delta_2 + \Delta_2^2 \le 0 \iff -\alpha\Delta_1 + \frac{\xi\Delta_1^2}{2d_2} \le \Delta_2 - \frac{\Delta_2^2}{2d_2}$$

$$\iff -\alpha\Delta_1\left(1 - \frac{\Delta_1}{2d_1}\right) \le \Delta_2\left(1 - \frac{\Delta_2}{2d_2}\right) \quad \text{(using } d_2 = \xi \cdot d_1/\alpha\text{)}$$

Thus, when using elitist selection, the mutant is accepted iff the last inequality holds. Note that whenever a mutation satisfying $-\alpha\Delta_1 > \Delta_2$ is accepted, then

$$1 - \frac{\Delta_1}{2d_1} < 1 - \frac{\Delta_2}{2d_2} \iff \frac{\Delta_1}{d_1} > \frac{\Delta_2}{d_2} \iff \Delta_1 > \frac{d_1}{d_2}\Delta_2 \iff \Delta_1 > \frac{\alpha}{\xi}\Delta_2,$$

implying that $\Delta_1 > 0$ and $\Delta_2 < 0$, and consequently, such a step surely results in $d_1''/d_2'' < d_1/d_2$, i.e. $\alpha'' < \alpha$. Hence, in the following we may concentrate on the accepted mutations for which $-\alpha\Delta_1 \leq \Delta_2$.

So, let us assume for a moment that the mutant replaces/becomes the current individual iff $-\alpha\Delta_1 \leq \Delta_2$. As Δ_{3-i}, $i \in \{1,2\}$, is random, $\mathsf{E}\big[\Delta_i \cdot \mathbb{1}_{\{-\alpha\Delta_1 \leq \Delta_2\}}\big]$ is a random variable taking the value $\mathsf{E}\big[\Delta_i \cdot \mathbb{1}_{\{-\alpha\Delta_1 \leq x\}}\big]$ whenever Δ_2 happens to take the value x. We are interested in $\mathsf{E}\big[\mathsf{E}\big[\Delta_i \cdot \mathbb{1}_{\{-\alpha\Delta_1 \leq \Delta_2\}}\big]\big] = \mathsf{E}[d_i - d_i'']$, the expected reduction of the distance from the optimum in S_i in a step, and $\mathsf{E}[d_1'']/\mathsf{E}[d_2''] \leq d_1/d_2$, i.e. we "expect" $\alpha'' \leq \alpha$, iff

$$\mathsf{E}\big[\mathsf{E}\big[\Delta_1 \cdot \mathbb{1}_{\{-\alpha\Delta_1 \leq \Delta_2\}}\big]\big]/d_1 \;\geq\; \mathsf{E}\big[\mathsf{E}\big[\Delta_2 \cdot \mathbb{1}_{\{-\alpha\Delta_1 \leq \Delta_2\}}\big]\big]/d_2$$
$$\Longleftrightarrow\; \xi \cdot \mathsf{E}\big[\mathsf{E}\big[\Delta_1 \cdot \mathbb{1}_{\{-\alpha\Delta_1 \leq \Delta_2\}}\big]\big] \;\geq\; \alpha \cdot \mathsf{E}\big[\mathsf{E}\big[\Delta_2 \cdot \mathbb{1}_{\{-\alpha\Delta_1 \leq \Delta_2\}}\big]\big].$$

In order to prove that this inequality holds for $\alpha = O(1)$, we aim at a lower bound on $\mathsf{E}\big[\mathsf{E}\big[\Delta_1 \cdot \mathbb{1}_{\{-\alpha\Delta_1 \leq \Delta_2\}}\big]\big]$ and an upper bound on $\mathsf{E}\big[\mathsf{E}\big[\Delta_2 \cdot \mathbb{1}_{\{-\alpha\Delta_1 \leq \Delta_2\}}\big]\big]$ in the following. Note that

$$
\begin{aligned}
\mathsf{E}\big[\mathsf{E}\big[\Delta_i \cdot \mathbb{1}_{\{-\alpha\Delta_1 \leq \Delta_2\}}\big]\big] \;=\;\; & \mathsf{E}\big[\mathsf{E}\big[\Delta_i \cdot \mathbb{1}_{\{-\alpha\Delta_1 \leq \Delta_2\}} \cdot \mathbb{1}_{\{\Delta_i<0\}}\big] \cdot \mathbb{1}_{\{\Delta_{3-i}<0\}}\big] + \\
& \mathsf{E}\big[\mathsf{E}\big[\Delta_i \cdot \mathbb{1}_{\{-\alpha\Delta_1 \leq \Delta_2\}} \cdot \mathbb{1}_{\{\Delta_i<0\}}\big] \cdot \mathbb{1}_{\{\Delta_{3-i}\geq 0\}}\big] + \\
& \mathsf{E}\big[\mathsf{E}\big[\Delta_i \cdot \mathbb{1}_{\{-\alpha\Delta_1 \leq \Delta_2\}} \cdot \mathbb{1}_{\{\Delta_i\geq 0\}}\big] \cdot \mathbb{1}_{\{\Delta_{3-i}<0\}}\big] + \\
& \mathsf{E}\big[\mathsf{E}\big[\Delta_i \cdot \mathbb{1}_{\{-\alpha\Delta_1 \leq \Delta_2\}} \cdot \mathbb{1}_{\{\Delta_i\geq 0\}}\big] \cdot \mathbb{1}_{\{\Delta_{3-i}\geq 0\}}\big]
\end{aligned}
$$

and that $\mathsf{E}\big[\mathsf{E}\big[\Delta_i \cdot \mathbb{1}_{\{-\alpha\Delta_1 \leq \Delta_2\}} \cdot \mathbb{1}_{\{\Delta_i<0\}}\big] \cdot \mathbb{1}_{\{\Delta_{3-i}<0\}}\big] = 0$ since the three indicator inequalities describe the empty set. Since $\Delta_1, \Delta_2 \geq 0$ implies $-\alpha\Delta_1 \leq \Delta_2$,

$$
\begin{aligned}
\mathsf{E}\big[\mathsf{E}\big[\Delta_i \mathbb{1}_{\{-\alpha\Delta_1 \leq \Delta_2\}} \mathbb{1}_{\{\Delta_i\geq 0\}}\big] \mathbb{1}_{\{\Delta_{3-i}\geq 0\}}\big] \;=\;\; & \mathsf{E}\big[\mathsf{E}\big[\Delta_i \mathbb{1}_{\{\Delta_i\geq 0\}}\big] \cdot \mathbb{1}_{\{\Delta_{3-i}\geq 0\}}\big] \\
=\;\; & \mathsf{E}\big[\Delta_i \mathbb{1}_{\{\Delta_i\geq 0\}}\big] \cdot \mathsf{P}\{\Delta_{3-i} \geq 0\}.
\end{aligned}
$$

As we need a lower bound on $\mathsf{E}\big[\mathsf{E}\big[\Delta_1 \cdot \mathbb{1}_{\{-\alpha\Delta_1 \leq \Delta_2\}}\big]\big]$, we may pessimistically assume that $\Delta_1 = -x/\alpha$ whenever Δ_2 happens to equal x. By this assumption,

$$
\begin{aligned}
& \mathsf{E}\big[\mathsf{E}\big[\Delta_1 \cdot \mathbb{1}_{\{-\alpha\Delta_1 \leq \Delta_2\}} \cdot \mathbb{1}_{\{\Delta_1<0\}}\big] \cdot \mathbb{1}_{\{\Delta_2\geq 0\}}\big] \\
\geq\;\; & -\mathsf{E}\big[\mathsf{E}\big[\Delta_2 \cdot \mathbb{1}_{\{-\alpha\Delta_1 \leq \Delta_2\}} \cdot \mathbb{1}_{\{\Delta_2\geq 0\}}\big] \cdot \mathbb{1}_{\{\Delta_1<0\}}\big]/\alpha,
\end{aligned}
$$

$$
\begin{aligned}
& \mathsf{E}\big[\mathsf{E}\big[\Delta_1 \cdot \mathbb{1}_{\{-\alpha\Delta_1 \leq \Delta_2\}} \cdot \mathbb{1}_{\{\Delta_1\geq 0\}}\big] \cdot \mathbb{1}_{\{\Delta_2<0\}}\big] \\
\geq\;\; & -\mathsf{E}\big[\mathsf{E}\big[\Delta_2 \cdot \mathbb{1}_{\{-\alpha\Delta_1 \leq \Delta_2\}} \cdot \mathbb{1}_{\{\Delta_2<0\}}\big] \cdot \mathbb{1}_{\{\Delta_1\geq 0\}}\big]/\alpha.
\end{aligned}
$$

All in all, we have

$$
\begin{aligned}
\mathsf{E}\big[\mathsf{E}\big[\Delta_1 \cdot \mathbb{1}_{\{-\alpha\Delta_1 \leq \Delta_2\}}\big]\big] \;\geq\;\; & \mathsf{E}\big[\Delta_1 \cdot \mathbb{1}_{\{\Delta_1\geq 0\}}\big] \cdot \mathsf{P}\{\Delta_2 \geq 0\} \\
& - \mathsf{E}\big[\mathsf{E}\big[\Delta_2 \cdot \mathbb{1}_{\{-\alpha\Delta_1 \leq \Delta_2\}} \cdot \mathbb{1}_{\{\Delta_2\geq 0\}}\big] \cdot \mathbb{1}_{\{\Delta_1<0\}}\big]/\alpha \\
& - \mathsf{E}\big[\mathsf{E}\big[\Delta_2 \cdot \mathbb{1}_{\{-\alpha\Delta_1 \leq \Delta_2\}} \cdot \mathbb{1}_{\{\Delta_2<0\}}\big] \cdot \mathbb{1}_{\{\Delta_1\geq 0\}}\big]/\alpha,
\end{aligned}
$$

$$E[E[\Delta_2 \cdot \mathbb{1}_{\{-\alpha\Delta_1 \le \Delta_2\}}]] = E[\Delta_2 \cdot \mathbb{1}_{\{\Delta_2 \ge 0\}}] \cdot P\{\Delta_1 \ge 0\}$$
$$+ E[E[\Delta_2 \cdot \mathbb{1}_{\{-\alpha\Delta_1 \le \Delta_2\}} \cdot \mathbb{1}_{\{\Delta_2 \ge 0\}}] \cdot \mathbb{1}_{\{\Delta_1 < 0\}}]$$
$$+ E[E[\Delta_2 \cdot \mathbb{1}_{\{-\alpha\Delta_1 \le \Delta_2\}} \cdot \mathbb{1}_{\{\Delta_2 < 0\}}] \cdot \mathbb{1}_{\{\Delta_1 \ge 0\}}].$$

Recall that we want to show that for some $\alpha = O(1)$

$$\xi \cdot E[E[\Delta_1 \cdot \mathbb{1}_{\{-\alpha\Delta_1 \le \Delta_2\}}]] \ge \alpha \cdot E[E[\Delta_2 \cdot \mathbb{1}_{\{-\alpha\Delta_1 \le \Delta_2\}}]],$$

and note that $E[\Delta_1 \cdot \mathbb{1}_{\{\Delta_1 \ge 0\}}] \cdot P\{\Delta_2 \ge 0\}$ and $E[\Delta_2 \cdot \mathbb{1}_{\{\Delta_2 \ge 0\}}] \cdot P\{\Delta_1 \ge 0\}$ are of the same order when $P\{\Delta_2 \ge 0\}$ and $P\{\Delta_1 \ge 0\}$ are $\Omega(1)$, respectively. Consequently, since $\xi = \omega(1)$, for the above inequality to hold for n large enough, it would be sufficient that

$$E[E[\Delta_2 \cdot \mathbb{1}_{\{-\alpha\Delta_1 \le \Delta_2\}} \cdot \mathbb{1}_{\{\Delta_2 \ge 0\}}] \cdot \mathbb{1}_{\{\Delta_1 < 0\}}]$$
$$+ E[E[\Delta_2 \cdot \mathbb{1}_{\{-\alpha\Delta_1 \le \Delta_2\}} \cdot \mathbb{1}_{\{\Delta_2 < 0\}}] \cdot \mathbb{1}_{\{\Delta_1 \ge 0\}}] \le 0 \tag{1}$$

because then we would have

$$E[E[\Delta_1 \cdot \mathbb{1}_{\{-\alpha\Delta_1 \le \Delta_2\}}]] \ge E[\Delta_1 \cdot \mathbb{1}_{\{\Delta_1 \ge 0\}}] \cdot P\{\Delta_2 \ge 0\} \quad \text{and}$$
$$E[E[\Delta_2 \cdot \mathbb{1}_{\{-\alpha\Delta_1 \le \Delta_2\}}]] \le E[\Delta_2 \cdot \mathbb{1}_{\{\Delta_2 \ge 0\}}] \cdot P\{\Delta_1 \ge 0\}.$$

Concerning the expected spatial gain in S_2, however, we are going to use the trivial upper bound $E[E[\Delta_2 \cdot \mathbb{1}_{\{-\alpha\Delta_1 \le \Delta_2\}}]] \le E[\Delta_2 \cdot \mathbb{1}_{\{\Delta_2 \ge 0\}}]$, and thus, we concentrate on a lower bound on the expected spatial gain in S_1 in the following. Therefore, we prove next that inequality (1) holds for $\alpha = O(1)$ at least if the actual length of m_2 differs by no more than a constant factor from $\bar{\ell}_1$, the expected length of m_1.

Lemma 2. *If $P\{\Delta_1 \ge 0\} = \Omega(1)$ and $|m_2| = \Theta(\bar{\ell}_1)$, there exists a constant α^* such that for n large enough inequality (1) on this page holds for all $\alpha \ge \alpha^*$.*

The proof can be found in Appendix A. Note that $\bar{\ell}_1 = \bar{\ell}_2$ in our scenario. We know (cf. Section 2) that

$$P\{||m_2| - \bar{\ell}_2| \ge (\sqrt{3} - 1) \cdot \bar{\ell}_2\} \le \left((\sqrt{3} - 1)^2 \cdot 2 \cdot (n-1)\right)^{-1} < (n-1)^{-1},$$

and thus, the condition "$|m_2| = \Theta(\bar{\ell}_1)$" is not met only with probability $O(1/n)$. Whether or not this condition is met, trivially $\Delta_1 \ge -|m_1|$, and consequently, $E[E[\Delta_1 \cdot \mathbb{1}_{\{-\alpha\Delta_1 \le \Delta_2\}}]] \ge -\bar{\ell}_1$. Applying this rough bound only in the case of $||m_2| - \bar{\ell}_1| > (\sqrt{3}-1) \cdot \bar{\ell}_1$ and $(\Delta_1, \Delta_2) \in \mathbb{R}_{<0} \times \mathbb{R}_{\ge 0} \cup \mathbb{R}_{\ge 0} \times \mathbb{R}_{<0}$, the preceding lemma reads: if $P\{\Delta_1 \ge 0\} = \Omega(1)$ then for $\alpha \ge \alpha^*$

$$E[E[\Delta_1 \cdot \mathbb{1}_{\{-\alpha\Delta_1 \le \Delta_2\}}]] \ge E[\Delta_1 \cdot \mathbb{1}_{\{\Delta_1 \ge 0\}}] \cdot P\{\Delta_2 \ge 0\} - \frac{\bar{\ell}_1}{n-1}.$$

Next we will see that this additive error term vanishes in situations that arise due to the 1/5-rule.

Lemma 3. *If* $P\{\Delta_1 \geq 0\}$ *and* $P\{\Delta_2 \geq 0\}$ *are* $\Omega(1)$, *respectively, there exists a constant* α^* *such that for* $\alpha \geq \alpha^*$ *and* n *large enough*

$$\mathsf{E}\big[\mathsf{E}[\Delta_1 \cdot \mathbb{1}_{\{f' \leq f\}}]\big] \;\geq\; \mathsf{E}\big[\Delta_1 \cdot \mathbb{1}_{\{\Delta_1 \geq 0\}}\big] \cdot P\{\Delta_2 \geq 0\}/\,2.$$

Proof. Recall that $f' \leq f \wedge -\alpha\Delta_1 > \Delta_2$ implies $\Delta_1 > 0 > \Delta_2$. Consequently, all (Δ_1, Δ_2)-tuples zeroed out by $\mathbb{1}_{\{-\alpha\Delta_1 \leq \Delta_2\}}$, but kept by $\mathbb{1}_{\{f' \leq f\}}$ are in $\mathbb{R}_{>0} \times \mathbb{R}_{<0}$. Analogously, $f' > f \wedge -\alpha\Delta_1 \leq \Delta_2$ implies $\Delta_1 < 0 < \Delta_2$ so that all (Δ_1, Δ_2)-tuples kept by $\mathbb{1}_{\{-\alpha\Delta_1 \leq \Delta_2\}}$, but zeroed out by $\mathbb{1}_{\{f' \leq f\}}$ are in $\mathbb{R}_{<0} \times \mathbb{R}_{>0}$. Hence,

$$\mathsf{E}\big[\mathsf{E}[\Delta_1 \cdot \mathbb{1}_{\{f' \leq f\}}]\big] \;\geq\; \mathsf{E}\big[\mathsf{E}[\Delta_1 \cdot \mathbb{1}_{\{-\alpha\Delta_1 \leq \Delta_2\}}]\big]$$
$$\big(\text{and } \mathsf{E}\big[\mathsf{E}[\Delta_2 \cdot \mathbb{1}_{\{f' \leq f\}}]\big] \;\leq\; \mathsf{E}\big[\mathsf{E}[\Delta_2 \cdot \mathbb{1}_{\{-\alpha\Delta_1 \leq \Delta_2\}}]\big]\big).$$

As $P\{\Delta_1 \geq 0\} = \Omega(1)$ implies $\mathsf{E}[\Delta_1 \cdot \mathbb{1}_{\{\Delta_1 \geq 0\}}] = \Omega(\bar{\ell}_1/\sqrt{n})$ (cf. the results restated in Section 2), the error term $\bar{\ell}_1/(n-1)$ is by an $O(1/\sqrt{n})$-factor smaller than $\mathsf{E}[\Delta_1 \cdot \mathbb{1}_{\{\Delta_1 \geq 0\}}] \cdot P\{\Delta_2 \geq 0\} = \Omega(\bar{\ell}_1/\sqrt{n}) \cdot \Omega(1)$. Finally, for n large enough $1 - O(1/\sqrt{n}) \geq 1/2$. ☐

Recall: we expect $\alpha'' = \alpha$ iff $\xi \cdot \mathsf{E}\big[\mathsf{E}[\Delta_1 \cdot \mathbb{1}_{\{f' \leq f\}}]\big] = \alpha \cdot \mathsf{E}\big[\mathsf{E}[\Delta_2 \cdot \mathbb{1}_{\{f' \leq f\}}]\big]$ or, equivalently, iff $\mathsf{E}\big[\mathsf{E}[\Delta_1 \cdot \mathbb{1}_{\{f' \leq f\}}]\big]/d_1 = \mathsf{E}\big[\mathsf{E}[\Delta_2 \cdot \mathbb{1}_{\{f' \leq f\}}]\big]/d_2$. Thus there exists a distinct α_0 such that there is no drift w.r.t. the ratio d_1/d_2, i.e., this ratio becomes steady-state. Then for $\alpha < \alpha_0$, α is more likely to increase than to decrease, and for $\alpha > \alpha_0$, α is more likely to decrease than to increase.

Since $\mathsf{E}\big[\mathsf{E}[\Delta_2 \cdot \mathbb{1}_{\{f' \leq f\}}]\big] \leq \mathsf{E}\big[\mathsf{E}[\Delta_2 \cdot \mathbb{1}_{\{-\alpha\Delta_1 \leq \Delta_2\}}]\big] \leq \mathsf{E}\big[\Delta_2 \cdot \mathbb{1}_{\{\Delta_2 \geq 0\}}\big]$ and $\xi = \omega(1)$, we have $\xi \cdot P\{\Delta_2 \geq 0\}/2 \geq \alpha^*$ for n large enough if $P\{\Delta_2 \geq 0\} = \Omega(1)$, and hence, $\alpha_0 \leq \alpha^* = O(1)$ under the conditions of Lemma 3. Besides, the 1/5-rule just ensures these conditions as long as $d_1 = O(d_2)$. For the same reasons, there exists $\alpha_\downarrow > \alpha_0$ such that $\xi \cdot \mathsf{E}\big[\mathsf{E}[\Delta_1 \cdot \mathbb{1}_{\{f' \leq f\}}]\big] \geq 2 \cdot \alpha \cdot \mathsf{E}\big[\mathsf{E}[\Delta_2 \cdot \mathbb{1}_{\{f' \leq f\}}]\big]$ (for n large enough) and $\alpha_\downarrow = O(1)$ again under the conditions of Lemma 3. Thus, when $\alpha \geq \alpha_\downarrow$ there is a drift towards smaller α; more formally:

Lemma 4. *Let the scaling factor* s *be fixed. If* $P\{\Delta_1 \geq 0\}$ *and* $P\{\Delta_2 \geq 0\}$ *are* $\Omega(1)$, *respectively, there exists a constant* $\alpha_\downarrow$ *such that for* n *large enough, if in the* i^{th} *step* $\alpha^{[i]} \geq \alpha_\downarrow$ *(yet* $\alpha^{[i]} = O(\xi)$), *then w.o.p. after at most* $n^{0.3}$ *steps the search is located at a point for which* $\alpha < \alpha^{[i]}$, *and furthermore, w.o.p.* $\alpha \leq \alpha^{[i]} + O(\alpha^{[i]}/n^{0.6})$ *in all intermediate steps.*

The proof can be found in Appendix B. Since the 1/5-rule keeps the scaling factor unchanged for n steps, we can virtually partition each such observation phase in $n/n^{0.3} = n^{0.7}$ sub-phases to each of which this lemma applies. Since $O(\alpha^{[i]}/n^{0.6}) \leq \alpha^{[i]}$ for n large enough, the preceding lemma tells us that, when starting at a point with $\alpha^{[0]} = O(1)$, i.e. $d_1^{[0]}/d_2^{[0]} = O(1/\xi)$, then α will be upper bounded by $2 \cdot \max\{\alpha^{[0]}, \alpha_\downarrow\} = O(1)$ w.o.p. for any polynomial number of steps. Incorporating these new insights into the argumentation for the 1/5-rule known from the analysis of SPHERE finally enables us to replace the objectionable condition "$d_1/d_2 = O(1/\xi)$ in the complete optimization process" in Theorem 1 by "$d_1/d_2 = O(1/\xi)$ for the initial search point" – yielding the main result on the rutime of the (1+1) ES on the quadratic forms considered:

Theorem 2. *If the initialization satisfies* $s = \Theta((\sqrt{f(c)}/n)/\xi)$ *and* $d_1/d_2 = O(1/\xi)$, *then w. o. p. the number of steps/f-evaluations to reduce the initial approximation error/f-value to a 2^{-t}-fraction, $t = poly(n)$, is $\Theta(t \cdot \xi \cdot n)$.*

Naturally, one might ask what happens if the optimization starts at a point for which d_1 is not $O(d_2/\xi)$. A closer look at the argumentation in the proof of the preceding lemma reveals that the same argumentation results in the proof of the existence of another constant $\alpha_{\Downarrow} > \alpha_{\downarrow}$ such that the drift towards smaller α is that strong when $\alpha \geq \alpha_{\Downarrow}$ that w. o. p. α drops by a constant fraction within at most n steps:

Lemma 5. *Let the scaling factor s be fixed. If $\mathsf{P}\{\Delta_1 \geq 0\}$, $1/2 - \mathsf{P}\{\Delta_1 \geq 0\}$, $\mathsf{P}\{\Delta_2 \geq 0\}$ are $\Omega(1)$, respectively, then there exists a constant $\alpha_{\Downarrow}$ such that for n large enough: if in the ith step $\alpha^{[i]} \geq \alpha_{\Downarrow}$ (yet $\alpha^{[i]} = O(\xi)$, i. e. $d_1 = O(d_2)$), then w. o. p. after at most n steps the search is located at a point with $\alpha \leq \alpha^{[i]} - \Omega(\alpha^{[i]})$.*

See Appendix C for the proof. Finally, this lemma shows that α drops very quickly if the lemma's conditions are met. Again utilizing the results for SPHERE, it is simple to check that these conditions are met when d_1 is $O(d_2)$ (and $\Omega(d_2/\xi)$, of course). If d_1 is not $O(d_2)$, for instance if we start at a point of steepest descent (w. r. t. all points of a fixed f-value), i. e. $d_2 = 0$ so that $f = \xi d_1^2$, then a simple argumentation using rough bounds on Δ_1 and Δ_2 yields that d_1/d_2 drops even faster than in situations covered by the preceding lemma – which is hardly surprising since the (expected) spatial gain of a step in S_1 (on the ξ-weighted SPHERE$_{n/2}$) is negative whereas the one in S_2 is positive.

5   Conclusion

Based on the results on how the $(1+1)$ES minimizes the well-known SPHERE-function, we have extended these results to a broader class of functions consisting of certain positive definite quadratic forms. The main insight of the results presented is that Gaussian mutations adapted by the 1/5-rule result in the optimization process to stabilize such that the trajectory of the evolving search point takes course very close to the gentlest descent of the ellipsoidal fitness landscape. However, more insight into how EAs for continuous optimization work is gained, contributing to building an algorithmic EA-theory for continuous search spaces.

Naturally, the results carry over to functions that are translations of the considered functions. Furthermore, the argumentation presented here yields that for arbitrary positive definite quadratic forms – which we may assume to be of the form $f_n(x) = \sum_{i=1}^{n} \xi_i \cdot x_i^2$ with $\xi_n \geq \cdots \geq \xi_1 > 0$ as we have seen – the number of steps to halve the function value is $O(n \cdot \xi_n/\xi_1)$. This is due to the maximum curvature being upper bounded by $(\xi_n/\xi_1)/\sqrt{f}$ so that the radius of the hypersphere S is at least $\sqrt{f} \cdot \xi_1/\xi_n$. As a direct consequence, we obtain a $\Theta(n)$-bound for functions where all the ξ_is are of the same order, i. e. $\xi_n = \Theta(\xi_1)$. This is the reason why ξ was chosen to be $\omega(1)$.

Acknowledgments

I would like to thank two of the three reviewers for their helpful comments, in particular for pointing out that applying differential geometry in the analysis of fitness landscapes was already proposed by Beyer (1994).

A   Proof of Lemma 2

"If $P\{\Delta_1 \geq 0\} = \Omega(1)$ and $|m_2| = \Theta(\bar{\ell}_1)$, there exists a constant α^* such that for n large enough inequality (1) on page 272 holds for all $\alpha \geq \alpha^*$."

Let us assume for a moment that the distribution of $|m_2|$ were concentrated at a certain ℓ_2, and let "$D\{\cdot\}$" denote the density of an event. Then

$$E\big[E\big[\Delta_2 \cdot \mathbb{1}_{\{-\alpha\Delta_1 \leq \Delta_2\}} \cdot \mathbb{1}_{\{\Delta_2 \geq 0\}}\big] \cdot \mathbb{1}_{\{\Delta_1 < 0\}}\big]$$

$$= \int_0^{\ell_2} x \cdot D\{\Delta_2 = x\} \cdot P\{-x/\alpha \leq \Delta_1 < 0\} \, dx \qquad \text{and}$$

$$E\big[E\big[\Delta_2 \cdot \mathbb{1}_{\{-\alpha\Delta_1 \leq \Delta_2\}} \cdot \mathbb{1}_{\{\Delta_2 < 0\}}\big] \cdot \mathbb{1}_{\{\Delta_1 \geq 0\}}\big]$$

$$= \int_{-\ell_2}^0 y \cdot D\{\Delta_2 = y\} \cdot P\{\Delta_1 \geq -y/\alpha\} \, dy$$

$$= \int_0^{\ell_2} -x \cdot D\{\Delta_2 = -x\} \cdot P\{\Delta_1 \geq x/\alpha\} \, dx.$$

We know from the analysis of SPHERE that for $x \in [0, \ell_2)$

$$D\{\Delta_2 = x\} < \frac{\Psi_n}{\ell_2} \cdot (1 - (x/\ell_2)^2)^{(n-3)/2} < D\{\Delta_2 = -x\}$$

(with $\Psi_n := \pi^{-1/2} \cdot \Gamma(n/2) / \Gamma(n/2 - 1/2) = \Theta(\sqrt{n})$, where "$\Gamma$" denotes the well-known Gamma function).

Thus, the LHS of (1) on page 272 is smaller than

$$\int_0^{\ell_2} x \cdot \frac{\Psi_n}{\ell_2} \cdot (1 - (x/\ell_2)^2)^{(n-3)/2} \cdot P\{-x/\alpha \leq \Delta_1 < 0\} \, dx$$

$$- \int_0^{\ell_2} x \cdot \frac{\Psi_n}{\ell_2} \cdot (1 - (x/\ell_2)^2)^{(n-3)/2} \cdot P\{\Delta_1 \geq x/\alpha\} \, dx$$

$$= \int_0^{\ell_2} x \frac{\Psi_n}{\ell_2} (1 - (x/\ell_2)^2)^{(n-3)/2} \big(P\{-x/\alpha \leq \Delta_1 < 0\} - P\{\Delta_1 \geq x/\alpha\}\big) \, dx.$$

Let $\Phi \colon [0, \ell_2] \to [-1, 1]$ be defined by $\Phi(y) := P\{-y \leq \Delta_1 < 0\} - P\{\Delta_1 \geq y\}$.

Hence,

$$\int_0^{\ell_2} x \cdot \frac{\Psi_n}{\ell_2} \cdot (1 - (x/\ell_2)^2)^{(n-3)/2} \cdot \Phi(x/\alpha)\,\mathrm{d}x \quad \leq \quad 0$$

implies the inequality (1). Note that, obviously, $\mathsf{P}\{-0 \leq \Delta_1 < 0\} = 0$ and, by assumption, $\mathsf{P}\{\Delta_1 \geq 0\} = \Omega(1)$. Since, $\mathsf{P}\{\Delta_1 \geq y\}$ decreases monotonically, whereas $\mathsf{P}\{-y \leq \Delta_1 < 0\}$ increases monotonically when y grows, $\Phi(y)$ is monotone increasing for $0 \leq y \leq \min\{\ell_1, \ell_2\}$ and equals $\mathsf{P}\{\Delta_1 < 0\}$ for $y \geq \ell_1$. Furthermore, if ε denotes an arbitrary constant with $0 < \varepsilon < \mathsf{P}\{\Delta_1 \geq 0\}$, then $\mathsf{P}\{\Delta_1 \geq y\} = \varepsilon$ implies $y = \Theta(\bar\ell_1/\sqrt{n})$. Analogously, if $0 < \varepsilon < \mathsf{P}\{\Delta_1 < 0\}$, then $\mathsf{P}\{-y \leq \Delta_1 < 0\} = \varepsilon$ implies $y = \Theta(\bar\ell_1/\sqrt{n})$. Thus, there exists $\breve{y} = \kappa \cdot \bar\ell_1/\sqrt{n-1}$ with $\kappa = \Theta(1)$ such that $\mathsf{P}\{\Delta_1 \geq \breve{y}\} = \mathsf{P}\{-\breve{y} \leq \Delta_1 < 0\}$, i.e., $\Phi(\breve{y}) = 0$, and hence, the inequality to be shown reads

$$-\frac{\Psi_n}{\ell_2} \int_0^{\alpha \cdot \breve{y}} x \cdot (1 - (x/\ell_2)^2)^{(n-3)/2} \cdot \Phi(x/\alpha)\,\mathrm{d}x$$

$$\geq \quad \frac{\Psi_n}{\ell_2} \int_{\alpha \cdot \breve{y}}^{\ell_2} x \cdot (1 - (x/\ell_2)^2)^{(n-3)/2} \cdot \Phi(x/\alpha)\,\mathrm{d}x. \tag{2}$$

For the RHS we have, using $(1 - a/(n-1))^{(n-1)/2} \leq e^{-a/2}$ for $n - 1 > a > 0$,

$$\int_{\alpha \cdot \breve{y}}^{\ell_2} x \cdot (1 - (x/\ell_2)^2)^{(n-3)/2} \cdot \Phi(x/\alpha)\,\mathrm{d}x$$

$$\leq \quad \int_{\alpha \cdot \breve{y}}^{\ell_2} x \cdot (1 - (x/\ell_2)^2)^{(n-3)/2} \cdot 1\,\mathrm{d}x$$

$$= \quad \left[\frac{-\ell_2^2}{2} \cdot \frac{(1 - (x/\ell_2)^2)^{(n-1)/2}}{(n-1)/2}\right]_{\alpha \cdot \breve{y}}^{\ell_2}$$

$$= \quad 0 - \left(\frac{-\ell_2^2}{n-1} \cdot (1 - (\alpha \cdot \breve{y}/\ell_2)^2)^{(n-1)/2}\right)$$

$$= \quad \frac{\ell_2^2}{n-1} \cdot (1 - (\alpha \cdot \breve{y}/\ell_2)^2)^{(n-1)/2}$$

$$\leq \quad \frac{\ell_2^2}{n-1} \cdot \left(1 - (\alpha \cdot \kappa \cdot \bar\ell_1/\ell_2)^2/(n-1)\right)^{(n-1)/2}$$

$$\leq \quad \frac{\ell_2^2}{n-1} \cdot e^{-(\alpha \cdot \kappa \cdot \bar\ell_1/\ell_2)^2/2} \quad \text{if } n - 1 > \left(\alpha \cdot \kappa \cdot \frac{\bar\ell_1}{\ell_2}\right)^2.$$

For the LHS of (2) note that, by the same arguments, there exists $\ddot{y} = \tau \cdot \bar\ell_1/\sqrt{n-1}$ with $\tau = \Theta(1)$ such that $\mathsf{P}\{\Delta_1 \geq \ddot{y}\} = 2 \cdot \mathsf{P}\{-\ddot{y} \leq \Delta_1 < 0\}$, and thus, for $0 \leq y \leq \ddot{y}$ we have $\mathsf{P}\{\Delta_1 \geq y\} \geq 2 \cdot \mathsf{P}\{-y \leq \Delta_1 < 0\}$, i.e., $-\Phi(y) \geq p := \mathsf{P}\{\Delta_1 \geq \ddot{y}\}/2 = \Omega(1)$. Hence,

$$-\int_0^{\alpha\cdot\ddot{y}} x\cdot(1-(x/\ell_2)^2)^{(n-3)/2}\cdot\Phi(x/\alpha)\,\mathrm{d}x$$

$$\geq\int_0^{\alpha\cdot\ddot{y}} x\cdot(1-(x/\ell_2)^2)^{(n-3)/2}\cdot p\,\mathrm{d}x$$

$$=p\cdot\left[\frac{-\ell_2^2}{2}\cdot\frac{(1-(x/\ell_2)^2)^{(n-1)/2}}{(n-1)/2}\right]_0^{\alpha\cdot\ddot{y}}$$

$$=p\cdot\frac{-\ell_2^2}{n-1}\cdot\left((1-(\alpha\cdot\ddot{y}/\ell_2)^2)^{(n-1)/2}-1\right)$$

$$=p\cdot\frac{\ell_2^2}{n-1}\cdot\left(1-\left(1-\frac{(\alpha\cdot\tau\cdot\bar{\ell}_1/\ell_2)^2}{n-1}\right)^{(n-1)/2}\right)$$

$$\geq p\cdot\frac{\ell_2^2}{n-1}\cdot\left(1-\mathrm{e}^{-(\alpha\cdot\tau\cdot\bar{\ell}_1/\ell_2)^2/2}\right)\qquad\text{if } n-1>\left(\alpha\cdot\tau\cdot\frac{\bar{\ell}_1}{\ell_2}\right)^2.$$

All in all, we have broken it down into the inequality

$$p\cdot\frac{\ell_2^2}{n-1}\cdot\left(1-\mathrm{e}^{-(\alpha\cdot\tau\cdot\bar{\ell}_1/\ell_2)^2/2}\right)\ \geq\ \frac{\ell_2^2}{n-1}\cdot\mathrm{e}^{-(\alpha\cdot\kappa\cdot\bar{\ell}_1/\ell_2)^2/2}.$$

Since p, τ, and κ are $\Theta(1)$, it is finally obvious that $\alpha=O(1)$ can be chosen large enough for this inequality to hold for n large enough if $\bar{\ell}_1/\ell_2=\Theta(1)$, i.e. $\ell_2=\Theta(\bar{\ell}_1)$.

B   Proof of Lemma 4

"Let the scaling factor s be fixed. If $\mathrm{P}\{\Delta_1\geq 0\}$ and $\mathrm{P}\{\Delta_2\geq 0\}$ are $\Omega(1)$, respectively, there exists a constant $\alpha_\downarrow$ such that for n large enough, if in the ith step $\alpha^{[i]}\geq\alpha_\downarrow$ (yet $\alpha^{[i]}=O(\xi)$), then w.o.p. after at most $n^{0.3}$ steps the search is located at a point for which $\alpha<\alpha^{[i]}$, and furthermore, w.o.p. $\alpha\leq\alpha^{[i]}+O(\alpha^{[i]}/n^{0.6})$ in all intermediate steps."

We begin by proving the second claim. Let us assume that, starting with the ith step, $\alpha\geq\alpha^{[i]}$ for $k\leq n^{0.3}$ steps. Recall that, due to elitist selection, the f-value is non-increasing. As $d_2>d_2^{[i]}$ and $f\leq f^{[i]}$ implies $d_1<d_1^{[i]}$, which again implies $\alpha/\xi=d_1/d_2<d_1^{[i]}/d_2^{[i]}=\alpha^{[i]}/\xi$, we have just proved that (surely) $d_2\leq d_2^{[i]}$ in these k steps, respectively. Since, irrespective of the adaptation of the length of an isotropic mutation, in a step w.o.p. $\Delta_2=O(d_2/n^{0.9})$, in all $k\leq n^{0.3}$ steps w.o.p. $d_2\geq d_2^{[i]}-k\cdot O(d_2^{[i]}/n^{0.9})\geq d_2^{[i]}-O(d_2^{[i]}/n^{0.6})$, i.e., $d_2=d_2^{[i]}(1-\psi)$ for some $\psi=O(n^{-0.6})$, respectively. Concerning an upper bound on d_1, we have

$$f\ =\ \xi d_1^2+d_2^2\ =\ \xi d_1^2+\left(d_2^{[i]}-\psi d_2^{[i]}\right)^2\ \leq\ f^{[i]}\ =\ \xi {d_1^{[i]}}^2+{d_2^{[i]}}^2,$$

and hence

$$\xi d_1^2 \;\leq\; \xi d_1^{[i]^2} + (2\psi - \psi^2)d_2^{[i]^2}$$

$$\Leftrightarrow \quad d_1^2 \;\leq\; d_1^{[i]^2} + (2\psi - \psi^2)\frac{d_2^{[i]^2}}{\xi} \;=\; d_1^{[i]^2} + (2\psi - \psi^2)\frac{d_1^{[i]^2}}{\alpha^{[i]}}$$

$$= d_1^{[i]^2}\left(1 + \frac{\psi(2-\psi)}{\alpha^{[i]}}\right)$$

Since $\psi(2-\psi)/\alpha^{[i]}$ is $O(\psi)$, i.e. $O(n^{-0.6})$, we finally get that in all k steps

$$\frac{\alpha}{\xi} = \frac{d_1}{d_2} \leq \frac{d_1^{[i]}}{d_2^{[i]}} \cdot \frac{\sqrt{1 + O(n^{-0.6})}}{1 - O(n^{-0.6})} = \frac{\alpha^{[i]}}{\xi} \cdot (1 + O(n^{-0.6})).$$

Now we are ready for the proof of the lemma's first claim. Therefore, assume that $\alpha \geq \alpha^{[i]} \geq \alpha_\downarrow$ for $n^{0.3}+1$ steps. We are going to show that the probability of observing such a sequence of steps is exponentially small. Note that, since w.o.p. $d_2 \geq d_2^{[i]}(1-\psi)$ as we have seen, our assumption implies that also w.o.p. $d_1 \geq d_1^{[i]}(1-\psi)$, i.e., w.o.p. $d_1 = d_1^{[i]} - O(d_1^{[i]}/n^{0.6})$ in all $n^{0.3}$ steps. Let $X_j^{[k]}$, $j \in \{1,2\}$, denote the RV $\Delta_j \cdot \mathbb{1}_{\{f' \leq f\}}$ in the $(i-1+k)^{\text{th}}$ step (so that $\mathsf{E}[X_j] = \mathsf{E}[\mathsf{E}[\Delta_j \cdot \mathbb{1}_{\{f' \leq f\}}]]$). Then, according to the arguments preceding the lemma, for $1 \leq k \leq n^{0.3}$, $\mathsf{E}\left[X_1^{[k]}\right]/d_1^{[k]} \geq 2 \cdot \mathsf{E}\left[X_2^{[k]}\right]/d_2^{[k]}$, i.e.,

$$\xi \cdot \mathsf{E}\left[X_1^{[k]}\right] \geq 2 \cdot \alpha^{[k]} \cdot \mathsf{E}\left[X_2^{[k]}\right] \geq 2 \cdot \alpha^{[i]} \cdot \mathsf{E}\left[X_2^{[k]}\right].$$

Let $S_j^{[k]} := X_j^{[1]} + \cdots + X_j^{[k]}$ denote the total gain of k steps w.r.t. to d_j. By linearity of expectation, $\mathsf{E}\left[S_1^{[k]}\right]/d_1^{[i]} \geq 2 \cdot \mathsf{E}\left[S_2^{[k]}\right]/d_2^{[i]}$ for $1 \leq k \leq n^{0.3}$; however, the goal is to show that $\mathsf{P}\left\{S_1^{[k]}/d_1^{[i]} \leq S_2^{[k]}/d_2^{[i]} \text{ for } 1 \leq k \leq n^{0.3}\right\}$ is exponentially small.

Therefore, we will assume the worst case (w.r.t. to the analysis, i.e. the best case w.r.t. the chance of observing such a sequence) that $\mathsf{E}\left[X_1^{[k]}\right]/d_1^{[i]} = 2 \cdot \mathsf{E}\left[X_2^{[k]}\right]/d_2^{[i]}$ in each step. To see that this is in fact the worst case consider a search point $\boldsymbol{x}$ for which $\alpha \geq \alpha^{[i]}$, i.e. $d_1/d_2 > d_1^{[i]}/d_2^{[i]}$, so that $\xi \cdot \mathsf{E}[X_1] > 2 \cdot \alpha \cdot \mathsf{E}[X_2]$. Now consider a search point $\widetilde{\boldsymbol{x}}$ with $f(\widetilde{\boldsymbol{x}}) = f(\boldsymbol{x})$ but $\widetilde{\alpha} < \alpha$, i.e., $\widetilde{d_1} < d_1$ and $\widetilde{d_2} > d_2$. Owing to the results on SPHERE we know that, for an isotropic mutation of an arbitrary fixed length ℓ_j, for any fixed $g \in (-\ell_j, \ell_j)$, $\mathsf{P}\{\Delta_j \geq g\}$ strictly increases with d_j (when $d_j > \ell_j$). Consequently, (independently of the distribution of $|\boldsymbol{m}|$) $\widetilde{\Delta}_1$ is stochastically dominated by Δ_1, whereas $\widetilde{\Delta}_2$ stochastically dominates Δ_2. This implies that X_1 dominates $\widetilde{X}_1$, whereas X_2 is dominated by $\widetilde{X}_2$ (in particular, we have $\mathsf{E}[X_1] < \mathsf{E}\left[\widetilde{X}_2\right]$ and $\mathsf{E}[X_2] > \mathsf{E}\left[\widetilde{X}_2\right]$).

As we have just seen, we may pessimistically assume that in each step the search is located at a point for which $\xi \cdot \mathsf{E}[X_1] = 2 \cdot \alpha \cdot \mathsf{E}[X_2]$. Hence, $\mathsf{E}\left[S_1^{[k]}\right]/d_1^{[i]} =$

$2 \cdot \mathsf{E}\left[S_2^{[k]}\right]/d_2^{[i]}$. Let $S_j := S_j^{[n^{0.3}]}$. Since $1.2/0.8 = 1.5 < 2$, it is sufficient to show that w. o. p. $S_1 \geq 0.8 \cdot \mathsf{E}[S_1]$ and w. o. p. $S_2 \leq 1.2 \cdot \mathsf{E}[S_2]$. The Hoeffding bounds (1963) (cf. Section 2.6.2 of (Hofri, 1987)) state that, for $X_j^{[k]} \in [a_j, b_j]$ and $t_j > 0$,

$$P\{S_1 - \mathsf{E}[S_1] \leq -n^{0.3} \cdot t_1\} \leq \exp\left(\frac{-2 \cdot n^{0.3} \cdot t_1^2}{(b_1 - a_1)^2}\right) \quad \text{and}$$

$$P\{S_2 - \mathsf{E}[S_2] \geq n^{0.3} \cdot t_2\} \leq \exp\left(\frac{-2 \cdot n^{0.3} \cdot t_2^2}{(b_2 - a_2)^2}\right).$$

For $t_j = 0.2 \cdot \mathsf{E}[S_j]/n^{0.3}$, both exponents equal

$$-0.08 \cdot n^{-0.3} \cdot \mathsf{E}[S_j]^2/(b_j - a_j)^2 = -\Omega(n^{-0.3}) \cdot \left(\frac{\mathsf{E}[S_j]}{b_j - a_j}\right)^2,$$

respectively. Therefore, our goal is to show that $\mathsf{E}[S_j]/(b_j - a_j) = \Omega(n^{0.2})$.

First we concentrate on $\mathsf{E}[S_1]$. Since S_1 is the sum of $n^{0.3}$ RVs $X_1^{[k]}$, it suffices to show that $\mathsf{E}\left[X_1^{[k]}\right]/(b_1 - a_1) = \Omega(n^{-0.1})$ for $1 \leq k \leq n^{0.3}$. In the following we assume that $d_1 = d_1^{[i]} \pm O(d_1^{[i]}/n^{0.6})$ and $d_2 \in \left[d_2^{[i]} - O(d_2^{[i]}/n^{0.6}), d_2^{[i]}\right]$ since we have seen (in the preceding proof of the second claim) that this happens w. o. p. Owing to the results for SPHERE, we know that $P\{\Delta_j \geq 0\} = \Omega(1)$ implies that the scaling factor s is $O(d_j/n)$, which results in $\bar{\ell}_j = O(d_j/\sqrt{n})$, and that, under these conditions, w. o. p. $|\Delta_j| = O(\bar{\ell}_j/n^{0.4})$. Recall that $\mathsf{E}\left[\Delta_1 \cdot \mathbb{1}_{\{f' \leq f\}}\right]$ is at least $\mathsf{E}\left[\Delta_1 \cdot \mathbb{1}_{\{\Delta_1 \geq 0\}}\right] \cdot P\{\Delta_2 \geq 0\}/2$. Since $P\{\Delta_2 \geq 0\} = \Omega(1)$ in i^{th} step and $d_2 \geq d_2^{[i]}(1 - O(n^{-0.6}))$ in all $n^{0.3}$ steps, in each of these steps $P\{\Delta_2 \geq 0\} = \Omega(1)$. Hence, $\mathsf{E}[X_1] = \Omega(\mathsf{E}\left[\Delta_1 \cdot \mathbb{1}_{\{\Delta_1 \geq 0\}}\right])$ in each of the $n^{0.3}$ steps. Owing to the results for SPHERE, we know that (since $\bar{\ell}_1 = O(d_1/\sqrt{n})$ as we have seen) $\mathsf{E}\left[\Delta_1 \cdot \mathbb{1}_{\{\Delta_1 \geq 0\}}\right] = \Theta(\bar{\ell}_1/\sqrt{n})$ so that $\mathsf{E}[X_1] = \Omega(\bar{\ell}_1/\sqrt{n})$. Thus, $\mathsf{E}[S_1] = n^{0.3} \cdot \Omega(\bar{\ell}_1/\sqrt{n}) = \Omega(\bar{\ell}_1/n^{0.2})$ and $b_1 - a_1 = O(\bar{\ell}_1/n^{0.4})$, i. e., $\mathsf{E}[S_1]/(b_1 - a_1) = \Omega(n^{0.2})$.

Concerning a lower bound on $\mathsf{E}[S_2]$, recall that $\mathsf{E}[S_1]/d_1^{[i]} = 2 \cdot \mathsf{E}[S_2]/d_2^{[i]}$, i. e., $\mathsf{E}[S_2] = \mathsf{E}[S_1] \cdot d_2^{[i]}/(2 \cdot d_1^{[i]}) = \Omega(\bar{\ell}_1/n^{0.2}) \cdot \Omega(\xi/\alpha^{[i]})$. As $\bar{\ell}_1 = \bar{\ell}_2$ and (by assumption) $\alpha^{[i]} = O(\xi)$, we have $\mathsf{E}[S_2] = \Omega(\bar{\ell}_2/n^{0.2})$. Since $b_2 - a_2 = O(\bar{\ell}_2/n^{0.4})$ (see above), $\mathsf{E}[S_2]/(b_2 - a_2) = \Omega(\bar{\ell}_2/n^{0.2})/O(\bar{\ell}_2/n^{0.4})$ is also $\Omega(n^{0.2})$.

All in all, our initial assumption that $\alpha \geq \alpha^{[i]} \geq \alpha_\downarrow$ for $n^{0.3} + 1$ steps implies that w. o. p. for the first $n^{0.3}$ steps $S_1/S_2 > \alpha^{[i]}/\xi$, i. e., that w. o. p. after at most $n^{0.3}$ steps α drops below $\alpha^{[i]}$ – showing that the sequence of steps we assumed to be observed happens only with an exponentially small probability.

C   Proof of Lemma 5

"Let the scaling factor s be fixed. If $P\{\Delta_1 \geq 0\}$, $1/2 - P\{\Delta_1 \geq 0\}$, $P\{\Delta_2 \geq 0\}$ are $\Omega(1)$, respectively, then there exists a constant $\alpha_\downarrow$ such that for n large enough: if in the ith step $\alpha^{[i]} \geq \alpha_\downarrow$ (yet $\alpha^{[i]} = O(\xi)$, i. e. $d_1 = O(d_2)$), then w. o. p. after at most n steps the search is located at a point with $\alpha \leq \alpha^{[i]} - \Omega(\alpha^{[i]})$."

By the same arguments used before, under the given assumptions there exists $\alpha' = O(1)$ such that for n large enough $\xi \cdot \mathsf{E}\big[\mathsf{E}[\Delta_1 \cdot \mathbb{1}_{\{f' \leq f\}}]\big] \geq 3 \cdot \alpha \cdot \mathsf{E}\big[\mathsf{E}[\Delta_2 \cdot \mathbb{1}_{\{f' \leq f\}}]\big]$. Let $\alpha_{\Downarrow} := 2 \cdot \alpha'$. Assume that $\alpha^{[i]} \geq \alpha_{\Downarrow}$ and $\alpha \geq \alpha_{\Downarrow}/2 = \alpha'$ for n steps (if α drops below $\alpha_{\Downarrow}/2$ within one of these n steps, there is nothing to show). Following the same argumentation used in the proof of the preceding lemma (except for S_j now being the sum of n (instead of $n^{0.3}$) RVs), we get that w.o.p. $S_1/S_2 > 2 \cdot \alpha^{[i]}/\xi$, and hence, after these n steps w.o.p.

$$
\begin{aligned}
\frac{d_1}{d_2} = \frac{d_1^{[i]} - S_1}{d_2^{[i]} - S_2} &< \frac{d_1^{[i]} - S_1}{d_2^{[i]} - S_1 \cdot \xi/(2 \cdot \alpha^{[i]})} = \frac{d_1^{[i]} - S_1}{d_1^{[i]} \cdot \xi/\alpha^{[i]} - S_1 \cdot \xi/(2 \cdot \alpha^{[i]})} \\
&= \frac{d_1^{[i]} - S_1}{d_1^{[i]} - S_1/2} \cdot \frac{\alpha^{[i]}}{\xi} = \left(1 - \frac{S_1/2}{d_1^{[i]} - S_1/2}\right) \cdot \frac{d_1^{[i]}}{d_2^{[i]}}.
\end{aligned}
$$

Thus, we must finally show that $S_1 = \Omega(d_1^{[i]})$. Recall that S_1 is the sum of n RVs $X_1^{[k]}$ ($\Delta_1 \cdot \mathbb{1}_{\{f' \leq f\}}$ in the $(i-1+k)^{\text{th}}$ step, respectively). In the following we consider the ith step. Our argumentation just bases on the fact that $\mathsf{E}[\Delta_1 \cdot \mathbb{1}_{\{f' \leq f\}}] \geq \mathsf{E}[\Delta_1 \cdot \mathbb{1}_{\{\Delta_1 \geq 0\}}] \cdot \mathsf{P}\{\Delta_2 \geq 0\}/2$ as we have seen, and since $\mathsf{P}\{\Delta_2 \geq 0\} = \Omega(1)$ by assumption, $\mathsf{E}[\Delta_1 \cdot \mathbb{1}_{\{f' \leq f\}}] = \Omega(\mathsf{E}[\Delta_1 \cdot \mathbb{1}_{\{\Delta_1 \geq 0\}}])$. Furthermore, since $\mathsf{P}\{\Delta_1 \geq 0\}$ as well as $1/2 - \mathsf{P}\{\Delta_1 \geq 0\}$ are $\Omega(1)$ by assumption, we know that $\mathsf{E}[\Delta_1 \cdot \mathbb{1}_{\{\Delta_1 \geq 0\}}] = \Theta(d_1/n)$ (cf. Section 2). Thus, the assumptions ensure $\mathsf{E}[\Delta_1 \cdot \mathbb{1}_{\{f' \leq f\}}] = \Omega(d_1/n)$, and hence, $\mathsf{E}[S_1] = n \cdot \Omega(d_1/n) = \Omega(d_1)$. Applying Hoeffding's bound just as in the proof of the preceding lemma, we immediately get that S_1 is $\Omega(\mathsf{E}[S_1])$, i.e. $\Omega(d_1^{[i]})$, w.o.p.

References

Beyer, H.-G. [1994]. Towards a theory of evolution strategies: Progress rates and quality gain for $(1\overset{+}{,}\lambda)$-strategies on (nearly) arbitrary fitness functions. *Proc. of Parallel Problem Solving from Nature 3 (PPSN)*, LNCS 866. Springer, 58–67.

Beyer, H.-G. [2001a]. On the performance of $(1, \lambda)$-evolution strategies for the ridge function class. *IEEE Transactions on Evolutionary Computation* 5(3): 218–235.

Beyer, H.-G. [2001b]. *The Theory of Evolution Strategies*. Springer.

Bienvenue, A. and Francois, O. [2003]. Global convergence for evolution strategies in spherical problems: Some simple proofs and difficulties. *Theoretical Computer Science* 306: 269–289.

Droste, S., Jansen, T., and Wegener, I. [2002]. On the analysis of the (1+1) evolutionary algorithm. *Theoretical Computer Science* 276: 51–82.

Giel, O. and Wegener, I. [2003]. Evolutionary algorithms and the maximum matching problem. *Proc. of the 20th Int'l Symposium on Theoretical Aspects of Computer Science (STACS)*, LNCS 2607. Springer, 415–426.

Greenwood, G. W. and Zhu, Q. J. [2001]. Convergence in evolutionary programs with self-adaptation. *Evolutionary Computation* 9(2): 147–157.

Hansen, N. and Ostermeier, A. [1996]. Adapting arbitrary normal mutation distributions in evolution strategies: The covariance matrix adaptation. *Proc. of the IEEE Int'l Conference on Evolutionary Computation (ICEC)*. 312–317.

Hoeffding, W. [1963]. Probability inequalities for sums of bounded random variables. *American Statistical Association Journal* 58(301): 13–30.

Hofri, M. [1987]. *Probabilistic Analysis of Algorithms*. Springer.

Jägersküpper, J. [2002]. Analysis of a simple evolutionary algorithm for the minimization in euclidean spaces. Technical Report CI-140/02, Univ. Dortmund, SFB 531. http://sfbci.uni-dortmund.de→Publications→Tech-Reports.

Jägersküpper, J. [2003]. Analysis of a simple evolutionary algorithm for minimization in Euclidean spaces. *Proc. of the 30th Int'l Colloquium on Automata, Languages and Programming (ICALP '03)*, LNCS 2719. Springer, 1068–1079.

Neumann, F. and Wegener, I. [2004]. Randomized local search, evolutionary algorithms, and the minimum spanning tree problem. *Proceedings of the Genetic and Evolutionary Computation Conference (GECCO)*, LNCS 3102. Springer, 713–724.

Rechenberg, I. [1973]. *Evolutionsstrategie*. Frommann-Holzboog, Stuttgart, Germany.

Rudolph, G. [1997]. *Convergence Properties of Evolutionary Algorithms*. Verlag Dr. Kovač, Hamburg.

Scharnow, J., Tinnefeld, K., and Wegener, I. [2002]. Fitness landscapes based on sorting and shortest paths problems. *Parallel Problem Solving from Nature 7 (PPSN)*, LNCS 2439. Springer, 54–63.

Schwefel, H.-P. [1995]. *Evolution and Optimum Seeking*. Wiley, New York.

Wegener, I. [2001]. Theoretical aspects of evolutionary algorithms. *Proc. of the 28th Int'l Colloquium on Automata, Languages and Programming (ICALP)*, LNCS 2076. Springer, 64–78.

Population Sizing of Dependency Detection by Fitness Difference Classification

Miwako Tsuji, Masaharu Munetomo, and Kiyoshi Akama

Hokkaido University, North 11 West 5, Sapporo, 060-0811, Japan
{m_tsuji,munetomo,akama}@cims.hokudai.ac.jp

Abstract. Recently, the linkage problem has attracted attention from researchers and users of genetic algorithms and many efforts have been undertaken to learn linkage. Especially, (1) perturbation methods (PMs) and (2) estimation of distribution algorithms (EDAs) are well known and frequently employed for linkage identification. In our previous work [TMA04], we have proposed a novel approach called Dependency Detection for Distribution Derived from df (D^5) which inherits characteristics from both EDAs and PMs. It detects dependencies of loci by estimating the distributions of strings classified according to fitness differences and can solve EDA difficult problems requiring a smaller number of fitness evaluations. In this paper, we estimate population size for the D^5 and its computation cost. The computation cost slightly exceeds $O(l)$, which is less than the PMs and some of EDAs.

1 Introduction

A set of loci tightly linked to form a building block is called a linkage set and encoding such loci loosely results building block disruptions. Several efforts have been undertaken to ensure appropriate building block processing without prior knowledge of problem. Two major methods of them are follows:

1. Perturbation Methods (PMs)
2. Estimation of Distribution Algorithms (EDAs)

PMs examine fitness differences by perturbations at loci to detect interdependency among them. They can recognize building blocks with lower marginal fitness contributions, but require a large number of extra fitness evaluations in addition to the usual fitness evaluations which are performed to select strings. For example, the LINC [MG99] requires $O(l^2)$ fitness evaluations for its linkage identification where l is string length. Heckendorn et al.[HW03] shows algorithm which uses the Walsh coefficients. This algorithm behaves similar to the LINC when it considers order-2 dependencies. But while the LINC guesses order-3 or more dependencies, it introduces higher-order perturbations (probes). EDAs like the BOA [PGCP99] employ probabilistic modeling of promising solutions to generate new solutions instead of the crossover and mutation operators of simple GAs. Some of EDAs are based on conditional probabilities to model dependency

A.H. Wright et al. (Eds.): FOGA 2005, LNCS 3469, pp. 282–299, 2005.
© Springer-Verlag Berlin Heidelberg 2005

of variables. They can construct their models without additional fitness evaluations, however, it is difficult for EDAs to recognize low scaling building blocks. For such problems, EDAs need more strings and generations, therefore, the total number of evaluations increases to that of the PMs.

If a problem is composed of variously scaled sub-problems, fitness of the problem is dominated by solutions of highly scaled sub-problems (i.e. important building blocks) and GAs should focus only on the sub-problems. The scale of sub-problems means the amount of contribution of each sub-problems. The highly scaled sub-problems gives large contribution to fitness, while lowly scaled sub-problems gives small contribution to fitness. Therefore, GAs solve sub-problems sequentially from those scaled larger to those scaled smaller, and sometimes, the low scaled building blocks are lost while they are on a waiting list. For example, consider maximizing a problem of 4 variables, $f(s_1, s_2, s_3, s_4) = g(s_1, s_3) + h(s_2, s_4)$ where $s_1, \cdots s_4$ are variables. If $g_{max}(s_1, s_3) = 5$ and $h_{max}(s_2, s_4) = 5$, $h(s_3, s_2)$ and $g(s_2, s_4)$ are searched in parallel. On the other hand, if $g_{max}(s_1, s_3) = 7$ and $h_{max}(s_2, s_4) = 3$, GAs focus on larger scaled sub-problem, $g(s_1, s_3)$ first and then give their eyes to $h(s_2, s_4)$. This sequential search procedure is referred as domino convergence [TGP98,LGP00].

In our previous work [TMA04], we have proposed another approach called the Dependency Detection for Distribution Derived from df (D^5) which combines both features of the previous methods. It detects dependencies of loci by estimating the distributions of strings classified according to fitness differences. Generally, EDAs estimate bias in selected sub-population and such bias come naturally from selection according to fitness. On the other hand, the D^5 makes sub-population biased artificially by perturbations. Therefore, the D^5 can solve EDA difficult problems using less computation cost than the PMs. The experiments showed that it reduces computation cost considerably and the number of evaluations is approximately $O(l)$ where l is string length.

In this paper, we estimate the growth of the required population size for the D^5 theoretically in order to understand scalability of the D^5. The number of strings is an important factor for linkage identification quality and computation cost. Resulting number of evaluations exceeds $O(l)$ slightly but still far less than PMs and even some of EDAs.

This paper is organized as follows. First, we show a brief introduction to the D^5. And its population sizing is discussed in section 4. After theoretical estimation of population size, some numerical experiments are performed to confirm the result in section 5 and finally this paper is concluded in section 7.

2   Background

Decomposability is one of the grounds for the advantage of genetic algorithm over random search. Additively decomposable functions are one of representations of the decomposable problem. An additively decomposable function is defined as follows:

$$f(s) = \sum_{\forall v \in V} f_v(s) \tag{1}$$

where s is a string, v is a set of loci that composing a sub-function (i.e a linkage set), $f_v(s)$ is a sub-function defined over v, and V is a set of disjoint sets of loci (i.e. a set of linkage sets). We consider only bit string as s. We consider only bit string as s. In addition, we assume that the sub-functions do not overlap each other, i.e. $v \cap v' = \emptyset$ $(v, v' \in V)$. The length of s is denoted by l and sum of the number of loci in each v is equal to l:

$$\sum_{v \in V} |v| = l \tag{2}$$

We limit maximum size of a linkage set v to k.

$$|v| \leq k \quad \forall v \in V \tag{3}$$

k is known as order of (sub-)problem or order of building block. The additively decomposable functions are known as order k delineable problems defined by Kargupta [Kar95]. In this paper, we consider the linkage set is the set of loci which are linked and are not separable. If a building blocks can be constructed by crossover, these loci should be separable that can be optimized separately.

3   Dependency Detection by Fitness Differences

In this section, we show a brief explanation of the Dependency Detection for Distribution Derived from df (D^5). The D^5 combines PMs and EDAs in order to obtain bias of sub-population to be estimated rapidly even if there are some low scaling sub-problems. It detects dependencies of loci by estimating the distributions of strings classified according to fitness differences.

EDAs learn problem structure from bias of sub-population and such bias is given by selection pressure based on fitness. Therefore, if problem is composed of variously scaled sub-problems, then modeling processes in EDAs focus only on highly-scaled sub-problems. On the other hand, the D^5 makes sub-population to be estimated biased artificially by perturbations. Therefore, the D^5 can detect such EDA-difficult sub-problems. Moreover, although the D^5 requires additional fitness evaluations due to calculate fitness differences, the number of evaluation is less than the PMs which generally perform pairwise perturbations because the D^5 employs estimation instead of higher order perturbations.

3.1   Algorithm

Genetic algorithm using the D^5 is composed of two parts (1) detecting linkage sets and (2) generating, increasing and combining building blocks. In this paper, we concentrate on the first part because if we know problem structures we can perform subsequent optimization processes easily and efficiently.

Fig. 1 shows the algorithm of the D^5. The algorithm consists of three parts: (1) calculating of fitness differences, (2) classifying of strings according to the differences and (3) estimating of the classified strings. After initializing population, following procedures are repeated for each locus i: At first, locus i in

---

1. initialize population with n strings
2. for each locus i
   (a) calculate fitness difference $df_i(s^p)$ by a perturbation at locus i in string s^p $(p = 1, 2, \cdots, n)$.
   (b) classify strings according to their fitness differences into sub-populations.
   (c) estimate sub-populations and construct linkage sets (see Fig. 2 for detail).

Fig. 1. Overall Algorithm of linkage identification in the D^5

---

1. for each sub-population p classified by the Classification Algorithm
   (a) initialize set of loci $v_1 = \{1, 2, \cdots, i - 1, i + 1, \cdots, l\}$ and $v_2 = \{i\}$
   (b) while $|v_2| < K$, where K is pre-defined problem complexity
       i. calculate a entropy $E_j = E(v_2 \cup \{j\})$ for all locus $j \in v_1$
       ii. $h = \arg \min_{j \in v_1} E_j$
       iii. update $v_1 = v_1 - \{h\}$ and $v_2 = v_2 \cup \{h\}$
   (c) $v_p = v_2$ and $E_p = E(v_2)$
2. select v_p with the smallest E_p as the linkage set for locus i

---

Fig. 2. Construct Linkage Set

each string s^p is perturbed and then fitness difference for the perturbation is calculated as follows:

$$df_i(s^p) = f(s^p) - f(s_i^p) \tag{4}$$

In the above equation, s_i^p is a string perturbed at locus i. Then, strings are classified into sub-populations according to their fitness differences $df_i(s^p)$. We employed a simple centroid method for classification, but other approaches like k-means can also be applied. In this method, the centroid of a cluster is determined by averaging $df_i(s^p)$ of all strings within that cluster. The distance between two clusters is defined as the distance between the centroids of the clusters. The pair of clusters with the smallest distance is merged until a termination criteria is satisfied. If the smallest distance of all the rest exceeds a threshold, the classification is terminated. The threshold should be small for problems which consist of independent sub-problems, while it should be large for those with interacted sub-problems.

Finally, the sub-populations are estimated in order to detect loci which depend on locus i.

Fig. 2 is the algorithm to construct a linkage set for locus i. First, a set v_2 is initialized as $\{i\}$. The locus j which gives the smallest entropy $E(v_2 \cup \{j\})$ is merged repeatedly until the size of linkage set exceeds problem complexity k. This defines the order of a sub-problem and is given by algorithm users. The order of a building block is equal to k because we assume the building block is the optimal solution of the sub-problem. The entropy measure is defined as

$$E(v_2) = -\sum_{x=1}^{2^{|v_2|}} p_x \log_2 p_x, \tag{5}$$

where p_x is the appearance ratio of each schema x and $2^{|v_2|}$ is the number of all possible schema defined by v_2. This procedure is applied to all sub-populations except those including small number of strings. The sub-populations having small number of strings are ignored because estimating distribution of small samples has risk of unreliable result. The population sizing in section 4 will give the threshold for the sampling size. After the estimation of all sub-populations, the linkage set v_2 which gives the smallest entropy $E(v_2)$ is selected as linkage set v_i of locus i.

3.2   Example

As example, we use sum of the order 3 deceptive problem which was used as an opponent of the messy GA[GKD89] and defined as follows:

$$f(s) = \sum_{\forall v \in V} f_v(s) \tag{6}$$

$$f_v(s) = 30 \text{ if } 111, \quad 0 \text{ if } 110, 101, 011,$$
$$14 \text{ if } 100, \quad 22 \text{ if } 010, \quad 26 \text{ if } 001, \quad 28 \text{ if } 000$$

where $V = \{\{1,2,3\},\{4,5,6\}\}$ and $f_v(s)$ is defined by each schema of $\{1,2,3\}$ and $\{4,5,6\}$. Fig. 3 shows the perturbation in the 1st locus. In this figure, strings having $df_1 = 30$ are belong to a sub-population. In the sub-population, linkage set $\{1,2,3\}$ has only schema 011 and $E(\{1,2,3\})$ should be zero. On the other hand, linkage set $\{1,4,5\}$ has schemata $010, 001, 010, 011$ and $E(\{1,4,5\})$ should be relatively large. Therefore the algorithm evaluates that a relationship between locus 1, 2, and 3 take place more likely than a relationship between locus 1, 4, and 5.

s	$f(s)$	s_1	$f(s_1)$	bias	$df_1(s)$	s	$f(s)$	s_1	$f(s_1)$	bias	$df_1(s)$
011100	14	111100	44	011***	30	110101	0	010101	27	110***	27
011010	27	111010	57	011***	30	110011	0	010011	27	110***	27
011101	0	111101	30	011***	30	110001	26	010001	53	110***	27
011110	0	111110	30	011***	30	101100	14	001100	40	101***	26
011110	0	111110	30	011***	30	101101	0	001101	26	101***	26
		$\cdots$						$\cdots$			

Fig. 3. Strings classified according to df_1

If problems are (quasi-)decomposable like $f(s) = \sum_{v \in V} f_v(s)$ then fitness differences for perturbation in locus i are calculated as

$$df_i(s) = f(s) - f(s_i)$$

$$= [f_{\hat{v}}(s) + \sum_{v \neq \hat{v}, v \in V} f_v(s)] - [f_{\hat{v}}(s_i) + \sum_{v \neq \hat{v}, v \in V} f_v(s_i)]$$
$$= f_{\hat{v}}(s) - f_{\hat{v}}(s_i).$$

where $\hat{v}$ is the sub-problem including locus i. It is clear that fitness differences depend only on the linkage set $\hat{v}$ and independent on loci $j \notin \hat{v}$. Therefore, we can obtain bias in sub-populations classified according to fitness differences and detecting such bias by minimizing the entropy measure we can learn the linkage set for locus i.

Our greedy search shown in Fig. 2 sometimes can not find k bit dependencies, because in some problems entropies for lower order linkage sets shows randomness even if those for higher order sets distribute unevenly. However, it is clear that strings can be divided into at least two sub-populations and there should be 2^{k-1} schemata for loci $\hat{v}$ in these sub-populations. On the other hand, loci $j \notin \hat{v}$ distribute perfectly random and for any k-bit combination of such loci, there should be all possible schemata (2^k schemata). For example, for $\hat{v}$ of 3-bit problem, a sub-population have schemata$\{111, 100, 010, 001\}$. In this sub-population, all order 1 and 2 loci have all possible schemata and only the order 3 loci has half of all. Such dependency can not be found by the greedy search. But if more sophisticated method is applied, it is not impossible. The refinement should require larger computation cost, but is worthy of consideration when fitness evaluation of a problem takes huge time.

4   Population Sizing

In this section, we calculate the number of strings required to detect correct dependencies by the D^5. This consists of two stages: sub-population sizing and overall population sizing. The above case defines the size enough to distinguish biased distribution of dependent loci and random distribution of independent loci in a sub-population. The overall population size must ensure the sufficient sub-population size when it is divided.

We calculate sub-population size and then define overall population size. But, first of all, we should simplify the problem to make population sizing easy.

4.1   Simplification

Sub-population should have enough number of strings to distinguish biased distribution of dependent loci and random distribution.

The level of bias changes with each fitness landscape of a sub-problem. Some of them give strong bias, others give weak bias. The strong bias makes distinction easy, while the weak bias makes it difficult. The former case means that the sub-population has a few unique schemata. Therefore, the D^5 can exploit a small part of original population. In the later case, the sub-population should have more schemata and it can use a larger part of the population. To make calculation easy, we consider the first case only. The resulting population size should not be upper bounds or precise predictions of population size, as a consequence of this simplicity. But it will help to understand how the number of strings grows as string length gets longer.

As mentioned in section 2, we consider additively decomposable functions only. In this type of functions, there are 2^k schemata for each sub function $f_v(s)$. One bit perturbation gives 2^k schemata changes such as

$$df_1(\underline{0}00...) = f(\underline{0}00...) - f(\underline{1}00...)$$
$$df_1(\underline{0}01...) = f(\underline{0}01...) - f(\underline{1}01...)$$
$$df_1(\underline{0}10...) = f(\underline{0}10...) - f(\underline{1}10...)$$
$$\cdots\cdots\cdots\cdots$$
$$df_1(\underline{1}11...) = f(\underline{1}11...) - f(\underline{0}11...)$$

It is clear that $df_i(s) = -df_i(s_i)$. For example,

$$df_1(\underline{0}00) = f(\underline{0}00...) - f(\underline{1}00...), \quad df_1(\underline{1}00) = f(\underline{1}00...) - f(\underline{0}00...).$$

Then

$$df_1(\underline{0}00) = f(\underline{0}00...) - f(\underline{1}00...) = -df_1(\underline{1}00).$$

We denote fitness difference of h-th schema by perturbation of i-th loci as df_i^h $(i = 1, 2, \cdots, l. \quad h = 1, 2, \cdots, 2^k)$.

The number of unique fitness differences by the perturbations in locus i varies from 2 to 2^k. The upper bound, 2^k, comes in the case where fitness changes of all schemata differ:

$$df_i^h \neq df_i^{h'} \quad (\forall h \neq h') \tag{7}$$

The lower bound, 2, takes place if the amount of fitness increase and decrease are always same:

$$|df_i^h| = |df_i^{h'}| \quad (\forall(h, h')) \tag{8}$$

In order to make calculation easy, we consider the first case. In this case, all 2^k schemata should be classified into different classes. One class has one schema for a linkage set. Loci in the linkage set of all strings in the class have same value. The number of strings must be enough to ensure that no single unlinked locus takes a same value in the class. To this end all classes does not have to have exactly one schema, but at least one of them must have exactly one schema. Therefore, we relax the equation (7) as follows:

$$\exists df_i^{h'} \in \{df_i^1, \cdots df_i^{2^k}\} \quad \text{that satisfies } df_i^{h'} \neq df_i^h \ (h \neq h') \tag{9}$$

If this condition is satisfied, the sub-population of $df_i^{h'}$ has only one unique schema. The entropy of the set of loci that depend on locus i should be zero. On the other hand, the entropy of the set of loci that do not depend on locus i is close to the number of loci in the set. Therefore, if there are enough strings in the sub-population, it is easy to identify linkage set correctly.

Despite such simplicity of linkage identification over a sub-population, whole population size for the function that satisfies the condition (9) should be large. The reason comes from the fact that we can exploit only a small part of whole population for the sub-population of $df_i^{h'}$ because h'-th schema should have $n/2^k$ copies where n is whole population size.

If fitness function does not satisfy (9), by contrast, there are two or more schemata in sub-populations. Consequently, entropies in the sub-populations should be larger than zero. However, they should be still smaller than that for random distributions. For example, the lower bound of fitness differences are 2, as mentioned earlier. In the case, there are 2^{k-1} unique schemata in each sub-population. Then the entropy should be less or equal to $k-1$. The signal difference of entropy increase from the above case, but the half part of original population can be used.

The precise population size for the D^5 is defined from both the string utilization ratio of original population and the signal difference of entropy between $\hat{v}$ and v. Both of those differ with respect to fitness functions. In the followings, to make population sizing easy, we consider the extremely biased case, the functions that satisfy (9) only. Although it does not give the precise population size, it at least should give how it grows with problem size.

After theoretical population sizing, we perform experiments with some classes of problems including the problem that does not satisfy the condition (9).

4.2   Sub-population Sizing

Let $C_{df_i^{h'}}$ a sub-population of $df_i^{h'}$ in (9), let n_1 size of the $C_{df_i^{h'}}$.

Locus $j \notin \hat{v}$ must have less certain distribution than locus $j \in \hat{v}$ to detect correct dependency for locus i.

If (9) is satisfied for additively decomposable functions (1), then there is only one unique schema of $\hat{v}$ in sub-population $C_{df_i^{h'}}$. Therefore, each locus of $\hat{v}$ takes the same gene value in the sub-population.

On the other hand, loci in $V - \{\hat{v}\}$ take 0 or 1 at random. If a locus $j \notin \hat{v}$ takes the same gene value in $C_{df_i^{h'}}$ accidentally, the dependency detection will fail. Therefore, we should employ enough sub-population size n_1 to avoid such undesirable coincidence for all $j \notin \hat{v}$.

The probability that every gene value of a locus $j \notin \hat{v}$ is same is

$$\left(\frac{1}{2}\right)^{n_1}.\tag{10}$$

The probability that it does not occur in all loci $j \in V - \{\hat{v}\}$ is

$$\left(1 - \left(\frac{1}{2}\right)^{n_1}\right)^{l-k}.\tag{11}$$

The probability P_1 that the previous condition hold true for all dependency detections for loci $i = 1, 2, \cdots, l$ is

$$P_1 = \left(\left(1 - \left(\frac{1}{2}\right)^{n_1}\right)^{l-k}\right)^l = \left(1 - \left(\frac{1}{2}\right)^{n_1}\right)^{l(l-k)}. \tag{12}$$

Randomly generated gene value in $j \notin \hat{v}$ can have same value in size n_1 sub-population $C_{df^{h'}}$ with probability $\left(\frac{1}{2}\right)^{n_1}$ and it can not occur with probability $\left(1 - \left(\frac{1}{2}\right)^{n_1}\right)$. This hold true for all $j \notin \hat{v}$ with probability $\left(1 - \left(\frac{1}{2}\right)^{n_1}\right)^{l-k}$ because $|\hat{v}| = k$ and then $|V - \hat{v}| = l - k$. Therefore the probability that we can avoid the undesirable biases for all $j \in \hat{v}$ in all l times perturbations is equation (12).

Rewriting (12), we can obtain sub-population size for expected success ratio P_1

$$n_1 = -\log(1 - P_1^{\frac{1}{l(l-k)}}). \tag{13}$$

Therefore, if an appropriate distribution is required with probability P_1, then sub-population must have more than $-\log(1 - P_1^{\frac{1}{l(l-k)}})$ strings.

The resulting sub-population size n_1 is also used as threshold for the estimation phase in the D^5. If a sub-populations has less than n_1 strings, it should not be used for dependency detection.

4.3　Overall Population Sizing

Now, we consider overall population size, n. It must be enough to obtain appropriate sub-population size, n_1. From (9), sub-population $C_{df^{h'}}$ has only one schema of possible 2^k schemata in $\hat{v}$. Because the original population is initialized by random coin toss, the distribution of the schema in the original population is the binomial distribution with the mean $\frac{n}{2^k}$ and the variance $n\frac{1}{2^k}\left(1 - \frac{1}{2^k}\right)$.

Therefore, the lower bound of expected sub-population size n_1 for large n is estimated as follows:

$$n_1 \geq \frac{n}{2^k} - \sqrt{n\frac{1}{2^k}\left(1 - \frac{1}{2^k}\right)} \tag{14}$$

Let $p = 1/2^k$ and $q = 1 - p$, we rewrite above equation as

$$n_1 \leq np - \sqrt{npq}. \tag{15}$$

Rewriting the inequality, required population size n is as follows:

$$n \geq \frac{(2n_1p + pq) + \sqrt{(2n_1p + pq)^2 - 4p^2n_1^2}}{2p^2}. \tag{16}$$

$$= \frac{1}{p}n_1 + \frac{q}{2p} + \sqrt{\frac{q}{p}n_1 + \frac{q^2}{4p^2}}. \tag{17}$$

Because p is constant if k is constant, the term $\frac{1}{p}n_1$ is dominate for population sizing, thus

$$n = O\left(\frac{1}{p}n_1\right) \tag{18}$$

$$= O(-2^k \log(1 - P_1^{\frac{1}{l(l-k)}})). \tag{19}$$

Using L'Hopital's rule,

$$\lim_{l\to\infty} \frac{1 - P_1^{1/l^2}}{1/l^2} = 1. \tag{20}$$

$$\lim_{l\to\infty} \frac{\ln(1 - P_1^{1/l^2})}{\ln l^2} = \lim_{l\to\infty} \frac{\ln(1 - P_1^{1/l^2})}{2\ln l} = \lim_{l\to\infty} \frac{-P_1^{1/l^2} \ln P_1}{(1 - P_1^{1/l^2})/(1/l^2)}. \tag{21}$$

The limit of the denominator is 1 from the equation (20), the limit of the numerator is $-\ln P_1$. Thus,

$$\lim_{l\to\infty} \frac{\ln(1 - P_1^{1/l^2})}{2\ln l} = -\ln P_1 \tag{22}$$

Therefore, approximating $\log(1 - P_1^{\frac{1}{l(l-k)}})$ to $\log(1 - P_1^{\frac{1}{l^2}})$

$$O(-2^k \log(1 - P_1^{\frac{1}{l(l-k)}})) \approx O(2^k \log l) \tag{23}$$

for large l.

4.4   Overall Complexity

In this section, we show the number of fitness evaluations for the D^5. Optimization using it consists of the dependency detection stage and the building block combination stage. However, if problem structure is revealed, the following evolution should be success using relatively small cost. Therefore, we consider computation cost for dependency detection * approximately equal to overall computation cost.

In dependency detection by the D^5, we should know original fitness of all strings and those after perturbations in all l loci of all strings. Therefore, the number of evaluations required to obtain appropriate linkage sets is $nl + n$ where l is string length and n is population size. substituting (16), the number of evaluations is

$$nl + n = \frac{(2n_1 p + pq) + \sqrt{(2n_1 p + pq)^2 - 4p^2 n_1^2}}{2p^2}(l + 1). \tag{24}$$

If the order of problem k is fixed, then $p = 1/2^k$ and $q = 1 - p$ are also fixed. From the (16) and (24), the number of evaluations is roughly

$$O(n_1 l) = O(-l \log(1 - P_1^{\frac{1}{l(l-k)}})) \tag{25}$$

for string length l.

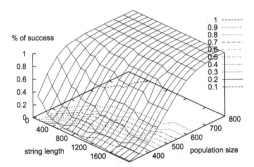

Fig. 4. Accuracy of linkage identification for trap function for various population size and string length

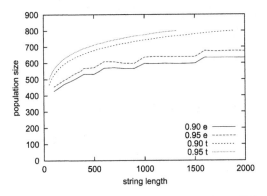

Fig. 5. Contours of fig. 4 and theoretical estimation in equation (16). In index, numbers like 0.95 or 0.90 mean contour levels, e or t means experimental result or theoretical result respectively

5 Experiments: Population Size, String Length and Success Ratio

In this section, we compare theoretical estimations in section 4 and experimental results.

Trap Function

Experiments in this section are performed on a deceptive trap function as follows:

$$f(s) = \sum_{i=1}^{m} \mathrm{trap}(u_i) \tag{26}$$

$$\mathrm{trap}(u_i) = \begin{cases} k & (u = k) \\ k - 1 - u & (\text{otherwise}) \end{cases} \tag{27}$$

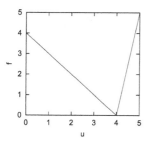

Fig. 6. Example of Trap Function

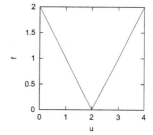

Fig. 7. Example of Valley Function

where m is the number of sub-functions and k is the order of a sub-function, u_i is the number of ones in each k-bit sub-string. This sub-function is called trap function.

Fig.6 shows the function. The x-axis of the figure shows the number of ones in a sub-string and y-axis is the contribution of each sub-string. The left side of the figure shows the contribution of all zeros sub-string and the right side shows that of all ones. From this function, it is clear that the fitness difference for 1-bit perturbation takes one of $\{k, 1, -1, -k\}$. Only all ones gives $-k$ (when $u = k$ to $k-1$ for perturbation , 11111 → 01111) and only sub-strings to be all ones by the perturbation gives k (when $u = k - 1$ to k for perturbation , 01111 → 11111). All the other 2^{k-1} sub-strings give -1 or 1 by $0 \rightarrow 1$ or $1 \rightarrow 0$ respectively. This function satisfies the assumption 1 because there are two schemata, all ones and that to be all ones, whose fitness differences differ from all the other fitness differences.

In these experiments, we try various string length $l = k \times m$ and population size n. The order of problem is fixed to $k = 5$. We record percent of linkage correctly identified for several (l, n) pairs. We perform 10 runs for each (l, n) pair and average success ratio of linkage identification.

Figure 4 shows the experimental result of accuracy of linkage identification for 5-bit trap function in various population size and string length. Figure 5 shows contours of the accuracy of experimental results and theoretical estimations in equation (16).

Because our theoretical population sizing is conservative – we assume one unique fitness difference but there are two unique fitness differences in the trap function –, the experimental result can archive a certain success ratio with smaller number of strings than the theoretical result. However traces of contour in experiments are follows that in experiment very well.

Valley Function

In this experiment, we employ a test function which does not satisfy (9). The function is defined as follows:

$$f(s) = \sum_{i=1}^{m} \text{valley}(u_i) \tag{28}$$

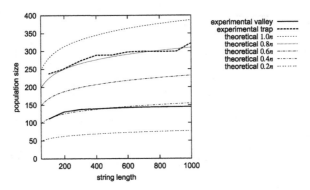

Fig. 8. D^5 for 4-bit valley function and 4-bit trap function. The lines mean threshold sizes for linkage identification with probability 0.95 for each string length. For reference, $c \times n$ where n c-s are constants and vary from 0.2 to 1.0 are added

$$\text{valley}(u) = |u - \frac{k}{2}| \tag{29}$$

where k is the order of a sub-function, m is the number of sub-functions, and u is the number of ones in each k-bit sub-string. Figure 7 shows this function. The valley function has no peak or needle like the optimal solution of the trap function. For all perturbations it gives only two types of fitness difference $\{-1, 1\}$ if k is even.

We fix k to 4 and try various (l, n) pairs to record ratio of correctly identified linkage sets. We perform 10 runs for each (l, n) pair and average success ratio of linkage identification.

Figure 8 shows contours of the accuracy of experimental results and theoretical estimations in equation (16). We also perform 4-bit trap function for comparison. And for reference, $c \times n$ where c-s are constants and vary from 0.2 to 1.0 are added. In this equation, n is the theoretical estimation and $n = -l \log(1 - 0.95^{1/(l(l-4))})$.

If many schemata have a same fitness difference, then all of them are classified into one sub-population. The entropy of linkage set in such sub-population is larger and closer to the entropy of random set of loci than the entropy of linkage set of a function which has various fitness differences like a trap function. On the other hand, because original population is divided into a few groups in this kind of problems, each sub-population size should be large and enough to detect the small signal difference of entropy.

For the 4-bit functions, there are $n_1^t = n^t/2^4 = n^t/16$ strings for the trap function and there are $n_1^v = n^v/2$ strings for the valley function where n^t is population size for trap function and n^v is population size for valley function. Because $n^v \approx n^t/2$ from the experiment, the sub-population size in valley function is $n_1^v = n^t/4 = 4n_1^t$. Sub-population sizes n_1^v and n_1^t are defined from the signal difference of entropy and population sizes n^v and n^t are defined to ensure sub-population sizes.

Although the number of strings for the valley function is smaller than for the trap function, the slope for the valley function is similar to that for the trap function. In fact, the population size for the valley function is similar to the $0.4 \times n$ and that for the trap function is similar to the $0.8 \times n$. These results show that the theoretical population size guides population sizings for problems which do not satisfy the condition (9).

6   Comparisons with Other Methods

In this section, we compare population sizings in existing literatures of GAs, PMs and EDAs. Depending on the strategies of each algorithm, their way of sizing differs. Please not that these population sizings do not address the same problem and the same purpose. However, they have some things in common and other things in contrast.

6.1   Population Sizing of Simple GAs

Various efforts have been focused on sizing population of genetic algorithms. One of the most accurate population size was calculated by Harik et al. [HCPGM97].

They estimated population size considering an initial supply of building blocks and a good decision making between competing building blocks using the gambler's ruin model. Resulting population size n that is enough for an optimal solution to take over population after several generations is

$$n = -2^{k-1} \ln(\alpha) \frac{\sigma_{bb} \sqrt{\pi(m-1)}}{d} \tag{30}$$

where k is the order of a building block, α is the probability of failure convergence in a sub-problem, σ_{bb} is the standard deviation (the square root of variance) of fitness of the building block, m is the number of building blocks (sub-problems) in a string, and d is the difference between the mean fitness of the best and the second building blocks. The term 2^{k-1} is required for the initial supply of building blocks and the other terms are for the decision making. The last term shows that if the population size increases as the average variance of the building blocks increases, as the problem size is grows and as the signal difference decreases.

If we assume that string length is approximately proportional to the number of building blocks, we obtain following result from equation (30):

$$n = O(2^{k-1}\sqrt{l}) \tag{31}$$

In their analysis, they assumed tight linkage and building block disruption was not considered. If such tight linkage is not ensured, SGA performs as random search and needs $O(l^2)$ strings for the worst case. In addition, some approximations are used in their calculation.

6.2 Population Sizing of PMs

For the LINC, the number of strings required to obtain correct linkage set was calculated by Munetomo et al [MG99]. The LINC identifies linkage by detecting the second order nonlinearity. It assumes that nonlinearity must exist within loci to form a building block.

This population sizing is differ from the population sizing of simple GAs because they do not concern the number of strings for optimal population convergence but for correct linkage identifications. However, if correct linkage sets are obtained it becomes easy to combine building blocks to find an optimal solution.

Their population sizing is based on the supply of building blocks in a linkage set because they assume that there is at least one schema which violates linear condition along the perturbations for the pair of loci belonging a same subproblem. They showed that if there are n strings then correct linkage sets is obtained with probability P as follows:

$$P = 1 - \left(1 - \left(\frac{1}{2^k}\right)\right)^n \tag{32}$$

where k is the order of a sub-problem. Therefore the population size required for a certain success probability P is

$$n = \frac{\log(1 - P)}{\log(1 - 1/2^k)} \simeq -2^k \log(1 - P). \tag{33}$$

From equation (33), the population size for the LINC depends only on the order of sub-problem.

Munetomo et al [MG99] calculate the number of strings which is enough to obtain pairs of loci that are linked with a given probability, P. Whereas, the population size in the D^5 is sized enough to obtain all sets of loci that linked. In addition, Heckendorn et al [HW03] define the number of strings to find all pairs of loci that linked with a global probability of success. The population size of their algorithm (the number of iterations in their algorithm) is $-2^k \ln(1 - \delta^{1/J})$ for order-2 linkage detection where δ is the probability of success and J is the number of order-2 relationships between loci (hyperedges). For the non-overlapping additively decomposable functions composed of order-k sub functions, $J = l/k \times_k C_2 = l(k - 1)/2$. Then the population size is $-2^k \ln(1 - \delta^{2/l(k-1)})$.

6.3 Population Sizing of EDAs

As an example of population sizing in EDAs, we show the population sizing for the BOA calculated by Pelikan et al.[PSG02] The BOA [PGCP99] exploits Bayesian network to represent conditional probabilities in order to encode dependency of variables in their models.

For uniformly scaled problem, the most important factor for population sizing is that the BIC metric can distinguish between the appropriate and the

Table 1. Comparison of population size and number of evaluations

method	population size	number of evaluations
D^5	$O(-2^k \log(1 - P_1^{\frac{1}{l(l-k))}}))$	$O(-2^k l \log(1 - P_1^{\frac{1}{l(l-k)}}))$
SGA	$O(2^{k-1}\sqrt{l}) \sim O(2^l)$	$O(2^{k-1}l) \sim O(2^{k-1}l\log l) \sim O(2^{k-1}l^2)$
LINC	$O(2^k)$	$O(2^k l^2)$
BOA	$O(2^k l^{1.05})$	$O(2^k l^{1.55})$

inappropriate dependencies and decision making between to add or to not add an edge from a variance to another variance. From the viewpoint, the required number of strings is

$$n = O(l^{1.05}). \tag{34}$$

This result is obtained using some approximations, however, it matches experimental results. The detail about this equation is available in the literature [PSG02].

Above equations were obtained for two bit dependency, but it can be extended for multiple dependencies as follows:

$$n = O(2^k l^{1.05}) \tag{35}$$

The term 2^k comes from the number of possible schemata and $l^{1.05}$ is the requirement for noise avoidance from contributions of other sub-solutions.

6.4   Discussions

Table 1 shows the comparison of population size in the D^5, SGA, the LINC and the BOA. Please note that the population size of SGA $O(2^{k-1}\sqrt{l})$ is for tight encoding strings. If such tight encoding is not ensured SGA requires an exhaustive search which needs exponential number of strings. In addition, the comparison is not completely fair because the population sizings for the LINC and the D^5 does not concern evolution of population and decision-making during the evolution. However, it should be true that if we know problem structure, then we can make decision easier than without such explicit information of problems.

The required number of evaluations is also shown in Table 1. Again, the numbers of the D^5 and the LINC are only for dependency detection and they need other evaluations for evolution phase but these are not dominant for overall computation cost. The number of evaluations for SGA and the BOA is simple multiply of population size and the number of generations for convergence. The numbers of evaluations for SGA are varied with respect to the selection methods [GD91] and is also an ideal case that tight linkage can be ensured from previous information of problems.

Streeter [Str04] improves the LINC from a traditional algorithm perspective. It uses binary search to detect specific loci j which depend on i. Therefore, it requires $O(2^k l \log l)$ fitness evaluations where $l \log l$ comes from binary search for each locus $i = 1, 2, \cdots, l$ and $O(2^k)$ is population size required to guarantee a certain success probability. This computation cost is similar to our result, $O(-2^k l \log(1 - P_1^{1/l(l-k)}))$. The main difference is that the log term of the D^5 comes from population size, $O(-2^k \log(1 - P_1^{1/l(l-k)}))$, while the Streeter's one comes from binary search.

All of population sizings have the factors from possible number of schemata, 2^k or $2^{k-1} = 2^{-1} \times 2^k$. This number, 2^k, is number of all possible schemata of order k. It should guarantee existence optimal schema which will be produced with probability $1/2^k$. SGA and BOA consider the effect from other sub-functions by $\sqrt{l}$ and $l^{1.05}$ respectively. The D^5 has also the term $\log(1 - P_1^{1/l(l-k)})$ for population sizing. This comes from requirement to avoid undesirable bias for all loci which do not belong a sub-problem including a perturbed locus. This requirement is approximately equal to the requirement that initial population should distribute enough randomly. The LINC and Streeter's algorithm, which are uses perturbations only for their dependency detections, have no term for string length in population sizing.

7   Conclusion

In this paper, we estimate the number of strings required for the D^5 under some assumptions. Estimated population size is $O(-2^k \log(1 - P_1^{\frac{1}{l(l-k)}}))$ where l is string length and k is the order of the sub-problem. This result shows the number of strings required to obtain correct linkage sets is defined mainly by the order of sub-problem and the D^5 can be scalable to large problem size. Validity of the population is also verified in experiments. The experimental population sizes follow $c \times -2^k \log(1 - P_1^{\frac{1}{l(l-k)}})$ where c is a constant.

References

[GD91]   David E. Goldberg and Kalyanmoy Deb. A comparative analysis of selection schemes used in genetic algorithms. In Gregory J.E. Rawlins, editor, *Foundations of Genetic Algorithms*, pages 69–93. Morgan Kaufmann Publishers, 1991.

[GKD89]   David E. Goldberg, Bradley Korb, and Kalyanmoy Deb. Messy genetic algorithms: Motivation, analysis, and first results. *Complex Systems*, 3(5):415–444, 1989.

[HCPGM97]   Georges Harik, Erick Cantú-Paz, David E. Goldberg, and Brad L. Miller. The gambler's ruin problem, genetic algorithms, and the sizing of populations. In *IEEECEP: Proceedings of The IEEE Conference on Evolutionary Computation, IEEE World Congress on Computational Intelligence*, pages 7–12, 1997.

[HW03]   Robert B. Heckendorn and Alden H. Wright. Efficient linkage discovery by limited probing. In *Proceedings of the 2003 Genetic and Evolutionary Computation Conference*, pages 1003–1014. Morgan Kaufmann Publishers, 12–16 July 2003.

[Kar95]   H. Kargupta. SEARCH, polynomial complexity, and the fast messy ge-
             netic algorithm. Technical Report 95008, University of Illinois at Urbana-
             Champaign, Urbana, IL, October 1995.

[LGP00]   Fernando G. Lobo, David E. Goldberg, and Martin Pelikan. Time com-
             plexity of genetic algorithms on exponentially scaled problems. In *Pro-
             ceedings of the 2000 Genetic and Evolutionary Computation Conference*,
             pages 151–158. Morgan Kaufmann Publishers, 10-12 July 2000.

[MG99]   Masaharu Munetomo and David E. Goldberg. Identifying linkage groups
             by nonlinearity/non-monotonicity detection. In *Proceedings of the 1999
             Genetic and Evolutionary Computation Conference*, pages 433–440, 7
             1999.

[PGCP99]   Martin Pelikan, David E. Goldberg, and Erick Cantú-Paz. BOA: The
             bayesian optimization algorithm. In *Proceedings of the 1999 Genetic and
             Evolutionary Computation Conference*, pages 525–532. Morgan Kaufmann
             Publishers, 1999.

[PSG02]   Martin Pelikan, Kumara Sastry, and David E. Goldberg. Scalability of the
             bayesian optimization algorithm. *International Journal of Approximate
             Reasoning*, 31(3):221–258, 2002.

[Str04]   Matthew J. Streeter. Upper bounds on the time and space complexity
             of optimizing additively separable functions. In *Proceedings of the 2004
             Genetic and Evolutionary Computation Conference*, pages 186–197, 2004.

[TGP98]   Dirk Thierens, David E. Goldberg, and Ângela G. Pereira. Domino con-
             vergence, drift and the temporalsalience structure of problems. In *Proceed-
             ings of the IEEE International Conference of Evolutionary Computation*,
             pages 535–540, 1998.

[TMA04]   Miwako Tsuji, Masaharu Munetomo, and Kiyoshi Akamae. Modeling de-
             pendencies of loci with string classification according to fitness differences.
             In *Proceedings of the 2004 Genetic and Evolutionary Computation Con-
             ference*, pages 246–257, 26-30 June 2004.

The Deceptive Degree of the Objective Function

Yun-qiang Li

Electronic Technique Institute,
Zhengzhou Information Engineering University,
450004 Zhengzhou, China
yunqiangli@126.com

Abstract. In this paper we present a novel quantitative measure metric for the "degree of deception" of a problem. We present a new definition for the deceptive degree of a function. We investigate the relationship between the best solution and the monomial coefficients of a function, and we give theorems that show the usefulness of the new definition. The new definition can be applied in three ways: it gives a quantitative measure of deception, it simplifies the evaluation of the GA difficulty, and it gives a relationship between the deceptive degree and the polynomial degree. Furthermore we use the deceptive degree of a function to discuss Goldberg's Minimal Deceptive Problem and derive the same result as Goldberg did. Finally, we make experiments with a class of fitness functions to verify the relation between the canonical GA difficulty and the deceptive degree of a function for this class of functions.

1 Introduction

What makes a function f difficult to optimize by a genetic algorithm (GA)? In the framework of applications of the GA as an optimization tool, this is one of the most importation questions, if not *the* most important question, to answer, but probably also the most difficult one. This question has become increasingly important as people have tried to apply the GA to ever more diverse types of problems. Much previous work on this question has studied the relationship between GA performance and the structure of a given fitness function. The introduction of the notion of *deception* by Goldberg in the GA appears to provide some insights into this question [3]. Subsequently, deception has come to be widely regarded as a central feature in the design of problems that are difficult for the GA.

Though a number of different definitions of deception as well as types of deception have been proposed in the GA literature (e.g., see [2,3,14,21,22]), the relationship between a deceptive problem and a difficult problem is still not easily explained. Some counterexamples in [10,11,12,13] show that deception is neither necessary nor sufficient to make a problem difficult for a GA. Since there is no generally accepted definition of deception, how to define deception is still a topic that deserves scrutiny.

In this paper we will present a new definition of deception according to the relation between the best string (or solution) and the monomial coefficients of a given function. Without loss of generality, we consider a pseudo-boolean function f: $\{0,1\}^n \rightarrow R$ as an objective function and aim to find the best string as defined by the given function. Every pseudo-boolean function can be expressed in polynomial form. The monomial coefficients in a polynomial function have an important effect on the

A.H. Wright et al. (Eds.): FOGA 2005, LNCS 3469, pp. 300–314, 2005.

best string. For example, for $f(x_1,x_2) = x_1 - 2x_2 + cx_1x_2$, where $x_1,x_2 \in \{0,1\}$, we can find that: if $c > 2$, the best string is 11; if $c < 2$, the best string is 10. We will see that 2 is a critical value of the monomial x_1x_2 with respect to x_2 in f. There is a close relation between the best string and the critical value. The relationships among the three factors, which are the best string, the critical values, and the monomial coefficients, should be taken into account in defining the deceptive degree of a given function.

The goal of the paper is to come up with a quantitative metric for the "degree of deception" of a problem. The remainder of the paper is organized as follows: Section 2 gives the concept of a critical value of a monomial with respect to a variable and gets some properties of a critical value. Section 3 defines the deceptive degree of a monomial and the deceptive degree of a function, which can be used to get a quantitative measure of deception, to simplify the evaluation of the GA difficulty of a given function and to get the relation between the deceptive degree and the polynomial degree. Section 4 discusses Goldberg's minimal deceptive problem by the new definition of deception and gets the same conclusion as Goldberg did. Section 5 describes experiments with a class of fitness functions to verify the relation between the canonical GA difficulty and the deceptive degree of functions. Section 6 summarizes the paper.

2 Definition and Properties of a Critical Value

2.1 Definition

Schemata have been widely studied in the field of GA, and are the basis of the deception analysis. We quickly remind the definition of a schema, which is also an important concept in the new definition of deception. A schema corresponds to a subset of the space $\Omega^n = \{0,1\}^n$ (the space of binary strings of length n for a GA using binary encoding), or more precisely a hyper-plane of Ω^n. An additional symbol "*" representing a wild card ("0" or "1") is used to represent a schema. For example, if n=4,the strings 0101 and 1101 are the two elements of the schema S=*101. Non-* positions in a schema are called defining positions. Defining positions and their corresponding values can be used to get another representation of a schema. For example, S[(2,1),(3,0),(4,1)] is also used to represent the schema S=*101. The latter representation of a schema is often used in this paper.

By convention, f will denote a pseudo-boolean function in polynomial form:

$$f(x) = a_0 + a_1x_1 + \cdots + a_nx_n + \cdots a_{i_1 \cdots i_k} x_{i_1} \cdots x_{i_k} + \cdots + a_{12 \ldots n} x_1 x_2 \cdots x_n$$

where $a_{i_1 \cdots i_k}$ is called the coefficient of the monomial $x_{i_1} \cdots x_{i_k}$, k is called the degree of the monomial $x_{i_1} \cdots x_{i_k}$, and the maximum of the degrees of all monomials is called the polynomial degree of f. $f_-(i_1, \cdots, i_k) = f - a_{i_1 i_2 \cdots i_k} x_{i_1} x_{i_2} \cdots x_{i_k}$ will denote f without the monomial $x_{i_1} \cdots x_{i_k}$. K will always represent a subset of

$\{1,2,\cdots,n\}$, and $f[(i_k,\theta_{i_k});i_k \in K]$ will denote f where $x_{i_k} = \theta_{i_k}$ for all $i_k \in K$ and $\theta_{i_k} \in \{0,1\}$. For example, let $f(x) = x_1 + x_2 + x_3 - x_1 x_2 x_3$, then $f_-(1,2,3) = x_1 + x_2 + x_3$, $f[(1,1),(2,0)] = 1 + 0 + x_3 - 1 \bullet 0 \bullet x_3 = 1 + x_3$. Now we define the concept of a critical value of a monomial with respect to a variable. A given function may have one or more best strings in Definition 1.

Definition 1. For a given function f, λ is called a critical value of the monomial $x_{i_1} \cdots x_{i_k}$ with respect to the variable x_{i_1} if for arbitrary $\varepsilon > 0$, all the best strings of $f_-(i_1,\cdots,i_k) + (\lambda+\varepsilon) x_{i_1} \cdots x_{i_k}$ are included in the schema $S[(i_1,1)]$ and one of the best strings of $f_-(i_1,\cdots,i_k) + (\lambda-\varepsilon) x_{i_1} \cdots x_{i_k}$ is included in the schema $S[(i_1,0)]$, which means that all the maximum of $f_-(i_1,\cdots,i_k) + (\lambda+\varepsilon) x_{i_1} \cdots x_{i_k}$ satisfy $x_{i_1} = 1$ and one of the maximum of $f_-(i_1,\cdots,i_k) + (\lambda-\varepsilon) x_{i_1} \cdots x_{i_k}$ satisfies $x_{i_1} = 0$.

In fact, the concept of **critical value** can reflect whether a monomial affects the position of the best string or not. If a critical value of a monomial with respect to a variable exists, a change of the monomial coefficient maybe lead to the change of the position of the best string. For a monomial of a given function, the best string of when the monomial coefficient is greater than the critical value is different from the best string of when the monomial coefficient is smaller than the critical value. For a function $f = f_-(i_1,\cdots,i_k) + a_{i_1 i_2 \cdots i_k} x_{i_1} x_{i_2} \cdots x_{i_k}$ where $a_{i_1 i_2 \cdots i_k} \neq 0$, according to Definition 1, if all the best strings of f satisfy $x_{i_1} = 1$, a critical value of the monomial $x_{i_1} x_{i_2} \cdots x_{i_k}$ with respect to the variable x_{i_1} is smaller than $a_{i_1 i_2 \cdots i_k}$. Conversely, if one of the best strings of f satisfies $x_{i_1} = 0$, a critical value of the monomial $x_{i_1} x_{i_2} \cdots x_{i_k}$ with respect to the variable x_{i_1} is greater than $a_{i_1 i_2 \cdots i_k}$. For example, a critical value of the monomial $x_1 x_2$ with respect to the variable x_2 in $f(x_1,x_2) = x_1 - 2x_2 + 3x_1 x_2$ is 2.

Now we consider the trap function $t(x)$ which is widely used. The definition of $t(x)$ is as follows:

$$t(x) = \begin{cases} n - bc(x) & \text{if } bc(x) \neq n \\ n+1 & \text{if } bc(x) = n \end{cases}$$

Here $bc(x) = \text{bitcount}(x) = $ number of ones in x. The trap function can be expressed in polynomial form:

$$t(x_1,x_2,\cdots,x_n) = n - x_1 - x_2 - \cdots - x_n + (n+1)x_1 x_2 \cdots x_n$$

According to Definition 1, a critical value of the monomial $x_1 x_2 \cdots x_n$ with respect to the variable x_i is n, a critical value of the monomial x_i with respect to the variable x_i is -2, where $i = 1,2,\cdots,n$.

2.2 Properties of a Critical Value

Theorem 1. *If there exists a critical value of the monomial* $x_{i_1} \cdots x_{i_k}$ *with respect to the variable* x_{i_1} *in f, then the critical value is unique.*

Proof. To prove uniqueness, suppose there are two distinct critical values of the monomial $x_{i_1} \cdots x_{i_k}$ with respect to the variable x_{i_1} in f, say λ^1 and λ^0 where $\lambda^1 > \lambda^0$. Let $\varepsilon = (\lambda^1 - \lambda^0)/2$, then $\lambda^1 > \lambda^0 + \varepsilon > \lambda^0$. For λ^0 is a critical value of the monomial $x_{i_1} \cdots x_{i_k}$ with respect to the variable x_{i_1}, by Definition 1, we get that all the best strings of $f_-(i_1, \cdots, i_k) + (\lambda^0 + \varepsilon) x_{i_1} \cdots x_{i_k}$ are included in the schema $S[(i_1, 1)]$. But λ^1 is also a critical value of the monomial $x_{i_1} \cdots x_{i_k}$ with respect to the variable x_{i_1}, we get that one of the best strings of $f_-(i_1, \cdots, i_k) + (\lambda^0 + \varepsilon) x_{i_1} \cdots x_{i_k}$ is included in the schema $S[(i_1, 0)]$. The above two results are contradictory, so the uniqueness is proved.

Theorem 2. *If one of the best strings of f is included in the schema* $S[(i_1, 0)]$, *then there exists a critical value of each monomial* $x_{i_1} \cdots x_{i_k}$ *with respect to the variable* x_{i_1}.

Proof. First we consider two functions about a variable λ as follows:

$$\max\{f_-(i_1, \cdots, i_k)[(i_1, 0)]\}, \ \max\{f_-(i_1, \cdots, i_k)[(i_1, 1)] + \lambda \, x_{i_2} \cdots x_{i_k}\}$$

Obviously the function $\max\{f_-(i_1, \cdots, i_k)[(i_1, 0)]\}$ is a constant and the other function $\max\{f_-(i_1, \cdots, i_k)[(i_1, 1)] + \lambda \, x_{i_1} \cdots x_{i_k}\}$ is an increasing function about the variable λ. If one of the best strings of f is included in the schema $S[(i_1, 0)]$, then

$$\max\{f_-(i_1, \cdots, i_k)[(i_1, 0)]\} \geq \max\{f_-(i_1, \cdots, i_k)[(i_1, 1)] + a_{i_1 \cdots i_k} \, x_{i_2} \cdots x_{i_k}\}$$

and moreover if $\lambda \to +\infty$, we get

$$\max\{f_-(i_1, \cdots, i_k)[(i_1, 1)] + \lambda \, x_{i_2} \cdots x_{i_k}\} \to +\infty.$$

So there must exist a $\lambda^0 \in R$, for arbitrary $\varepsilon > 0$, all the best strings of $f_-(i_1, \cdots, i_k) + (\lambda^0 + \varepsilon) x_{i_1} \cdots x_{i_k}$ are included in the schema $S[(i_1, 1)]$ and one of the best strings of $f_-(i_1, \cdots, i_k) + (\lambda^0 - \varepsilon) x_{i_1} \cdots x_{i_k}$ is included in the schema $S[(i_1, 0)]$. Therefore, by Definition 1, λ^0 is a critical value of the monomial $x_{i_1} \cdots x_{i_k}$ with respect to the variable x_{i_1} and the theorem is proved.

Corollary 1. *If there doesn't exist a critical value of the monomial* $x_{i_1} \cdots x_{i_k}$ *with respect to the variable* x_{i_1} *, then all the best strings of f must be included in the schema* $S[(i_1, 1)]$.

Theorem 3. *Let* $K = \{i_1, \cdots, i_k\}$ *be a subset of* $\{1, 2, \cdots n\}$. *If there exists a string* $(x_1, \ldots, x_n)$, *which satisfies that* $(x_1, \ldots, x_n)$ *is included in the schema* $S[(i_1, \theta_{i_1}), \cdots, (i_k, \theta_{i_k})]$ *where* $\theta_{i_1} = 1$ *and the other* θ_{i_j} *aren't all equal to* 1, *moreover* $f(x_1, \cdots, x_n) > \max\{f[(i_1, 0)]\}$, *then there doesn't exist a critical value of the monomial* $x_{i_1} \cdots x_{i_k}$ *with respect to the variable* x_{i_1} *.*

Proof. Let $f_\lambda = f_-(i_1, \cdots, i_k) + \lambda x_{i_1} \cdots x_{i_k}$. By the conditions of the theorem, we get

$$\max\{f_\lambda\} \geq f(x_1, \cdots, x_n) > \max\{f[(i_1, 0)]\} = \max\{f_\lambda[(i_1, 0)]\} .$$

For an arbitrary real number λ, all the best strings of f_λ must be included in the schema $S[(i_1, 1)]$. According to Definition 1, there doesn't exist a critical value of the monomial $x_{i_1} \cdots x_{i_k}$ with respect to the variable x_{i_1} .

Example. For $f(x_1, x_2) = x_1 - 2x_2 + 3x_1 x_2$, there exists the string 10 which is included in the schema $S[(1,1), (2,0)]$, moreover

$$f(1,0) = 1 > 0 = \max\{-2x_2\} = \max\{f[(1,0)]\}$$

By Theorem 3, there doesn't exist a critical value of the monomial $x_1 x_2$ with respect to the variable x_1 .

Theorem 4. *If two critical values of the same monomial* $x_i x_j x_{i_1} \cdots x_{i_k}$ *with respect to variables* x_i, x_j *exist, then they are equal.*

Proof. Let f be in the form

$$f = f_0 + x_i f_i + x_j f_j + x_i x_j f_{ij} + a_{ij} x_i x_j x_{i_1} \cdots x_{i_k}$$

where f_0 denotes f with monomials which include neither x_i nor x_j, $x_i f_i$ denotes f with monomials which include x_i but not x_j, $x_j f_j$ denotes f with monomials which includes x_j but not x_i, $x_i x_j f_{ij}$ denotes f with monomials which includes not only x_i but also x_j, where monomial $x_i x_j x_{i_1} \cdots x_{i_k}$ isn't included in $x_i x_j f_{ij}$.

Let λ_i, λ_j respectively be critical values of the same monomial $x_i x_j x_{i_1} \cdots x_{i_k}$ with respect to variables x_i, x_j , then we will prove that they are equal by contradiction. Suppose they are unequal. Let $\lambda_i > \lambda_j$, $\varepsilon = (\lambda_i - \lambda_j)/2$. According to Definition 1, we have

$$\max\{f_0 + x_j f_j\} \geq \max\{f_0 + f_i + x_j f_j + x_j f_{ij} + (\lambda_j + \varepsilon)x_j x_{i_1} \cdots x_{i_k}\} \qquad (1)$$

$$\max\{f_0 + x_j f_j\} < \max\{f_0 + f_i + x_j f_j + x_j f_{ij} + (\lambda_j + \varepsilon) x_j x_{i_1} \cdots x_{i_k}\} \qquad (2)$$

$$\max\{f_0 + x_i f_i\} \geq \max\{f_0 + f_j + x_i f_i + x_i f_{ij} + (\lambda_j - \varepsilon)x_i x_{i_1} \cdots x_{i_k}\} \qquad (3)$$

$$\max\{f_0 + x_i f_i\} < \max\{f_0 + f_j + x_i f_i + x_i f_{ij} + (\lambda_j + \varepsilon)x_i x_{i_1} \cdots x_{i_k}\} \qquad (4)$$

According to whether the best string of $f_0 + x_j f_j$ is included in the schema $S[(j,0)]$ or not, we discuss two cases.

Case I. If the best string of $f_0 + x_j f_j$ is included in the schema $S[(j,0)]$, then $\max\{f_0\} = \max\{f_0 + x_j f_j\}$. Applying (1) we get

$$\max\{f_0\} \geq \max\{f_0 + f_i + f_j + f_{ij} + (\lambda_j + \varepsilon)x_{i_1} \cdots x_{i_k}\}$$

Furthermore, applying (4) we get

$$\max\{f_0\} < \max\{f_0 + f_j\}.$$

This contradicts condition of Case I.

Case II. If the best string of $f_0 + x_j f_j$ isn't included in the schema $S[(j,0)]$, then $\max\{f_0 + f_j\} = \max\{f_0 + x_j f_j\}$. Applying (1) we get

$$\max\{f_0 + f_j\} \geq \max\{f_0 + f_i + f_j + f_{ij} + (\lambda_j + \varepsilon)x_{i_1} \cdots x_{i_k}\}$$

Furthermore, applying (3) we get

$$\max\{f_0 + x_i f_i\} \geq \max\{f_0 + f_j\}$$
$$\geq \max\{f_0 + f_i + f_j + f_{ij} + (\lambda_j + \varepsilon)x_{i_1} \cdots x_{i_k}\}$$

This contradicts the result of (4).

Since both cases lead to contradictions, our supposition is false and the theorem is proved.

Corollary 2. *If multiple critical values of the same monomial with respect to all variables exist, if exist, they are equal.*

Theorem 5. *If a critical value of the monomial $x_{i_1} \cdots x_{i_k}$ with respect to the variable x_{i_1} of f is 0, then*

$$\max\{f_-(i_1,\cdots,i_k)[(i_1,0)]\} = \max\{f_-(i_1,\cdots,i_k)[(i_1,1)]\}$$

Proof. Suppose the equality is not true. Now we discuss two cases.

Case I. Suppose

$$\max\{f_-(i_1,\cdots,i_k)[(i_1,0)]\} > \max\{f_-(i_1,\cdots,i_k)[(i_1,1)]\}$$

Let ε satisfy

$$0 < \varepsilon < \max\{f_-(i_1,\cdots,i_k)[(i_1,0)]\} - \max\{f_-(i_1,\cdots,i_k)[(i_1,1)]\}$$

then

$$\begin{aligned}
&\max\{f_-(i_1,\cdots,i_k)[(i_1,0)]\}\\
&> \max\{f_-(i_1,\cdots,i_k)[(i_1,1)]\} + \varepsilon\\
&= \max\{f_-(i_1,\cdots,i_k)[(i_1,1)]\} + \max\{\varepsilon \, x_{i_1}\cdots x_{i_k}[(i_1,1)]\}\\
&\geq \max\{(f_-(i_1,\cdots,i_k) + \varepsilon \, x_{i_1}\cdots x_{i_k})[(i_1,1)]\}
\end{aligned}$$

According to Definition 1, a critical value of the monomial $x_{i_1}\cdots x_{i_k}$ with respect to the variable x_{i_1} in f must be greater than 0, which contradicts the condition of the theorem.

Case II. Suppose

$$\max\{f_-(i_1,\cdots,i_k)[(i_1,0)]\} < \max\{f_-(i_1,\cdots,i_k)[(i_1,1)]\}.$$

Let ε satisfy

$$0 < \varepsilon < \max\{f_-(i_1,\cdots,i_k)[(i_1,1)]\} - \max\{f_-(i_1,\cdots,i_k)[(i_1,0)]\}$$

then

$$\begin{aligned}
&\max\{f_-(i_1,\cdots,i_k)[(i_1,0)]\}\\
&< \max\{f_-(i_1,\cdots,i_k)[(i_1,1)]\} - \varepsilon\\
&= \max\{f_-(i_1,\cdots,i_k)[(i_1,1)]\} - \max\{\varepsilon \, x_{i_1}\cdots x_{i_k}[(i_1,1)]\}\\
&\leq \max\{(f_-(i_1,\cdots,i_k) - \varepsilon \, x_{i_1}\cdots x_{i_k})[(i_1,1)]\}
\end{aligned}$$

According to Definition 1, a critical value of the monomial $x_{i_1}\cdots x_{i_k}$ with respect to the variable x_{i_1} in f must be less than 0, which also contradicts the condition of the theorem.

From results of both cases, we get

$$\max\{f_-(i_1,\cdots,i_k)[(i_1,0)]\} = \max\{f_-(i_1,\cdots,i_k)[(i_1,1)]\}$$

and the theorem is proved.

If there doesn't exist a critical value of the monomial $x_{i_1}\cdots x_{i_k}$ with respect to the variable x_{i_1} in f, we say the critical value is $-\infty$ where $-\infty$ denotes a negative infinite number. The following result is easy to prove.

Theorem 6. *If the coefficient of the monomial* $x_{i_1}\cdots x_{i_k}$ *is greater than critical values of the monomial* $x_{i_1}\cdots x_{i_k}$ *with respect to all variables, then the best string must be* *included in* $S[(i_1,1),(i_2,1),\cdots,(i_k,1)]$. *If the coefficient of the monomial* $x_{i_1}\cdots x_{i_k}$ *is less than one of critical values of the monomial* $x_{i_1}\cdots x_{i_k}$ *with respect to all variables, then the best string mustn't be included in* $S[(i_1,1),(i_2,1),\cdots,(i_k,1)]$.

Corollary 3. *If the coefficient of each monomial which includes* x_i *is greater than a critical value of the corresponding monomial with respect to the variable* x_i, *then the best string must be included in* $S[(i,1)]$.

In the following, we discuss the deception of a given function according to the relation between a monomial coefficient and a critical value.

3 The Deceptive Degree of a Function

In the polynomial representation of f, a monomial coefficient may determine the position of the best string of f. When a monomial coefficient changes, the best string maybe change from one to another. For a monomial $x_{i_1}\cdots x_{i_k}$, different bits between the best string of f and the best string of $f_-(i_1,\cdots,i_k)$ can be used to depict the GA difficulty. If it is easy for a GA to find the best string of f which is included in the schema $S[(i_1,\theta_{i_1}),\cdots,(i_k,\theta_{i_k})]$, it may be not easy for the GA to find the best string of $f_-(i_1,\cdots,i_k)$ only because of these different bits, and vice visa. So when we want to find the best string of a given function, a monomial coefficient maybe make it easy for a GA, another monomial coefficient maybe make it difficult for the GA which lead to deception. In order to depict the properties, we introduce the definition of a monomial has deception about a variable.

3.1 A Monomial Having Deception About a Variable

Definition 2. *If a variable in a monomial satisfies one of the following three conditions, we say that the monomial has deception about the variable.*

(1) If a critical value of the monomial with respect to the variable is 0;

(2) If a critical value of the monomial with respect to the variable is positive and less than the monomial coefficient;

(3) If a critical value of the monomial with respect to the variable is negative and greater than the monomial coefficient.

Obviously these three conditions contradict each other and at most one can be satisfied.

Example. For the function $x_1 - 2x_2 + 3x_1x_2$, the monomial x_1x_2 has deception about the variable x_2, but the monomial x_1x_2 doesn't have deception about the variable x_1. For the trap function $t(x_1,x_2,\cdots,x_n)$, the monomial $x_1x_2\cdots x_n$ has deception about every variable in it, the monomial x_i doesn't have deception about the variable x_i, where $i = 1,2,\cdots,n$.

Theorem 7. *Let f have only one best string, then the monomial* $x_{i_1} \cdots x_{i_k}$ *doesn't have deception about every variable in it if and only if the best string of* $f_-(i_1,\cdots,i_k)$ *is the same as the best string of f.*

Proof. First the sufficiency is proved. Suppose the monomial $x_{i_1} \cdots x_{i_k}$ has deception about the variable x_{i_1}. In the following, we will discuss three cases.

Case I. If a critical value of the monomial $x_{i_1} \cdots x_{i_k}$ with respect to the variable x_{i_1} in f is 0, Theorem 5 shows that:

$$\max\{f_-(i_1,\cdots,i_k)[(i_1,0)]\} = \max\{f_-(i_1,\cdots,i_k)[(i_1,1)]\},$$

So $f_-(i_1,\cdots,i_k)$ has at least two best strings. For the best string of $f_-(i_1,\cdots,i_k)$ is the same as the best string of f, f has at least two best strings, which contradicts the fact that f has only one best string.

Case II. If a critical value of the monomial $x_{i_1} \cdots x_{i_k}$ with respect to the variable x_{i_1} is positive and less than the coefficient $a_{i_1 \cdots i_k}$, then the best string of f must be included in the schema $S[(i_1,1)]$. Moreover according to Definition 1, we get

$$\max\{f_-(i_1,\cdots,i_k)[(i_1,0)]\} \geq \max\{f_-(i_1,\cdots,i_k)[(i_1,1)]\},$$

Thus either $f_-(i_1,\cdots,i_k)$ has only one best string and the best string is included in the schema $S[(i_1,0)]$, or $f_-(i_1,\cdots,i_k)$ has more than one best strings, some of them are included in $S[(i_1,1)]$ and some of them are included in $S[(i_1,0)]$, the two results both contradict the condition that the best string of $f_-(i_1,\cdots,i_k)$ is the same as the best string of f.

Case III. If a critical value of the monomial $x_{i_1} \cdots x_{i_k}$ with respect to the variable x_{i_1} is negative and greater than the coefficient $a_{i_1 \cdots i_k}$, moreover f has only one best string, then we get that the best string of f must be included in the schema $S[(i_1,0)]$, but the best string of $f_-(i_1,\cdots,i_k)$ must be included in the schema $S[(i_1,1)]$, this contradicts the fact that the best string of $f_-(i_1,\cdots,i_k)$ is the same as the best string of f.

From results of three cases, we get the monomial $x_{i_1} \cdots x_{i_k}$ does not have deception about every variable in it and the sufficient condition is proved.

Conversely, suppose the best string of f is included in the schema $S[(i_1,\theta_{i_1}),\cdots,(i_k,\theta_{i_k})]$. If there exists a $\theta_{i_t}=0\,(1\leq t\leq k)$, then the best string of $f_-(i_1,\cdots,i_k)$ must be included in the schema $S[(i_1,\theta_{i_1}),\cdots,(i_k,\theta_{i_k})]$; If all

$\theta_{i_t} = 1 (1 \le t \le k)$ and one of the best strings of $f_-(i_1, \cdots, i_k)$ is not included in the schema $S[(i_1, \theta_{i_1}), \cdots, (i_k, \theta_{i_k})]$, then there must exist a $i_t (1 \le t \le k)$ and the best string of $f_-(i_1, \cdots, i_k)$ is included in the schema $S[(i_t, 0)]$, thus a critical value of the monomial $x_{i_1} \cdots x_{i_k}$ with respect to the variable x_{i_t} is positive and less than the coefficient $a_{i_1 \cdots i_k}$, hence the monomial $x_{i_1} \cdots x_{i_k}$ has deception about the variable x_{i_t}, a contradiction. So the best string of $f_-(i_1, \cdots, i_k)$ is the same as the best string of f and the necessary condition is proved.

3.2 The Deceptive Degree of a Function

In fact, for a monomial $x_{i_1} \cdots x_{i_k}$, if the best string of f is different from the best string of $f_-(i_1, \cdots, i_k)$, one of the two best strings must be included in the schema $S[(i_1, 1), (i_2, 1), \cdots, (i_k, 1)]$. When we assume that the best string of f is included in the schema $S[(i_1, 1), (i_2, 1), \cdots, (i_k, 1)]$, then the number of variables for which the monomial has deception can reflect the difficulty of a GA evolving to the schema $S[(i_1, 1), (i_2, 1), \cdots, (i_k, 1)]$. Furthermore, if the best string of f is $11 \ldots 1$, we can reflect the evaluation of the GA difficulty of f by the number of variables about which the monomial has deception.

For simplicity, in the next part of the paper, we only discuss the deceptive degree of a function that has only one best string.

Definition 3. *Let the best string of a function be* $11 \ldots 1$, *for a monomial whose coefficient isn't equal to 0, the number of variables about which the monomial has deception is called the deceptive degree of the monomial. The maximum of deceptive degrees of monomials of the function is called the deceptive degree of the function.*

Let $\oplus$ denote the binary XOR operation on bits and the bitwise XOR operation on strings.

Proposition 1. *If* $g(x_{i_1}, x_{i_2}, \cdots, x_{i_n}) = f(x_{i_1} \oplus 1, x_{i_2}, \cdots, x_{i_n})$, *then the GA difficulties of g and f are similar.*

According to Proposition 1, the deceptive degree of a function, which has only one best string, can be defined as follows.

Definition 4. *Let the best string of f be* $11 \ldots 1$, *and*

$$g(x_{i_1}, x_{i_2}, \cdots, x_{i_n}) = f(x_{i_1} \oplus 1, x_{i_2}, \cdots, x_{i_n}),$$

then we define the deceptive degree of g to be equal to the deceptive degree of f.

Example. The deceptive degree of the function $x_1 - 2x_2 + 3x_1 x_2$ is 1. The deceptive degree of the trap function $t(x_1, x_2, \cdots, x_n)$ is n.

Obviously when the best string of a function is 11...1, it is impossible that the third condition in Definition 2 comes into existence. So we may discuss two conditions for the deceptive of a monomial, and then get the deceptive degree of a function. From Definition 3 and 4, the deceptive degrees of functions range from 1 to n. We may class functions according to their deceptive degrees.

3.3 Relations of the Deceptive Degrees of Functions

According to the definition of the deceptive degree of a function, we easily get some relationships among the deceptive degrees of functions.

Theorem 8. *For two functions* $f, g : Z_2^n \to R$, *if there exists a permutation* $i_1, i_2, \cdots, i_n$ *of the integers* $1, 2, \cdots, n$, $g(x_{i_1}, x_{i_2}, \cdots, x_{i_n}) = f(x_1, x_2, \cdots, x_n)$, *then the deceptive degrees of* f, g *are equal.*

Theorem 9. *For an arbitrary positive constant* $C \in R$, *the deceptive degrees of* $Cf(x), f(x) + C$ *both are equal to the deceptive degree of* $f(x)$.

Theorem 10. *For an arbitrary constant* $C \in \{0,1\}^n$, *the deceptive degree of* $f(x \oplus C)$ *is equal to the deceptive degree of* $f(x)$.

3.4 Applications of the Deceptive Degree

The above definition of the deceptive degree can be applied in three aspects. First we get a quantitative measure of deception for a function that has only one best string. The deceptive degrees of functions range from 1 to n. Second we can get the deceptive degree of a complex function by the deceptive degree of a relative simpler function. Usually the GA hardness of a function in simpler form is easier to evaluate. From the properties of the coefficients of a function whose best string is $11 \cdots 1$, we may simplify the function as follows:

(1) If all critical values of a monomial with respect to variables in it are negative and less than the monomial coefficient, we use 0 instead of the monomial coefficient.
(2) If all critical values of a monomial with respect to variables in it are positive and greater than the monomial coefficient, we use 0 instead of the monomial coefficient.
(3) If all critical values of a monomial with respect to variables in it are 0 and the monomial coefficient is positive, we use an arbitrary positive ε instead of the monomial coefficient. If all critical values of a monomial with respect to variables in it are 0 and the monomial coefficient is negative, we use an arbitrary negative ε instead of the monomial coefficient.

If GA difficulty is measured by the expected time to the first occurrence of the best string, then the simpler function in the above conditions has the same GA difficulty as the original function. For example, the function

$$f(x_1, \cdots, x_n) = x_1 + x_2 + \cdots + x_n$$

has the same GA difficulty as the function

$$f(x_1, \cdots, x_n) = x_1 + x_2 + \cdots + x_n + 2^n x_1 x_2 \cdots x_n.$$

The function

$$f(x_1, \cdots, x_n) = 1 + \varepsilon \quad x_1 x_2 \cdots x_n$$

has the same GA difficulty as the function

$$f(x_1, \cdots, x_n) = 1 + 2^n x_1 x_2 \cdots x_n,$$

where ε is an arbitrary positive number.

The third application is that we get the relation between the deceptive degree and the polynomial degree. By Theorem 11, if the polynomial degree of a function is small, then deceptive degree of the function is small.

Theorem 11. *The deceptive degree of a function isn't greater than the polynomial degree of the function.*

Proof. For an arbitrary constant $C \in \{0,1\}^n$, $f(x \oplus C)$ and $f(x)$ have the same polynomial degrees and the same deceptive degrees. According to Definition 4, we may consider the relation between the polynomial degree and the deceptive degree of a function whose best string is only $11 \cdots 1$. Furthermore the deceptive degree of a function is defined by the number of variables about which a monomial has deception, thus the deceptive degree of f isn't greater than the polynomial degree of f.

4 Goldberg's Minimal Deceptive Problem (MDP)

We can explain Goldberg's Minimal Deceptive Problem (MDP) easily by the definition of the deceptive degree of a function. Goldberg discussed the MDP with a function of 2 bits, say f, whose best string is 11. If $[f(00) + f(01)]/2 > [f(10) + f(11)]/2$, then deception exists. Let $c = f(01) / f(00)$, Goldberg called $c > 1$ a Type I deceptive problem and called $c \leq 1$ a Type II deceptive problem. For a GA, to optimize a Type II deceptive problem is more difficult than to optimize a Type I deceptive problem.

Now we discuss these two types of deceptive problems again by the definition of the deceptive degree of a function. Let the best string of the function $f(x_1, x_2) = a_0 + a_1 x_1 + a_2 x_2 + a_3 x_1 x_2$ be 11. According to the requirements for a fitness function of a GA, we suppose all functions values are positive, which implies that $a_0 > 0$. Since the best string of the function is 11, we get

$$a_1 + a_2 + a_3 > \max\{a_1, a_2\}. \tag{5}$$

Furthermore for

$$[f(00) + f(01)]/2 > [f(10) + f(11)]/2,$$

we get

$$2a_1 + a_3 < 0 \tag{6}$$

Case I. If $c > 1$, then $a_2 > 0$ since $a_0 > 0$. Because of inequalities (5) and (6), we get

$$a_1 < 0, \quad a_2 > 0, \quad a_3 > |a_1|$$

By Definition 3, the deceptive degree of the function is 1.

Case II. If $c \le 1$, then $a_2 \le 0$ since $a_0 > 0$. Because of inequalities (5) and (6), we get

$$a_1 < 0, \quad a_2 \le 0, \quad a_3 > \max\{|a_1| + |a_2|\}$$

By Definition 3, the deceptive degree of the function is 2. We get the same result as Goldberg, namely to optimize a Type II deceptive problem is more difficult than to optimize a Type I deceptive problem for a GA.

5 Experimental Results

In order to verify the relation between the canonical GA (i.e. with proportionate selection (roulette wheel selection), one-point crossover and bit inversion mutation) difficulty and the deceptive degree of a function, we have done experiments with a class of functions f_k, where f_k is defined as:

$$f_k(x_1, x_2, \cdots, x_n) = x_1 + \cdots + x_{n-k} - x_{n-k+1} - \cdots - x_n + (k+1)x_{n-k+1}x_{n-k+2}\cdots x_n$$

Obviously the deceptive degree of f_k is k.

Parameters of the canonical GA which was used to optimized f_k were given as follows: the maximal generation was 1000, the population size and the chromosome length both were 100, an initial population of bit strings was generated randomly, the selection operator was roulette wheel selection, the crossover operator was one-point crossover and the crossover probability was 0.8, the mutation operator was bit inversion mutation and the mutation probability was 0.01. We did 100 tests for every f_k: a success test was that the test found the best string of f_k, a success probability is the percentage of success tests. The average success generations were the average number of generations of that a successful test used to find the best string of f_k. Figure 1 shows different experimental results for different deceptive degrees k.

From the left of Figure 1 we see the relation between the GA difficulty and the deceptive degree of a function: the greater the deceptive degree of a function is, the smaller the success probability is. From the right of Figure 1 we get the change of average success generations: the greater the deceptive degree of a function is, the stronger the change of average success generations is. On the one hand, the stronger fluctuations are due to random initial populations and few success tests, on the other hand, the strong fluctuations are due to the greater variance of average success generations for the greater deceptive degree. To summarize, the deceptive degree of a function is closely related to the canonical GA difficulty for this class of functions.

6 Summary

This paper presents a new definition for the deceptive degree of a function and many theorems that show the usefulness of the new definition. The new definition can be applied in three aspects: it gives a quantitative measure of deception, it simplifies the

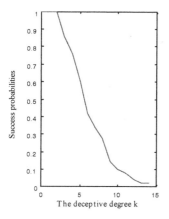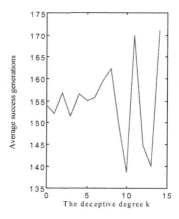

Fig. 1. Success probabilities and Average success generations for different deceptive degree k

evolution of the GA difficulty, and it gives the relationship between the deceptive degree and the polynomial degree. Furthermore, this paper uses the new definition to discuss Goldberg's Minimal Deceptive Problem and derives the same result as Goldberg did. Experiments with a class of functions show the relation between the canonical GA difficulty and the deceptive degree of a function for this class of functions.

Many of the criticisms of Goldberg's definition of deception relate to the fact that it is based on a *static* analysis of the function. The new definition of deception is also based on a *static* analysis of the properties of the function, and thus it may be subject to some of the same criticisms. What is the rigorous connection between deception and GA difficulty is still the main difficulty of the new definition. Whether the new definition can play an important role in the connection between the GA performance and the structure of a given function deserves further consideration. The relation between the new definition and the old ones should also be taken into account for future research.

Acknowledgements

The work presented in this paper was supported by the fund of National Laboratory for Modern Communications (China) grant number 51436020203JB0602. Thanks to professor Alden H. Wright and the reviewers for their help and suggestions to improve the presentation of this paper.

References

1. J. H. Holland.: Adaptation in Natural and Artificial Systems. The University of Michigan Press. (1976)
2. D. Bethke.: Genetic Algorithm as Function Optimizations. Doctoral Dissertation, University of Michigan. Dissertation abstracts International 41(9), 3503B. University Microfilms No.8106101. (1981).
3. Goldberg, D. E.: Simple genetic algorithms and the minimal deceptive problem. In L. Davis(Ed.), Genetic algorithms and simulated annealing (pp.74-88). London: Pitman. 1987.

4. Goldberg D E.: Genetic Algorithm and Walsh Functions: Part I, A Genetic Introduction. Complex Systems (3) (1989): 129-152.
5. Goldberg D E.: Genetic Algorithm and Walsh Functions: Part II, Deceptive and its Analysis. Complex Systems (3) (1989): 129-152.
6. W. E. Hart, R. K Belew.: Optimizing an Arbitrary Function is Hard for the Genetic Algorithm. Proceedings of the Fourth International Conference on Genetic Algorithm. University of California, San Diego July (1991)13-16.
7. W. Spears.: Evolutionary Algorithms: The Role of Mutation and Recombination. 2000, Springer-Verlag.
8. K. DeJong, W. Spears, and D. Gordon.: Using Markov Chains to Analyze GAFOs. FOGA-3, Morgan Kaufmann, 1995, pp 115-137.
9. Das and Whitley.: The Only Challenging Problems are Deceptive: Global Search by Solving Order-1 Hyperplanes. ICGA'91. pp166-173
10. J. J. Grefenstette.: Deception considered harmful. in: L. Darrell Whitley(Ed.), FOGA-2, Morgan Kaufmann, 1993, pp. 75-91.
11. J. J. Grefenstette and J. E. Baker.: How Genetic Algorithms Work: A Critical Look at Implicit Parallelism. Proceedings of the Third International Conference on Genetic Algorithms, J. D. Schaffer, Editor, San Mateo, CA, Morgan Kaufmann, 1989.
12. M. Mitchell, S. Forrest, and J. H. Holland.: The Royal Road Genetic Algorithms: Fitness Landscapes and GA Performance. Proceedings of the First European Conference and Artificial Life, Cambridge, MA, MIT Press/Branford Books, 1992.
13. Stephanie Forrest and Melanie Mitchell.: What Makes a Problem Hard for Genetic Algorithm? Some Anomalous Results in the Explanation. Machine Learning, 13, pages 285-319, 1993.
14. Y. Davidor.: Epistasis variance: a viewpoint on GA-hardness. In G.J.E. Rawlins, editor, Foundations of Genetic Algorithms, pages 23-35. Morgan Kaufmann, 1991.
15. C.R.Reeves and C. C. Wright.: Epistasis in genetic algorithms: an experimental design perspective. In L. J. Eshelman (Ed.) (1995)Proceedings of the 6 th International Conference on Genetic Algorithms, Morgan Kaufmann, San Mateo, CA,217-224.
16. Naudts, B., Suys, D., Verschoren, A.: Epistasis as a Basic Concept in Formal Landscape Analysis. In: Th. Baeck, ed., Proceedings of the 7th International Conference on Genetic Algorithms, Morgan Kaufmann, San Francisco, Ca.,pp. 65--72.
17. B. Leblanc and E. Lutton.: Bitwise regularity and GA-hardness. ICEC 98, May 5 Anchorage, Alaska, 1998.
18. A Homaifar, Qi Xiaoyun, and John Fost.: Analysis and Design of a General GA Deceptive Problem. Proceedings of the Fourth International Conference on Genetic Algorithm University of California, San Diego July(1991)13-16.
19. Xiaoping, C. Liming.: Genetic Algorithm—Theory, Application and Software Implementation. Xi'an University of Communication Publishers. (2002) (in Chinese)
20. Liepins, G. E. and Vose, M. D.: Representational Issues in Genetic Optimization. Journal of Experimental and Theoretical Artificial Intelligence. 1990(2):101-115.
21. Whitley, L D.: Fundamental Principles of Deception in Genetic Search. In G. Rawlins (ED.). Foundations of Genetic Algorithms. San Mateo, CA:Morgan Kaufmann(1991).

Author Index